Governance for Mediterranean Silvopastoral Systems

This book is about the resilience of silvopastoral systems now, and in the future. As such, it is about people. The goal is to fill the gap in the knowledge on silvopastoral systems and their changing trends, by adding the *human dimension*, with enough detail to draw inferences about the new governance solutions that are needed to address the multiple challenges faced by silvopastoral systems. As such, the book provides knowledge applicable to current and future silvopastoral territories in other regions across the world.

The volume is divided into three sections: people and institutions, the institutional framework and governance models. Each section, composed of several chapters, draws on empirical work about the Iberian montado and dehesa as well as from other similar systems in the Mediterranean, both on the northern and on the southern sides, in order to broaden its scope and cover a wider range of situations and examples. Some of the chapters rely more strongly on empirical findings and current experiences, others on a literature review and reflection by the authors over many years working with these systems. The conclusion sums up the most relevant findings from each chapter and discusses how research can progress so that new scientific approaches and evidence can support better adapted governance models of silvopastoral systems to face future challenges.

This text will be highly valuable to university and research institute libraries, academics, policy officials and stakeholder groups, such as NGOs and sectoral organizations, who wish to better understand the relevance of the human factor and use this knowledge to find sustainable solutions. It will be a central reading for postgraduate students enrolled in rural planning, landscape management and governance, agronomy and forestry, as well as geography and socio-ecology programmes, that have a focus on sustainable land use management and supporting mixed farming systems.

Teresa Pinto-Correia is a Portuguese geographer, with a record of publications on the dynamics and change of European rural landscapes at multiple scales, with a particular focus on silvopastoral systems of Iberia. She is a full professor at the University of Évora and the director of MED (www.med.uevora.pt), an R&I unit of 180 researchers, with a systemic and interdisciplinary perspective on Mediterranean agriculture and environment.

Maria Helena Guimarães is a Portuguese environmental scientist, a researcher at MED www.med.uevora.pt, and coordinator of the thematic line: Transdisciplinarity and co-construction of knowledge of LABscape – Mediterranean Landscape Systems Lab and Co-coordinator of MED thematic Line: Governance and Rural Dynamics.

Gerardo Moreno is a Spanish biologist, specialized on soil science and agroforestry, focused on the study of the functioning, management and provision of ecosystem services by Iberian dehesas. He is a full professor at University of Extremadura and a member of INDEHSA (Institute for the study of Iberian Dehesas; www.indehesa.unex.es).

Rufino Acosta-Naranjo is a Spanish anthropologist and professor of the University of Seville, specialized in Ecological Anthropology and Rural Societies focusing on traditional agroecosystems, in particular the dehesa, as well as on Agroecology, and biodiversity. He is currently working on initiatives against rural depopulation.

Perspectives on Rural Policy and Planning

Series Editors: Andrew Gilg and Mark Lapping

This well-established series offers a forum for the discussion and debate of the often-conflicting needs of rural communities and how best they might be served. Offering a range of high-quality research monographs and edited volumes, the titles in the series explore topics directly related to planning strategy and the implementation of policy in the countryside. Global in scope, contributions include theoretical treatments as well as empirical studies from around the world and tackle issues such as rural development, agriculture, governance, age and gender.

Globalization and Europe's Rural Regions
Edited by John McDonagh, Birte Nienaber and Michael Woods

Service Provision and Rural Sustainability
Infrastructure and Innovation
Edited by Greg Halseth, Sean Markey and Laura Ryser

The Changing World of Farming in Brexit UK
Edited by Matt Lobley, Michael Winter and Rebecca Wheeler

Rural Gerontology
Towards Critical Perspectives on Rural Ageing
Edited by Mark Skinner, Rachel Winterton and Kieran Walsh

Tourism and Socio-Economic Transformation of Rural Area
Evidence from Poland
Edited by Joanna Kosmaczewska and Walenty Poczta

Governance for Mediterranean Silvopastoral Systems
Lessons from the Iberian Dehesas and Montados
Edited by Teresa Pinto-Correia, Maria Helena Guimarães, Gerardo Moreno and Rufino Acosta-Naranjo

For more information about this series, please visit: www.routledge.com/Perspectives-on-Rural-Policy-and-Planning/book-series/ASHSER-1035

Governance for Mediterranean Silvopastoral Systems

Lessons from the Iberian Dehesas and Montados

Edited by Teresa Pinto-Correia,
Maria Helena Guimarães,
Gerardo Moreno
and Rufino Acosta-Naranjo

Routledge
Taylor & Francis Group

LONDON AND NEW YORK

First published 2022
by Routledge
2 Park Square, Milton Park, Abingdon, Oxon OX14 4RN

and by Routledge
605 Third Avenue, New York, NY 10158

Routledge is an imprint of the Taylor & Francis Group, an informa business

British Library Cataloguing-in-Publication Data
A catalogue record for this book is available from the British Library

Library of Congress Cataloging-in-Publication Data
Names: Pinto-Correia, Teresa, editor.
Title: Governance for Mediterranean silvo-pastoral systems : lessons from the Iberian dehesas and montados / edited by Teresa Pinto-Correia, Helena Guimarães, Gerardo Moreno, and Rufino Acosta Naranjo.
Description: Milton Park, Abingdon, Oxon ; New York, NY : Routledge, 2022. | Series: Perspectives on rural policy and planning | Includes bibliographical references and index.
Identifiers: LCCN 2021018030 (print) | LCCN 2021018031 (ebook) | ISBN 9780367463571 (hardback) | ISBN 9781032073354 (paperback) | ISBN 9781003028437 (ebook)
Subjects: LCSH: Silvopastoral systems--Government policy--Iberian Peninsula. | Forest management--Iberian Peninsula. | Pastoral systems--Iberian Peninsula.
Classification: LCC S494.5.A47 G68 2022 (print) | LCC S494.5.A47 (ebook) | DDC 634.9/209366--dc23
LC record available at https://lccn.loc.gov/2021018030
LC ebook record available at https://lccn.loc.gov/2021018031

ISBN: 978-0-367-46357-1 (hbk)
ISBN: 978-1-032-07335-4 (pbk)
ISBN: 978-1-003-02843-7 (ebk)

DOI: 10.4324/9781003028437

Typeset in Bembo
by KnowledgeWorks Global Ltd.

Contents

Contributors

José Muñoz-Rojas is a Rural Geographer and Landscape Planner, currently working as a Research Scientist in Rural Landscape Dynamics at the Mediterranean Institute for Agriculture, Environment and Development of the University of Évora (Portugal), where he heads the Landscape Dynamics and Management research group. Having worked across diverse rural settings over the past decade, he has researched and published extensively in the complexity of governance frameworks and regimes for the sustainability and resilience of agroforestry and other farming systems.

Isabel Loupa-Ramos is an Assistant Professor at the University of Lisbon, where she teaches courses across a range of fields within the landscape and urban planning domain. She is trained as a Landscape Architect, holds a master's degree in Human Geography and a PhD in Environmental Engineering. Her main interests are at the interface between cultural and natural processes, and urban and rural landscapes using scenario development and transdisciplinary approaches. Her research is on the integration of concepts as landscape preferences, expectations towards the future and landscape identity into spatial planning.

José Ramón Guzmán-Álvarez is a Spanish Forest and Agricultural Engineer with a degree in Geography and History. He is interested in the historical relationship between humans and nature and its reflection on the landscape. He has been involved in several research and management projects and studies related to main land-use systems of southern Spain such as dehesas, olive groves, natural and implanted forests, steppes and others. His published work includes a number of monographies, contributions to collective works, scientific papers and divulgation books and articles about these matters.

Laura Amores-Lemus is a Spanish anthropologist, PhD student and researcher at the Department of Social Anthropology, University of Seville. She has worked with food anthropology, rural studies and local development since she became a member of the Andalusian multidisciplinary Research Group "GICED" (Culture, Ecology and Development of Small Territories). Currently, her thesis deals with the evolution and innovation of food practices in southwestern Spanish society. Her previous publications also address rural depopulation issues in those regions.

Francisco M. Parejo-Moruno holds a PhD in Applied Economics and is an Assistant Professor of Economic History and Institutions at the University of Extremadura. He is the author of many research papers on agrarian history and, more specifically, on the history of cork business to which he has dedicated most part of his career. He is currently working on regional economic issues and is involved in the study of microeconomic aspects of rural development from a historical perspective.

Esteban Cruz-Hidalgo holds a PhD in Economics and Business and is a Professor of Foundations of Economic Analysis at the University of Extremadura. He is the author of multiple scholarly articles focused on macroeconomics, regional development and political economy, with a focus on the design of functional institutional mechanisms for the mobilization of real resources for sustainable and inclusive development.

Antonio M. Linares-Luján holds a PhD in History and is an Associate Professor of Economic History and Economic Institutions at the University of Extremadura. He is the author of many research papers on agrarian history and, more specifically, on forest history. Among them, the works dedicated to the analysis of the long-term changes experienced in the dehesa-systems stand out, fundamentally the changes in ownership, management and agro-silvopastoral exploitation models.

Amélia Branco holds a PhD in Economic and Social History and is an Assistant Professor at ISEG from the University of Lisboa. She is a researcher at the Economic and Social History Centre that integrates the Consortium Research in Social Sciences and Management, where she is a part of the research group Sustainability and Policy. Her main research interests are forestry value chain along the 19th and 20th centuries, where she has several publications.

José Francisco Rangel-Preciado holds a PhD in Economics and Business and is a Professor of History and Economic Institutions in the Department of Economics at the University of Extremadura. He is the author of numerous scientific articles in national and international journals on agricultural and forestry economy, business history and regional economics, although he has focused its attention on the cork business.

Elisa Oteros-Rozas is a trained Biologist and is currently a Sustainability Scientist, developing her research on social-ecological perspectives in pastoralism and agroecology. She has mostly worked in the Mediterranean ecoregion, with a particular interest in transhumance and local ecological knowledge. She is currently "Juan de la Cierva – Incorporación" Postdoctoral Fellow at the Chair on Agroecology and Food Systems of the University of Vic. She is also a member of FRACTAL, Feminist Researchers in Action for Transformation.

María E. Fernández-Giménez leads the Rangeland Social-Ecological Systems Lab in the Department of Forest and Rangeland Stewardship at the Colorado State University. Her research focuses on pastoralist decision-making at the individual, household and community scales; community-based resource management and collaborative governance; traditional ecological knowledge and knowledge integration; and participatory and transdisciplinary research. Her geographic focal areas are the western United States, Spain and Mongolia.

Ignacio García-Pereda is a member of the Research Centre CIUHCT-Universidade de Lisboa. His interests are focused on forest history and the agrarian products industry. He holds a master's degree (2006) in Forestry and a PhD (2018) in Science History. He currently works on the project "Horto Aquam Salutarem – water-wise management in gardens in the early modern period" (2018–2021).

Federica Ravera is a Ramón y Cajal Senior Researcher at the University of Girona. She holds a PhD in Environmental Science and has mainly developed her research experience in Ecological Economics and Political Ecology. Her line of research focuses on the analysis of socio-institutional innovations, collective actions and the role of traditional and local knowledge in an adaptation of global environmental, cultural and socioeconomic changes, especially in agro-pastoral systems of the Mediterranean context and high mountain regions (Pyrenees, Andes and Himalayas). Recently her research interest has focused on gender and power dynamics issues in global environmental change studies. Since October 2017, she is also a member of FRACTAL, Feminist Researchers in Action for Transformation.

Victoria Quintero-Morón is a Senior Lecturer in Social Anthropology at the University of Pablo de Olavide (Spain). Her research areas are focused on natural and cultural heritage and she has developed different projects in landscapes, environmental anthropology, tourism and social participation. Currently, she has two lines of research opened with a gender perspective: the first one on the logics of governance and the habitation of intangible heritage and the second one on local knowledge and narratives of the Mediterranean dehesas.

Diana Surová worked for 12 years in research dealing with landscapes in southern Portugal and their perception by different society members. Her publications focus mainly on landscape perception, preferences and values attributed to rural landscapes by various groups of people. She is currently working as a Researcher in the Department of Humanities at the Czech University of Life Sciences in Prague.

Pedro Herrera Calvo is a Spanish biologist, specializing in land-related issues, with long experience both as consultant and researcher, as freelance and part of small consultancies. In 2011 he was part of the birth of Entretantos Foundation, an NGO devoted to public participation in sustainable development. Currently, he works there as Project Director. His main research work is related to participation, conflict and governance in land use, pastoralism and extensive farming.

Fernando Pulido is a Professor of Biology and Forest Conservation at the University of Extremadura (Spain) since 1999. He has 25 years of expertise working with the ecology, biodiversity, management and conservation of Mediterranean forests. His research focuses on alternatives for improving the management of working forest landscapes including the use of agro-silvopastoralism for fire prevention. He is the Director of the Institute for Dehesa Research (INDEHESA).

Isabel Ferraz-de-Oliveira has a PhD in Animal Nutrition and is an Assistant Professor at the University of Évora in Portugal. She is an integrated member of the Mediterranean Institute for Agriculture, Environment and Development (MED) and her research has been mostly devoted to animal feeding and nutrition in extensive production systems. More recently she has been working in High Nature Value farming systems and in the co-construction of outcome-based agri-environmental schemes to improve the sustainability of silvopastoral systems.

Ali Chebil is an Agricultural Economist currently working as a Senior Researcher at INRGREF in Tunisia. He has more than 15 years of experience in research, teaching at the university level and supervision of graduate students. Ali's research interests include impact assessment of technologies, economics of natural resources, productivity analysis and economic impact of climate change. He has published more than 40 publications.

Mariem Khalfaoui is a Tunisian Agricultural Economist working on forest economics and management. She is concerned about climate change impacts and environmental issues and has participated in several national and international workshops and conferences. She also has vast experience in participating and managing international research projects.

Hamed Daly-Hassen is a Tunisian Professor in Rural Economics at INRAT, with a record of publications on forest's economics, especially on the economic valuation of forest ecosystems. He taught in different institutions at both national and international levels and contributed to international and European research projects. He is currently the Director General of the National Observatory of Agriculture.

Aymen Frija is an Agricultural Economist with ICARDA's Social, Economic and Policy Research Team. He is also the coordinator of ICARDA's activities in Tunisia and Algeria. His current research interests focus on farm modelling, farm efficiency and productivity analysis, agricultural water management instruments, institutional performance analysis and the economics of conservation agriculture. Earlier in his career, Dr. Frija was a post-doctoral researcher at Ghent University in Belgium, specializing in agricultural water policy analysis in developing countries. He was also an Assistant Professor and Researcher at the College of Agriculture of Mograne, Carthage University in Tunisia.

Mariem Sghaier is currently a PhD student at Avignon University in France, member of the research team at Arid Regions Institute (IRA) of Medenine, Laboratory of Economy and Rural Societies and the International Center for Agricultural Research in the Dry Areas – ICARDA. She got her master's degree in Economic Intelligence from the University of Poitiers France. Her areas of interest and research include territorial development and biostatistics.

Mondher Fetoui is an Associate Professor (Agro-Socio-Economy, Geography) at the Arid Regions Institute (IRA) of Medenine, Laboratory of Economy and Rural Societies. He is distinguished for his research and teaching on natural resources management, geographic information systems (GIS) and spatial analysis, environmental information systems and models, development of decision-making tools for rural development and assessment/monitoring of desertification risks, analysis of agropastoral systems vulnerability and sustainability, analysis of livelihood vulnerability, value chains, social networks analysis and stakeholders analysis, social innovation platforms and inclusive development. He has contributed to several international research projects since 2000. He has authored and co-authored more than 40 peer-reviewed journal articles and book chapters. He received his PhD in Geography and Development from Paul-Valéry III University, Montpellier, France.

Boubaker Dhehibi is an Agricultural Resource Economist at the International Center for Agricultural Research in the Dry Areas – ICARDA. He is distinguished for his research and teaching on production economics, economics of climate change, economics of natural resources management, value chain analysis, economics of development and competitiveness and productive analysis of the agricultural sector in the MENA region. Prior to joining ICARDA, he had worked at the National Agricultural Research Institute of Tunisia. Dr. Dhehibi has authored more than 60 peer-reviewed journal articles and book chapters. He received his PhD in Economics from the University of Zaragoza, Spain.

Mongi Sghaier is a Specialist in Agricultural and Natural Resources Economics. He is the founder and former Chief of the Laboratory "Rural Economy and Societies in arid regions LESOR" and he coordinates and takes part in several programs and international research projects in the field of natural resources management and local development. He has published many publications in peer-reviewed journals and books.

Athanasios Ragkos is an Agricultural Economist, Associate Researcher at the Agricultural Economics Research Institute. His main fields of expertise are non-market valuation techniques, optimization models, efficiency analysis and economics of rural development. He has published more than 80 papers in peer-reviewed journals and at international conferences and has participated in more than 30 projects concerning agricultural and livestock production economics, rural development and environmental economics.

Stavriani Koutsou is a Professor at the International Hellenic University. She is an Agricultural Economist specializing in human geography. Her scientific interests include collective actions in agriculture, rural development and the evolution of rural societies. She has authored numerous papers in peer-reviewed journals and conferences and has coordinated research projects regarding social capital in Greece and collaboration in rural areas.

Claudio Porqueddu holds a degree in Agricultural Sciences and is a Senior Researcher at the Italian National Research Council (CNR) – Institute for Animal Production System in Mediterranean Environment (ISPAAM). His research is focused on the agronomy of pasture and forage crops in Mediterranean areas, evaluation and multiplication of native pasture species, sustainability of extensive farming systems and cover crops management. He is currently the Coordinator of the FAO-CIHEAM Network on Pastures and Fodder Crops. He is a member of the Scientific Advisory Board of the European Grassland Federation.

Antonello Franca holds a PhD in Crop Productivity, is a researcher at the National Research Council and conducts research at the Institute for Animal Production in the Mediterranean Environment. His scientific interests concern the ecophysiology of Mediterranean pastures, the sustainability of silvopastoral systems and the multiple roles of pastoralism for the sustainable use of rural territories. He's a member of the Italian Network on Pastoralism.

Giovanna Seddaiu is an Associate Professor of Agronomy and Crop Science at the University of Sassari, Italy. Her research interests concern the analysis of the interrelations between Mediterranean cropping systems, the associated ecosystem services and their environmental impacts in terms of nitrate pollution and GHG emissions, with a special focus on large-scale grazing systems. She coordinated the project PASCUUM on ecosystem services of Mediterranean agro-silvopastoral systems. She is a member of the European Agroforestry Federation.

Pier Paolo Roggero holds a PhD in Agronomy, is Full Professor at the Faculty of Agriculture of the University of Sassari, Italy and Director of the Desertification Research Centre. His research focus includes agroecology, agro-silvopastoral systems and climate change adaptation.

Acknowledgements

Nick Parrott of TextualHealing.eu revised the texts of all the chapters in this book, and showed diligence, competence and professionalism in securing the quality of the language and a consistency of style across the book.

Introduction

Teresa Pinto-Correia, Rufino Acosta-Naranjo,
Gerardo Moreno and Maria Helena Guimarães

I.1 Why a book about silvopastoral systems?

In this first quarter of the 21st century, there is a lively debate in both policy and scientific circles about new pathways for a transition to sustainable farming that respects the need to preserve the planet's natural resources while also achieving food security for a growing world population (Pe'er et al. 2020; Rasmussen et al. 2018; Rockström et al. 2017). This debate and the urgency of solutions are reinforced by the growing evidence of climate change which is affecting all terrestrial ecosystems. The scenarios are extremely worrying for some regions, in particular in the Mediterranean basin, where land degradation and water scarcity are already causing concern and worsening every year (Garcia et al. 2014; Leal et al. 2019; Pecl et al. 2017) The European Environmental Agency noted that 25% of land in Southern and Eastern Europe was at high or very high risk of desertification in 2017 – and predicted an increase of 11% in desertification in just 10 years (E.E.A. 2019). Several of the UN Sustainable Development Goals (UN 2015) call for a transformative change of agricultural production, in particular making sustainable use of ecosystems (SGD15), achieving sustainable production and consumption (SDG12) and taking action to combat climate change and its impacts (SDG13). In Europe, the Green Deal (European Commission 2019), and the Farm to Fork Strategy as well as the Biodiversity Strategy, define ambitious goals for a transition to sustainability in farming, a process which the Commission has already started implementing (E.C. 2020a, 2020b). This reflects the recent paradigm change in European policy, from 'land sparing' to 'land sharing': a shift from the decades-long policies that favoured the abandonment of agriculture on less productive land and focused on areas of 'higher potential'. There is now growing interest in conserving farming practices on all types of land, since it recognized that on less productive land, HNV (High Nature Value) farming can support the achievement of environmental objectives.

Two polarized versions of the way forward have been under debate (Marsden 2012): on one side there is the bio-economy, which is the dominant paradigm of agricultural modernization and arguably espouses a weaker

DOI: 10.4324/9781003028437-1

conception of sustainability (Rasmussen et al. 2018); on the other side, there is the eco-economy, which embraces a holistic, stronger view of sustainability that is firmly embedded in agroecological knowledge (Biely, Maes, and Van Passel 2018). Agroecology involves respecting and integrating ecological processes in production systems, so as to improve efficiency in the use of production factors and secure ecosystem functions – which are having to adapt to the effects of climate change (González de Molina 2011; Nichols et al. 2015; van der Ploeg et al. 2019). Eco-economy strategies and practices seek to intensify the ecological processes that underpin long-term agricultural productivity, enhance soil fertility, and conserve water and organic matter. The eco-economy paradigm seeks to integrate local knowledge about local agro-ecologies in processes of scientific knowledge co-production in order to provide solutions for local and regional food security (or as many would argue food sovereignty). This relies more on the ecological intensification of food production than on agrochemical intensification (Bommarco, Kleijn, and Potts 2013). Despite the many claims about the multiple benefits of agroecological approaches (see for example van der Ploeg et al. 2019), it is the intensive and specialized production systems that continue to be the focus of research and research funding, leading to the development of increasingly sophisticated analytical tools (Vanloqueren and Baret 2009). Extensive, multifunctional production systems, with (seemingly) lower productivity levels, have remained the poor cousins, in terms of dedicated research development and funding.

Nevertheless, these mixed farming systems occupy large areas of the world under natural conditions which would not support intensive specialized uses. They have been developed through the strong adaptive capacities of communities, especially in regions facing severe biophysical constraints, mainly in arid or semi-arid climates. In such regions, multiple-use systems have developed as a strategy to cope with scarcity and as a source of resilience. Silvopastoral systems are examples of such multifunctional production systems (Ferraz-de-Oliveira et al. 2016; Pinto-Correia et al. 2011; Torralba et al. 2018). Many are shining examples of the possibilities of maintaining and developing the links between ecology, the economy and community development that could, if given the attention they merit, substantially shift the balance of current farming paradigms towards more integrative approaches (Pe'er et al. 2020). They are also living examples of humans' adaptive capacity to the environmental constraints of their environment. The long-term view of sustainability, or stronger sustainability, creates grounds for renewed interest in the viability of such systems (Biely et al. 2018).

The silvopastoral systems of Iberia, and around the Mediterranean basin, are interesting to study as they are paradigmatic examples of European and world silvopastoral systems. This is due to several reasons:

- Unlike many other of Europe's silvopastoral systems, they persist as economically viable and relevant land-use systems;

- They occupy a large share of the land in the regions where they are found (in total, circa 4 M ha);
- They include a large variability in soils and morphology, as well as in structure and composition, e.g. tree density, vertical vegetation structure and state of conservation, as a result, they are hotspots of Mediterranean biodiversity (Moreno et al. 2016; Torralba et al. 2018);
- They maintain production on marginal agricultural land, with an extreme scarcity of natural resources, typically with shallow soils with a low organic matter content and a dry Mediterranean climate (Guerra and Pinto-Correia 2016);
- They coexist on private, public and communal land;
- In recent years montados and dehesas have proven to be resilient to the impact of known and anticipated climate change scenarios, to a level that cannot be expected in many other land use systems and;
- In the long term, these systems have shown themselves to be highly resilient to the impact of extreme natural or socio-economic phenomena and long-term changes in the agricultural and forestry sectors (Jepsen et al. 2015; Pinto-Correia and Fonseca 2009).

Studying the montados and dehesas thus allows us to examine the key challenges and resilience strategies in what can safely be said to be the best performing silvopastoral and the most extensive high nature value and cultural agroforestry system in Europe (Rolo et al. 2020).We find the silvopastoral systems of Iberia to be not only a living example of resilient farming on the margins, but also treasures that deserve to be better understood, in order to be better managed for the future. They are also a potential source of knowledge that could be applicable to current (and future) silvopastoral territories in other regions across the world.

I.2 Why a book about governance?

The preservation of the multifunctional production systems of silvopastoral systems is increasingly threatened by the effects of global economic forces and socio-cultural changes all over the planet and is reinforced by the impact of climate change (Moreno et al. 2014; (Pinto-Correia and Azeda 2017; Pinto-Correia, Primdahl, and Pedroli 2018). Agricultural intensification and specialization are increasingly driven by external financial interests and global market forces, and consequently the patterns of farm ownership, employment and production are changing at an unprecedented scale (Pinto-Correia et al. 2018; Silveira et al. 2018).

In southern Europe, the processes of agricultural intensification and specialization have come later than in western and northern Europe but are ongoing today in a dramatically strong and unprecedented rhythm (Exposito et al. 2020; Ortiz-Miranda et al. 2013). This is exacerbated by a pronounced and accelerating degree of appropriation of territorial and natural capital, resulting in *de-territorialization* (van der Ploeg et al. 2015; Rodríguez-Cohard et al. 2018). By de-territorialization we mean a *multiple decoupling process* between economic activities, the community and the territory:

- *Ecological decoupling*, as new technical and scientific knowledge, appear to make baseline environmental conditions in each context less relevant;
- *Socio-cultural decoupling*, as progressive linkages to urban and global networks, stimulate the dissemination of urban values and discourage those associated with rural ways of life and local contexts, and;
- *Institutional decoupling*, in which a command-and-control administration system devalues or dismantles proximity-based decision-making mechanisms (Ferrão 2016; Silveira et al. 2018).

Farm structures are also changing, as family-farming is being replaced by large corporate enterprises. The same is true of employment patterns, with local labour being replaced by external, and often, seasonal labour. New owners and intensive uses make the exploitation of the externalities to farming (mushrooms, hunting, bee-keeping) more difficult for local people. Links to local communities are being weakened. The homogenization of the landscape and the spread of fencing affect the well-being and the attachment of local communities to places and reduce the potential for tourism and leisure. As a consequence the *territorial capital* of the region is depleted (Berbel et al. 2019; Corbera et al. 2019).

Territorial capital refers to all the geographically bounded assets of a territorial nature on which the competitiveness potential of regions and places is based. This can be broken down into sub-dimensions: the economic, human and labour markets, social, institutional, environmental, cultural and symbolic aspects (Camagni 2019). Within these, *social capital* is particularly relevant as it is key in development: social capital refers to the intangible resources that reside in relationships that enable the creation of value, achieving goals and getting things done (Rivera et al. 2019). The environmental sub-dimension corresponds to *natural capital*, and encompasses natural assets which play the role of providing natural resource inputs and environmental services for economic production: natural resources stocks (water, biodiversity), land, and ecosystems (Corbera et al. 2019; Rasmussen et al. 2018). Territorial capital, as the interactive sum of natural and social capitals, is thus crucial for the development and resilience of a region and its capacity to progress towards sustainability.

In face of these pressures, finding new pathways for farming strategies and practices is highly challenging. The interaction between societal and political actors changes over time as the challenges faced by society change. In the case of silvopastoral systems, besides their inherent qualities and the trends, they are subject to, there are now also new societal demands as they can provide multiple ecosystem services and public goods. This demand increases the complexity of their management, as new actors and expectations need to be integrated.

As Elinor Ostrom (2005) argues, simple designs such as private property, government ownership or community organizations are not the solution to the governance of the complexity of problems we face today. Hodge (2013) adds that relying on the role of governments, self-governing networks or market relations does not seem to be the proper pathway. Learning how to live with uncertainty and finding new governance mechanisms and institutional arrangements can perhaps increase resilience. Despite the importance of various combinations of networks and market relations in governance, a wider range of interactions, aimed at securing collective interests, need to be taken into account, including interactions between public and private actors (Kooiman 2003). In this perspective, attention needs to be paid to the art of steering interactions and establishing the foundations for the complex set of relationships that emerge from governance models. Governance mechanisms can foster innovation and facilitate the processes of transition towards sustainability that need to occur. The contribution from social sciences is thus of uppermost importance.

Surprisingly, the social and institutional drivers of the changes taking place and also the possible solutions for launching a sustainability transition, remain almost unseen, by science and by decision-makers at different governance scales. The existing scientific literature on dehesa/montado pays little attention to its social dimension. While there is already a tradition of scientific research on the ecology and productive capacity of silvopastoral systems, texts on the social fabric of the dehesa are very limited, especially in the last two decades (Fagerholm et al. 2016; Garrido et al. 2017). Social sciences seem to have moved away from the dehesa/montado, perhaps with the exception of geography's focus on the subject of landscape values (Pinto-Correia et al. 2019).

By studying governance this book compiles knowledge and initiatives as experiments and learning opportunities that can be incorporated into the future management of silvopastoral (and other) systems, increasing their capacity to face up and adapt, to changes.

I.3 What are silvopastoral systems?

Silvopastoral systems are land-use systems that combine open tree cover, in varying densities, with grazing in the under-cover, often with pastures and dispersed patches of shrub (Figure I.1). They are also called wood pastures, or rangelands. However, these two last terms do not fully cover the complexity of the silvopastoral systems and their management. Grazing can be by different types of livestock. Occasionally some crops are grown in the under-cover, as a complementary, though secondary, farming practice. These systems have developed as adaptive systems to the scarcity of natural conditions and are lately gaining prominence as a sustainable and climate-resilient livestock production system (Moreno and Rolo 2019). The agricultural component, in rotation with grazing, has generally been stronger in the past and

Figure I.1 Montado close to Évora, Portugal.

tends to be minimized by the intensification of crop production in other regions of the world, as well as the growth of global markets and global trade in fodder. The forestry component needs to be articulated with the livestock grazing and the pastures' carrying capacity, in order to secure the balance of the system and maintain the productivity of both components (Figure I.2) (Godinho et al., 2016; Moreno et al. 2014).

Silvopastoral systems also have the ability to produce high-quality food (meat, milk, cheese) together with recreational and cultural services, which can act as important drivers for their economic revitalization. These systems support multiple activities (firewood, hunting, bee-keeping, gathering mushrooms and wild plants, conservation of local breeds, bird-watching and recreation) and bring together private and public uses. In addition, they support exceptionally high levels of biodiversity and have an outstanding

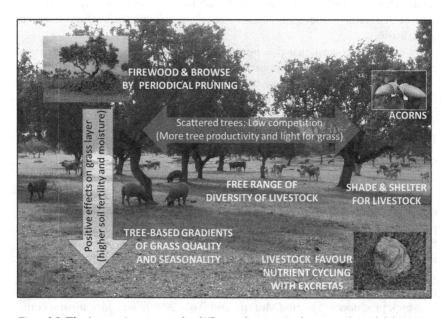

Figure I.2 The interaction among the different elements in the montado and dehesa.

heritage value, providing unique landscapes that are the source of local and regional cultural identities, making silvopastoral systems unique cultural icons (Fagerholm et al. 2016; Surová and Pinto-Correia 2016). Further, these systems also provide an effective synergy between climate change adaptation and mitigation, by directly contributing to the sequestration of greenhouse gases while at the same time buffering livelihoods against variability and recurrent drought as the sheltering effect of trees and because trees are less affected by variability in climate than grass pastures.

Silvopastoral systems are one particular type of a larger category generally referred to as agroforestry systems. Agroforestry can be defined as integrated land-use systems in which trees are grown in combination with agriculture on the same land or as the practice of deliberately integrating woody vegetation (trees or shrubs) with crop and/or animal systems to benefit from the resulting ecological and economic interactions and providing agricultural products (den Herder et al. 2017). There are several categories of common agroforestry practices worldwide, including wood pastures, hedgerows, windbreaks, riparian buffer strips, intercropped and grazed orchards, grazed forests, forest farming, silvoarable and silvopastoral practices, alley cropping, alley coppices and woodland chickens (Mosquera-Losada et al. 2012)

Silvopastoral systems can have different compositions and structures which can be found in many regions of the world (Aronson, Santos-Pereira, and Pausas 2009). They are long-established, although some modern silvopastoral systems, frequently based on high-quality timber species, have appeared in recent decades (Cubbage et al. 2012; Pantera et al. 2018). They are mostly extensive systems and the multiple habitats that they provide lead them to be considered as High Nature Value Farming systems (Lomba et al. 2020; Moreno et al. 2018; Oppermann et al. 2012; Pinto-Correia, et al. 2018)

Silvopastoral systems occupy land considered marginal to agriculture, in southern Europe, as well its northern and eastern peripheries. They are particularly evident in the Mediterranean region and others with a seasonally dry climate (Campos et al. 2013; Pinto-Correia and Vos 2004; Hartel and Plieninger 2014). According to den Herder et al (2017), grazed wood pastures cover about 15.1 million hectares in the EU, corresponding to about 3.5% of the EU's total landmass and 15% of European grasslands, and up to 35% of actually grazed land. The largest areas of grazed wood pastures are concentrated in Spain, Portugal and other Mediterranean countries (den Herder et al. 2017). Silvopastoral systems are also widespread in other Mediterranean regions, such as 'WANA' region (Western Asia and Northern Africa) as well as other semiarid and sub-humid regions in Africa, India and the Americas. Globally they account for an estimated 450 million ha (Nair 2012, Moreno and Rolo, 2019).

In this book, we focus on the montado (Portugal) and dehesa (Spain) as highly paradigmatic silvopastoral systems of the Mediterranean basin, which occupy 4 M ha (Ferraz-de-Oliveira, Azeda, and Pinto-Correia 2016; Guerra,

Pinto-Correia, and Metzger 2014; Moreno et al. 2014; Pinto-Correia, Ribeiro, and Sá-Sousa 2011).

In montado and dehesa, the tree cover is composed of cork and holm oaks (*Quercus suber* and *Quercus Rotundifolia*). Cork oaks are more frequent in areas with relatively higher precipitation and humidity, and holm oaks in dryer environments. In some areas with higher humidity, there are rare patches of mountain oak (*Quercus pyrenaica*). The tree layer has a variety of spatial configurations, with fuzzy boundaries between patches of different densities, resulting in a highly heterogeneous pattern at the landscape level. The shrub layer includes a combination of serial and climax shrub species. At the lower level, natural pastures are the most common, although these are sometimes improved and sometimes cultivated (in rotation). Livestock grazing has, in the past, been dominated by sheep and the Iberian pig, but cattle are increasingly dominant. As a response to market drivers and the homogenization of production models, autochthonous cattle breeds, small and less demanding in feed, are increasingly being replaced by heavier, but also more demanding breeds, imported from northern Europe. Goats were used when shepherds were common in montado farm units – but today are increasingly rare and concentrated in small intensive goat farms.

The main products of the Iberian silvopastoral systems are cork and meat. Cork is only important when the biophysical conditions are favourable for cork trees. In holm oak areas, in Spain and also Portugal, the production of Iberian pig is more characteristic, with high added value due to the resulting food products. Otherwise, the meat is from cattle and sheep. In some regions, goats and sheep are kept for their milk. In addition, there is the collection of wild edible plants, specially asparagus (Acosta-Naranjo, Guzmán-Troncoso & Gómez-Melara 2020) aromatics, mushrooms and bee-keeping, as well as hunting activities, all of which may bring extra income, some to landowners and some to other local users.

Silvopastoral systems have become economically marginal in many other countries, and the recent interest in these systems comes from their nature and cultural values, which are becoming progressively important (Moreno et al., 2018).

I.4 The *threats* to silvopastoral systems

Many traditional silvopastoral systems face a range of threats that compromise their long-term persistence, because of changes in, or a loss of, multiple management practices that concern both the grazing schemes and forestry practices – such as thinning, pruning and shrub clearing. Unfortunately, most of the traditional practices have largely vanished in more developed countries: firewood has been replaced by fossil fuels, fodder browsing and forages by commercial concentrates, herders by fences, transhumance by continuous grazing, local cattle breeds by imported breeds.

In Iberia, the analysis of different data sources shows a marked decrease in the total area or in the density of trees. The total extent of the montado land cover is decreasing every year, through a reduction of the total area covered, combined with a decrease in tree density and an increase in the size of tree openings. For instance, in southern Portugal, since the early nineties, there has been a reduction of approximately 5000ha of tree cover per year (Godinho et al. 2016). In Spain, while the total surface of dehesa remains quite constant, there has been a slight loss of tree cover (Plieninger, Rolo, and Moreno 2010). We suspect that the actually managed (grazed) area is decreasing, but no data are available. There is thus an issue not just of area shrinking but also a lack of tree renewal in the remaining area, compromising the natural renovation of the system (Godinho et al. 2016; Pinto-Correia and Godinho 2013; Plieninger, Rolo, and Moreno 2010). Indeed, some authors have hypothesized that silvopastoral systems inevitably evolve to treeless pastures in the long term, as grazing is not compatible with tree regeneration in wood pastures unless it is seasonally and/or periodically restricted to allow regeneration. In the meantime, the abandonment of silvopastoral practices results in new risks and costs, such as wildfire, loss of biodiversity, and even the reduction of blue water yield (Varela et al. 2020).

A major challenge that silvopastoral systems will face in the coming decades is the need to provide sustainable pasture yields while conserving ecosystem services (Moreno and Rolo 2019). But, there are many other major challenges. So far, the acknowledgement of the outstanding potential of silvopastoral systems has only led to the sustainable management of montados and dehesas in limited and specific cases.

I.5 Why we focus on the governance of silvopastoral systems?

The *direct decisions* on silvopastoral management are taken by the managers, primarily landowners, sometimes others who are in charge. The sustainability of these systems can be measured by biophysical and structural indicators, but the drivers that explain how they evolve are social and institutional. *Decisions are complex* and depend not only on the individual profile of the land manager but also on the market and public policies, and on the family context and social norms and practices, e.g. institutions and institutional arrangements (Pinto-Correia et al. 2019; Pinto-Correia and Azeda 2017). Thus, there are *other people involved*, at the very local scale and in multi-scalar governance mechanisms, who influence the way management decisions are taken.

Furthermore, increased societal expectations on silvopastoral systems lead to a growing protagonism of other actors. As a result, understanding the management practices and motivations of landowners and producers, as well as the ways in which the many other stakeholders (individual and institutional) seek to shape the management of these silvopastoral systems is absolutely central to understanding what is at stake in securing their future. Competing interests can not only exist between different interest groups but also within the same interest group. These conflicting positions and the resultant alliances and power relations can be very fluid (Primdahl 2018). There is also a variety of fundamentally different institutional arrangements and combinations of collective and individual decision-making that drive the practice and use of the silvopastoral systems, with a potential to affect the landscape (Hodge 2016). The governance of these systems has become more complex, with a need to integrate multiple actors and interests, different sectoral and spatial policies and the boundaries between public and private sectors, which have become progressively more blurred. New management solutions are needed.

The importance of the human factor in maintaining these systems, the decisions taken (and the underlying reasons for these decisions), the institutional arrangements and the multi-scalar governance mechanisms are rarely addressed in the literature. In order to understand what is at stake today and to safeguard the future of these agro-ecosystems, it is essential to address the management practices and motivations and the way in which the many individuals and institutions involved influence the management and functioning of these systems.

By looking at current governance questions in the silvopastoral systems of Iberia and Mediterranean Europe, we intend to reveal evidence and discuss the issues that are of relevance to all types of silvopastoral systems around the world, albeit in different contexts.

I.6 The main concepts used throughout this book

The analytical perspective in this book is strongly inspired by diverse and complementary approaches to agency and governance in farming and the countryside, governance and institutional arrangements (Hodge 2016) understanding institutional diversity (Ostrom 2005), farmers' profiles and how they relate to the wider countryside (Primdahl et al. 2018; Primdahl and Kristensen 2016) the farming systems perspective, farmers' motivations, strategies and resilience (Darnhofer et al. 2016), the specific characteristics of Mediterranean agriculture (Ortiz-Miranda et al., 2013), and the spatial and temporal continuum among agriculture and forest lands that feature silvopastoral systems (Fonseca et al., 2019; Hartel and Plieninger, 2014; Pinto-Correia et al., 2018).

Governance is an emergent concept in the socio-political sphere. The interaction between societal and political actors has modified over time as the challenges society faces become more complex (Ostrom 2005, 2010). Governance can be approached from different angles; therefore, we need to define the way the concept is used in this book. Governance is about self-organizing networks (Hodge 2013, 2016). Governance means steering multiple agencies, institutions, and systems that are autonomous from each other in terms of operations but structurally coupled through diverse interdependencies. It is more about steering and negotiating than commanding and controlling, and is about networks and reciprocal relationships. Focusing on governance means a broad perspective to conflict resolution and decision-making about resources, a territory or a strategy.

Actors are all those who participate in an action or process which directly affects silvopastoral system. *Stakeholders*, by contrast, are all those who have a stake in the system, even if they cannot directly influence the system (Pinto-Correia and Kristensen 2013). They are people or organizations with an interest or concern in something, in this case, a land-use system and the resulting land cover, its composite elements and the resulting landscapes.

Institutions are the rules through which groups, such as families, villages, communities, companies, and states, regularize and channel individual actions and interactions (Ostrom 2005, 2010). Institutions may be formal rules, or they may be informal norms and conventions. Many are informal, defining the boundaries of what is an acceptable behaviour. They may be written, or just known and accepted. Institutions limit the behaviour of each individual but they also support action and coordination. They shape our collective understanding of natural and social processes and the values we attribute to them (Vatn 2005). An institution is a much broader and different concept from an organization. Institutions are dynamic. Formal and informal institutions change in response to changing incomes, preferences and values, technology and social conventions, composition, and influences. Yet they are held steady by kinship ties, traditions, etc. The way institutions interact with each other are *institutional arrangements*: formal and informal cooperation structures that support and link public and private institutions and or organizations.

Organizations are well-defined units, an organized group of people with a particular purpose, such as a business or government department.

In order to understand what 'goes on' in the countryside, and thus on silvopastoral systems, it is necessary to focus on the institutions that influence the use and management of the land. In this instance property rights are central. *Property rights* define how people can benefit from and can make changes to, the resources available, and are legally defendable rights that define how resources may be used and how their benefits and costs may be allocated.

The notion of *resilience* was originally conceived of in relation to ecological systems and the adaptive capacity of such systems to persist in the face of significant natural changes such as fires or floods: "ecological resilience refers to a system's capacity to reorganize under change to reach a new equilibrium while retaining the same essential functions" (Robinson and Carson 2015). Resilience is in line with the vision of strong sustainability which goes beyond the original social, economic, environmental and institutional requirements of the concept as it also has social-ecological aspects (Folk et al. 2016). The concept has been evolving to also include the dynamics of the social sphere, and the notion of 'social-ecological resilience allows for the conceptualization of interactions between the socio-economic and ecological domains. The interaction between the social and the ecological are seen as dynamic, with any resultant resilience emerging from the configuration of relations between the two domains, that may incorporate both human and non-human elements (Darnhofer et al. 2016).

By *ecosystem services* we mean the services provided and the benefits people derive from these services, both at the ecosystem and the landscape scale, including public goods related to the wider ecosystem functioning and society well-being (Haines-Young and Potschin 2018; MA 2005).

I.7 The objectives and structure of this book

This book is about the resilience of silvopastoral systems now and in the future. As such, it is about people. The goal is to fill the gap in the knowledge on silvopastoral systems and their changing trends, by adding the *human dimension*, with enough detail to draw inferences about the new governance solutions that are needed to address the multiple challenges faced by silvopastoral systems.

The management arrangements for silvopastoral systems currently applied are largely outdated and either not functioning or inadequate for keeping the balance between the multiple components. Understanding the practices and motivations of landowners and other actors, as well as the ways in which the many other stakeholders (individual and institutional) seek to shape the management of these silvopastoral systems is absolutely central to understanding what is at stake today and in securing the future of these systems. This led us to first of all (in Section A) examine and analyze the *many actors* who play in these systems, their motivations, and goals. We also look at how these actors are *organized and interact*, and the subsequent conflicts and synergies, in Section B. Finally, in Section C, we discuss the *collective actions* and *governance mechanisms* that are in place and discuss innovative solutions that could enhance the resilience of these systems and better respond to the new societal demands for ecosystem services. All in all, we look at ways to enhance such systems' contribution to ensuring that these systems can become examples of farming practices that are adapted to new societal demands on ecosystem services.

While the chapters are written by different authors, they are interrelated and complementary. We have tried to create coherence in this book through meetings with the authors, at which we discussed the book's structure, the goals and guiding lines, as well as the content of the chapters. The editors have jointly discussed and commented on all chapters several times throughout the book's preparation. We hope in this way to offer the reader a coherent book, in which the knowledge provided in each chapter is complementary to the knowledge of the other chapters.

The structure of the book reflects these goals.

Following this introduction, this book is divided in three sections: (A) people and institutions; (B) the institutional framework; (C) governance models. Each section, composed of several chapters, draws on empirical work about the Iberian montado and dehesa as well as from other similar systems in the Mediterranean, both on the northern and the southern sides, in order to broaden its scope and cover a wider range of situations and examples. Some of the chapters rely more strongly on empirical findings and current experiences, others on a literature review and reflection by the authors over many years working with these systems. The conclusions sum up the most relevant findings from each chapter and discuss how research can progress so that new scientific approaches and evidence can support better-adapted governance models of silvopastoral systems in order to face future challenges.

More specifically, Section A contains 6 chapters. Chapter 1 provides an overview of the landowners in montado and dehesa (i.e. private, state, organizations) and how different typologies of ownership influence management at the property and landscape levels. Chapter 2 details the actors making everyday decisions at the farm level and discusses the multiple factors influencing farm management. Chapter 3 focuses on all the other actors who do not own the land or directly manage it but whose practices, motivations and perspectives influence the current management of these systems. Chapter 4 gives a special focus to the large private companies that commercialize cork, a key product for the systems' perpetuation, and evolution. Society also plays an increasing role in agriculture. Chapter 5 explores the issue of gender and the role of women in the management and governance of silvopastoral systems. Such a focus is important due to the increasing acknowledgment of the influence that gender issues can have on the resilience of such systems. Closing this section, Chapter 6 explores the societal values on the silvopastoral systems today, and their actual and potential ability to influence policies and practice.

Section B consists of 4 chapters. It starts by detailing how montados and dehesas are defined, in legislative and regulatory contexts. We did not envisage such a chapter at the start but finally understood that the many contradictions and conflicts identified at different governance scales result from inconsistencies in the definitions, and therefore decided on the need for such a chapter. Chapter 8 is about property rights and rights of use. This chapter explores the different combinations of rights that exist and that are being modified by the entrance of new owners, aspirations, and business models. The relevance of public policies is explored in Chapters 9 and 10 which consider the distinct political context of Mediterranean countries within and outside Europe, Chapter 9 details how public intervention is structured in the Maghrebian context while Chapter 10 discusses the different EU and national policy frameworks influencing silvopastoral systems in the montado and dehesas while arguing for the need of coordination and a customized policy framework for such systems.

Section C contains six chapters. Considering the importance of the interactions between public and private actors, Chapter 11 focuses on the conflicts between the actors described in Section A, as well as, the policies described in Section B. The attention given to conflicts is explained by the current challenges that silvopastoral systems face and the key role that such tensions play. Nonetheless, collaboration and collective actions exist and are perhaps pathways towards the resilience of these systems. Chapter 11 discusses the conflicts that exist or potentially may develop in these systems. Chapter 12 focuses on examples of collective governance that are in place in the Maghreb. Chapter 13 explores the structures that are in place to promote dialogue, active participation and coordination. Chapter 14 provides an overview of collective initiatives that foster the adoption of innovation since innovation and transitions towards adaptation play a key role. This section ends by revisiting silvopastoral management practices that have almost disappeared and that are being readopted due to their relevance for resilience. Chapter 15 discusses transhumance and Chapter 16 pastoralism, both of which play a role in and influence territorial management.

References

Acosta-Naranjo, R., A. Jesus, and J. Gómez-Melara. 2020. "The Persistence of Wild Edible Plants in Agroforestry Systems: The Case of Wild Asparagus in Southern Extremadura (Spain)." *Agroforestry Systems* 0123456789:2391–2400.

Aronson, J., J. Santos-Pereira, and J. Pausas. 2009. *Cork Oaks Woodlands in the Edge:Ecology, Adaptive Management and Restoration.* 1st ed. edited by J. Aronson, J. Santos-Pereira, and J. Pausas. Washington D.C.: Island Press.

Berbel, J., A. Expósito, and M. M. Borrego-Marín. 2019. "Conciliation of Competing Uses and Stakeholder Rights to Groundwater: An Evaluation of Fuencaliente Aquifer (Spain)." *International Journal of Water Resources Development* 35(5):830–46.

Biely, K., D. Maes, and S. Van Passel. 2018. "The Idea of Weak Sustainability Is Illegitimate." *Environment, Development and Sustainability* 20(1):223–32.

Bommarco, R., D. Kleijn, and S. G. Potts. 2013. "Ecological Intensification : Harnessing Ecosystem Services for Food Security." *Trends in Ecology & Evolution* 28(4):230–38.

Camagni, R. 2019. "Territorial Capital and Regional Development: Theoretical Insights and Appropriate Policies." P. 688 in *Handbook of Regional Growth and Development Theories,* edited by R. Capello and P. Nijkamp. Handbook of Regional Growth and Development Theories.

Campos, P., J. L. Oviedo, P. Starrs, M. Díaz, R. Standiford, and G. Montero. 2013. *Mediterranean Oak Woodland Working Landscapes. Dehesas of Spain and Ranchlands of California.* 1st ed. edited by P. Campos, J. L. Oviedo, P. Starrs, M. Díaz, R. Standiford, and G. Montero. Dordrecht,NL: Springer Netherlands.

Corbera, E., D. Roth, and C. Work. 2019. "Climate Change Policies, Natural Resources and Conflict : Implications for Development." *Climate Policy* 3062(19):51–57.

Cubbage, F., G. Balmelli, A. Bussoni, E. Noellemeyer, A. N. Pachas, H. Fassola, L. Colcombet, B. Rossner, G. Frey, F. Dube, M. L. de Silva, H. Stevenson, J. Hamilton, and W. Hubbard. 2012. "Comparing Silvopastoral Systems and Prospects in Eight Regions of the World." *Agroforestry Systems* 86(3):303–14.

Darnhofer I., C. Lamine, A. Strauss, and M. Navarrete. 2016. "The Resilience of Family Farms : Towards a Relational Approach." *Journal of Rural Studies* 44:111–22.

E.C. 2020a. *Farm to Fork Strategy.* Brussels.

E.C. 2020b. "Just Transition Mechanism." *European Commission.* Retrieved (https://ec.europa.eu/info/strategy/priorities-2019-2024/european-green-deal/actions-being-taken-eu/just-transition-).

E.E.A. 2019. *Climate Change Adaptation in the Agriculture Sector in Europe.* Luxembourg.

European Commission. 2019. *The European Gree Deal, COM(2019) 640 Final.* Bruxels.

Exposito, A., F. Beier, and J. Berbel. 2020. "Hydro-Economic Modelling for Water-Policy Assessment Under Climate Change at a River Basin Scale : A Review." *Water* 12(1559).

Fagerholm, N., C. M. Oteros-Rozas, E. Raymond, M. Torralba, G. Moreno, and T. Plieninger. 2016. "Assessing Linkages between Ecosystem Services, Land-Use and Wellbeing in an Agroforestry Landscape Using Public Participation GIS." *Applied Geography* 74:30–46.

Ferrão, J. 2016. "Ruralidades e Território No Capitalismo Contemporâneo: Uma Visão de Longa Duração Sobre Portugal." Pp. 229–45 in *Sociologia e Sociedades*, edited by F.. Machado, A. Nunes de Almeida, and A. F. Costa. Lisbon: Ed. Mundos Sociais.

Ferraz-de-Oliveira, I., C. Azeda, and T. Pinto-Correia. 2016. "Management of Montados and Dehesas for High Nature Value: An Interdisciplinary Pathway." *Agroforestry Systems* 90(1):1–6.

Folk, C., R. Biggs, V. Norstrom, B. Reyers, and J. Rockstrom. 2016. "Social-Ecological Resilience and Biosphere-Based Sustainability Science." *Ecology and Society* 21(3):41–50

Fonseca A. M., C. A. F. Marques, T. Pinto-Correia, N-Guiomar and D. E. Campbell. 2019. "Energy Evaluation for Decision-Making in Complex Multifunctional Farming Systems." *Agricultural Systems* 171(December 2018):1–12.

Garcia RA, Cabeza M, Rahbek C, and Araujo MB. 2014. "Multiple Dimensions of Climate Change and Their Implications for Biodiversity." *Science* 344(6183):486.

Garrido, P., M. Elbakidze, P. Angelstam, T. Plieninger, F. Pulido and G. Moreno. 2017. "Stakeholder Perspectives of Wood-Pasture Ecosystem Services: A Case Study from Iberian Dehesas." *Land Use Policy* 60:324–33.

Godinho, S., A. Gil, N. Guiomar, N. Neves, and T. Pinto-Correia. 2016. "A Remote Sensing-Based Approach to Estimating Montado Canopy Density Using the FCD Model: A Contribution to Identifying HNV Farmlands in Southern Portugal." *Agroforestry Systems* 90(1):23–34.

González de Molina, M. 2011. *Introducción a La Agroecología*. Madrid: SEAE.

Guerra, C. A. and T. Pinto-Correia. 2016. "Linking Farm Management and Ecosystem Service Provision: Challenges and Opportunities for Soil Erosion Prevention in Mediterranean Silvopastoral Systems." *Land Use Policy* 51.

Guerra, C. A., T. Pinto-Correia, and M. J. Metzger. 2014. "Mapping Soil Erosion Prevention Using an Ecosystem Service Modeling Framework for Integrated Land Management and Policy." *Ecosystems* 17(5): 878–889.

Haines-Young, R. and Potchin, M. (2018). Common International Classification of Ecosystem Services (CICES) V5.1. Guidance on the Application of the Revised Structure. https://cices.eu/content/uploads/sites/8/2018/01/Guidance-V51-01012018.pdf

Hartel, T. and T. Plieninger. 2014. *European Wood-Pastures in Transition. A Socio-Ecological Approach*. edited by T. Hartel and T. Plieninger. London: Routledge.

den Herder, N., G. Moreno, R. Mosquera-Rosada, J. Palma, A. Sidiropoulou, J. J. Freijanes, J. Crous-Duran, J. Paulo, M. Tomé, A. Pantera, K. Mantzanas, P. Pachana, A. Papadopoulos, T. Plieninger, and P. Burgess. 2017. "Current Extent and Stratification of Agroforestry in the European Union." *Agriculture, Ecosystems and the Environment* 241:121–32.

Hodge, J. 2013. "The Governance of Rural Landscapes: Properpy, Complexity, and Policy." Pp. 88–100 in *The Economic value of landscapes*, edited by M. van der Heide and W. Heijman. London: Routledge.

Hodge, J. 2016. *The Governance of the Countryside. Property, Planning and Policy*. 1st ed. Cambridge, UK: Cambridge University Press.

Jepsen, M. R., T. Kuemmerle, D. Müller, K. Erb, P. H. Verburg, H. Haberl, J. P. Vesterager, M. Andrič, M. Antrop, G. Austrheim, I. Björn, A. Bondeau, M. Bürgi, J. Bryson, G. Caspar, L. F. Cassar, E. Conrad, P. Chromý, V. Daugirdas, V. Van Eetvelde, R. Elena-Rosselló, U. Gimmi, Z. Izakovicova, V. Jančák, U. Jansson, D. Kladnik, J. Kozak, E. Konkoly-Gyuró, F. Krausmann, Ü. Mander, J. McDonagh, J. Pärn, M. Niedertscheider, O. Nikodemus, K. Ostapowicz, M. Pérez-Soba, T. Pinto-Correia, G. Ribokas, M. Rounsevell, D. Schistou, C. Schmit, T. S. Terkenli, A. M. Tretvik, P. Trzepacz, A. Vadineanu, A. Walz, E. Zhllima, and A. Reenberg. 2015. "Transitions in European Land-Management Regimes between 1800 and 2010." *Land Use Policy* 49.

Kooiman, J. 2003. *Governing as Governance*. London, UK: Sage.

Leal, A. I., R. A. Correia, J. M. Palmeirim, and M. N. Bugalho. 2019. "Is Research Supporting Sustainable Management in a Changing World? Insights from a Mediterranean Silvopastoral System." *Agroforestry Systems* 93(1):355–68.

Lomba, A., F. Moreira, S. Klimek, R. H. G. Jongman, C. Sullivan, J. Moran, X. Poux, J. P. Honrado, T. Pinto-Correia, T. Plieninger, and D. I. McCracken. 2020. "Back to the Future: Rethinking Socioecological Systems Underlying High Nature Value Farmlands." *Frontiers in Ecology and the Environment* 18(1):36–42.

Marsden, T. 2012. "Towards a Real Sustainable Agri-Food Security and Food Policy: Beyond the Ecological Fallacies ?" *The Political Quarterly* 83:139–45.

Moreno, G., S. Aviron, S. Berg, J. Crous-Duran, A. Franca, S. García de Jalón, T. Hartel, J. Mirck, A. Pantera, J. H. N. Palma, J. A. Paulo, G. A. Re, F. Sanna, C. Thenail, A. Varga, V. Viaud, and P. J. Burgess. 2018. "Agroforestry Systems of High Nature and Cultural Value in Europe: Provision of Commercial Goods and Other Ecosystem Services." *Agroforestry Systems* 92(4):877–91.

Moreno, G., A. Franca, T. Pinto-Correia, and S. Godinho. 2014. "Multifunctionality and Dynamics of Silvopastoral Systems." *Options Méditerranéennes* A(109):421–36.

Moreno, G., G. Gonzalez-Bornay, F. Pulido, M. Lopez-Diaz, M. Bertomeu, E. Juárez, and M. Diaz. 2016. "Exploring the Causes of High Biodiversity of Iberian Dehesas: The Importance of Wood Pastures and Marginal Habitats." *Agroforestry Systems* 90(1):87–105.

Moreno, G. and V. Rolo. 2019. "Agroforestry Practices:Silvopastoralism." in *Agroforestry for sustainable agriculture*, edited by M. R. Mosquera-Rosada and R. Prabhu. Cambridge, UK: Burleigh Dodds Science Publishing.

Mosquera-Losada, M. R., G. Moreno, A. Pardini, J. H. McAdam, V. Papanastasis, P. J. Burgess, and A. Rigueiro-Rodríguez. 2012. "Past, Present and Future of Agroforestry Systems in Europe." Pp. 285–312 in *Agroforestry-The Future of Global Land Use*. Dordrecht, NL: Springer Netherlands.

Nair, P. K. R. 2012. "Climate Change Mitigation: A Low-Hanging Fruit of Agroforestry." in *Agroforestry - The Future of Global Land Use. Advances in Agroforestry*, edited by P. K. R. Nair and D. Garrity. Dordrecht: Springer.

Nichols, C. I., A. Henaro, and M. A. Altieri. 2015. "Agroecología y El Diseño de Sistemas Agrícolas Resilientes Al Cambio Climático." *Agroecologia* (10).

Oppermann, R., G. Beaufoy, and G. Jones (eds.). 2012. *High Nature Value Farming in Europe*. 1st ed. Ubstadr-Weiher: Verlag Regionalkultur.

Ortiz-Miranda, D., A. Moragues-Faus, and E. Arnalte-Alegra. 2013. *Agriculture in Mediterranean Europe. Between Old and New Paardigms*. 1st ed. Emerald.

Ostrom, E. 2005. *Understanding Institutional Diversity*. P. U. Press. Princeton, NJ.

Ostrom, E. 2010. "A Gerenal Framework for Analysing Sustainability of Socio-Ecological Systems." *Science* 325(2009):419–22.

Pantera, A., P. J. Burgess, R. Mosquera Losada, G. Moreno, M. L. López-Díaz, N. Corroyer, J. McAdam, A. Rosati, A. M. Papadopoulos, A. Graves, A. Rigueiro Rodríguez, N. Ferreiro-Domínguez, J. L. Fernández Lorenzo, M. P. González-Hernández, V. P. Papanastasis, K. Mantzanas, P. Van Lerberghe, and N. Malignier. 2018. "Agroforestry for High Value Tree Systems in Europe." *Agroforestry Systems* 92(4):945–59.

Pe'er, G., A. Bonn, H. Bruelheide, P. Dieker, N. Eisenhauer, P. H. Feindt, G. Hagedorn, B. Hansjürgens, I. Herzon, Â. Lomba, and E. Marquard. 2020. "Action Needed for the EU Common Agricultural Policy to Address Sustainability Challenges." (November 2019):305–16.

Pecl, G.T., M.B. Araújo, J.D. Bell J.D., Blanchard J., Bonebrake T.C., Chen I.C., Clark T.D., Colwell R.K., Danielsen F., Evengård B., Falconi L, Ferrier S, Frusher S, Garcia RA, Griffis RB, Hobday AJ, Janion-Scheepers C, Jarzyna MA, Jennings S, Lenoir JL, and Williams SE. 2017. "Biodiversity Redistribution under Climate Change: Impacts on Ecosystems and Human Well-Being." *Science* 355(6332):31.

Pinto-Correia, T. and C. Azeda. 2017. "Public Policies Creating Tensions in Montado Management Models: Insights from Farmers' Representations." *Land Use Policy* 64.

Pinto-Correia, T. and A. Fonseca. 2009. "Historical Perspective of Montados: The Example of Évora." Pp. 49–54 in *Cork Oak Woodlands on the Edge: Ecology, Adaptive Management, and Restoration*, edited by J. Aronson, J. Santos Pereira, and J. Pausas. Island Press.

Pinto-Correia, T. and S. Godinho. 2013. *Changing Agriculture-Changing Landscapes: What Is Going on in the High Valued Montado Landscapes of Southern Portugal?* In:Ortiz-Miranda, D., Moragues-Faus, A.M., Arnalte-Alegre, E. (eds.), Agriculture in Mediterranean Europe Between old and new paradigms. Research in Rural Sociology and Development, vol. 19, Emerald, pp 75–90.

Pinto-Correia, T., N. Guiomar, M. I. Ferraz-de-Oliveira, E. Sales-Baptista, J. Rabaça, C. Godinho, N. Ribeiro, P. Sá Sousa, P. Santos, C. Santos-Silva, M. P. Simões, A. D. F. Belo, L. Catarino, P. Costa, E. Fonseca, S. Godinho, C. Azeda, M. Almeida, L. Gomes, J. Lopes de Castro, R. Louro, M. Silvestre, and M. Vaz. 2018. "Progress in Identifying High Nature Value Montados: Impacts of Grazing on Hardwood Rangeland Biodiversity." *Rangeland Ecology & Management*.

Pinto-Correia, T. and L. Kristensen. 2013. "Linking Research to Practice: The Landscape as the Basis for Integrating Social and Ecological Perspectives of the Rural." *Landscape and Urban Planning* 120:248–56.

Pinto-Correia, T., J. Muñoz-Rojas, M. H. Thorsøe, and E. B. Noe. 2019. "Governance Discourses Reflecting Tensions in a Multifunctional Land Use System in Decay; Tradition versus Modernity in the Portuguese Montado." *Sustainability (Switzerland)* 11(12).

Pinto-Correia, T., J. Primdahl, and B. Pedroli. 2018. *European Landscapes in Transition: Implications for Policy and Practice*. 1st ed. Cambridge, UK: Cambridge University Press.

Pinto-Correia, T., N. Ribeiro, and P. Sá-Sousa. 2011. "Introducing the Montado, the Cork and Holm Oak Agroforestry System of Southern Portugal." *Agroforestry Systems* 82(2).

Pinto-Correia, T. and W. Vos. 2004. "Multifunctionality in Mediterranean Landscapes – Past and Future." Pp. 135–64 in *The New Dimensions of the European Landscape*, edited by R. H. G. Jongman. Dordrecht,NL: Springer Netherlands.

Plieninger, T., V. Rolo, and G. Moreno. 2010. "Large-Scale Patterns of Quercus Ilex, Quercus Suber, and Quercus Pyrenaica Regeneration in Central-Western Spain." *Ecosystems* 13(5):644–60.

van der Ploeg, J. D., J. C. Franco, and S. M. Borras. 2015. "Land Concentration and Land Grabbing in Europe: A Preliminary Analysis." *Canadian Journal of DEvelopment Studies* 36(2):147–62.

van der P., J. Douwe, D. Barjolle, J. Bruil, G. Brunori, L. M. C. Madureira, J. Dessein, Z. Drąg, A. Fink-Kessler, P. Gasselin, M. G. de Molina, K. Gorlach, K. Jürgens, J. Kinsella, J. Kirwan, K. Knickel, V. Lucas, T. Marsden, D. Maye, P. Migliorini, P. Milone, E. Noe, P. Nowak, N. Parrott, A. Peeters, A. Rossi, M. Schermer, F. Ventura, M. Visser, and A. Wezel. 2019. "The Economic Potential of Agroecology: Empirical Evidence from Europe." *Journal of Rural Studies*, 71:46–61

Primdahl, J. 2018. "The Contested Nature of the Farmed Landscape." Pp. 1468–87 in *The handbook of Nature*, edited by T. Marsden. SAGE.

Primdahl, J. and L. S. Kristensen. 2016. "Landscape Strategy Making and Landscape Characterisation — Experiences from Danish Experimental Planning Processes." 6397(March).

Primdahl, J., L. S. Kristensen, F. Arler, P. Angelstam, A. A. Christensen, and M. Elbakidse. 2018. "Rural Landscape Governance and Expertise:On Landscape Agents and Democracy." Pp. 153–64 in *Defining Landscape Democracy. A Path to social justice*, edited by S. Egoz, K. Jorgensen, and D. Ruggeri. Edward Elgar.

Rasmussen, L. V., B. Coolsaet, A. Martin, O. Mertz, U. Pascual, E. Corbera, N. Dawson, J. A. Fisher, P. Franks, and C. M. Ryan. 2018. "Social-Ecological Outcomes of Agricultural Intensification." *Nature Sustainability* 1:275–82.

Rivera, M., K. Knickel, J. M. Díaz-Puente, and A. Afonso. 2019. "The Role of Social Capital in Agricultural and Rural Development: Lessons Learnt from Case Studies in Seven Countries." *Sociologia Ruralis* 59(1):66–91.

Robinson, G. and D. Carson. 2015. "Resilient Communities: Transitions, Pathways and Resourcefulness." *The Geographical Journal*.

Rockström, J., J. Williams, G. Daily, A. Noble, N. Matthews, L. Gordon, H. Wetterstrand, F. DeClerck, M. Shah, P. Steduto, C. de Fraiture, N. Hatibu, O. Unver, J. Bird, L. Sibanda, and J. Smith. 2017. "Sustainable Intensification of Agriculture for Human Prosperity and Global Sustainability." *Ambio* 46(1):4–17.

Rodríguez-Cohard, C. J., J. D. Sánchez-Martínez, and V. J. Gallego-Simón. 2018. "Olive Crops and Rural Development: Capital, Knowledge and Tradition." *Regional Science Policy & Practice* (November 2017):1–15.

Rolo, V., T. Hartel, S. Aviron, S. Berg, J. Crous-Duran, A. Franca, J. Mirck, J. H. N. Palma, A. Pantera, J. A. Paulo, F. J. Pulido, G. Seddaiu, C. Thenail, A. Varga, V. Viaud, P. J. Burgess, and G. Moreno. 2020. "Challenges and Innovations for Improving the Sustainability of European Agroforestry Systems of High Nature and Cultural Value: Stakeholder Perspectives." *Sustainability Science* 15(5):1301–15.

Silveira, A., J. Ferrão, J. Munoz-Rojas, T. Pinto-Correia, M. H. Guimarães, and M. L. Schmidt. 2018. "The Sustainability of Agricultural Intensification in the Early 21st Century: Insights from the Olive Oil Sector in Alentejo (Southern Portugal)." Pp. 258–85 in *The Diverse Worlds of Sustainability*, edited by A. Delicado, N. Domingos, and L. Sousa. Lisbon: ICS.

Stefano, Lucia De and M. R.. Llamas. 2012. *Water, Agriculture and the Environment in Spain : Can We Square the Circle ?* 1st ed. edited by C. Press. London, UK.

Surová, D. and T. Pinto-Correia. 2016. "A Landscape Menu to Please Them All: Relating Users' Preferences to Land Cover Classes in the Mediterranean Region of Alentejo, Southern Portugal." *Land Use Policy* 54.

Torralba, M., N. Fagerholm, T. Hartel, G. Moreno, and T. Plieninger. 2018. "A Social-Ecological Analysis of Ecosystem Services Supply and Trade-Offs in European Wood-Pastures." *Science Advances* 4:1–13.

UN. 2015. United Nations Sustainable Development Goals. https://sdgs.un.org/goals

Vanloqueren, G. and P. Baret. 2009. "How Agricultural Research Systems Shape a Technological Regime That Develops Genetic Engineering but Locks out Agroecological Innovations." *Research Policy* 38:971–83.

Varela, E., F. Pulido, G. Moreno, and M. Zavala. 2020. "Targeted Policy Proposals for Managing Spontaneous Forest Expansion in the Mediterranean." *Journal of Applied Ecology* accepted.

Vatn, A. 2005. "Rationality, Institutions and Environmental Policy." *Ecological Economics* 69:203–17.

Section A

People

1 Landowners

Teresa Pinto-Correia, Pedro M. Herrera
and Rufino Acosta-Naranjo

1.1 Introduction

Montado and dehesa of southern Portugal and southern Spain are found mainly in large estates. In Spain there are approximately 15,000 *dehesa*-properties, with an average surface around 250 ha, although about 10% of them are bigger than 500 ha (Pulido and Picardo, 2010; MAPA, 2008). In southern Portugal, 5,300 farms were identified as being partly covered with montado in 2018 (Guimarães et al., 2018). We refer to the very large ones (in both countries), with more than 500 ha, as estates, or large-scale farms. These are quite rare. More common are dehesa and montado farms of between 100 and 500 ha. Smaller scale farms of less than 100 ha are also less frequent. However, it is not known how many montado landowners there are, as the same person may own or be in charge of different farms (either through family members or leasing). Montado and dehesa farms are mostly private properties, although in Spain quite a significant share of dehesa (approximately 25% of them, irregularly distributed) is public land or under different systems of common ownership, whether owned by a group of neighbours or a municipality (Pulido and Picardo, 2010). Municipal ownership is specifically represented in the *dehesas boyales*, a public dehesa commonly managed to feed working animals and lately readapted to different land uses and regimes (Herrera, 2016).

Historically, large estates have been the property in possession of large and established families, belonging to the highly homogeneous rural elite (Cutileiro, 1971; Seita-Coelho, 1996; Acosta-Naranjo, 2002, 2008). These landlord families have a social culture and universe that is oriented exclusively towards their own social class and, in many cases, aspires to urban bourgeois values. Most estates have passed through the same family through generations via inheritance, or changed from one family to the other, within the same social group, through marriage. Despite periods of instability, the relative social stability, both in Spain and in Portugal, until the end of the 20th century also resulted in stability in the farm structure (Pinto-Correia and Fonseca, 2009; Martínez Alier, 1968). There is thus a customary approach to montado and dehesa management that is strongly linked to this specific social group of landowners, in both Iberian countries (Picão, 1983; Acosta-Naranjo, 2002).In recent decades, some changes have emerged due to shift in

DOI: 10.4324/9781003028437-2

social and economic strategies. Historically, there was a trend for intermarriage within the landowners, regrouping the divided heritages through marriage, while strengthening the social and political relationships to maintain property, power, status and privileges. However, the agriculture modernization process that started in the second half of 20th century (although somewhat later in Iberia) changed this system. Big landowners became dependent on agribusiness and lost their central status and power in rural governance systems as local society became more fully integrated within urban areas and increasingly global systems. New activities and relationships with urban areas were established, changing their economic profile and redefining marriage strategies so that they incorporated outsiders (Acosta-Naranjo, 2008).

At the same an increasing number of estates are being bought up by new owners. While the tradition of generation long family ownership still persists, there are new individual and corporate entrants into this farming sector, investing in large estates as a business opportunity or to become an amenity provider (leisure, hunting), or even just for social status (Barroso and Pinto-Correia, 2014). This has an impact on management practices as these new owners may be influenced by different ideas, often from their experiences of a globalized world and thus more prone to innovative business models (Pinto-Correia et al., 2019).

Beside the landowners of large estates, we can also find small landowners, especially in Castilla (Spain) and in some municipalities in the littoral of Alentejo (southern Portugal). As depicted in Chapter 8, despite being land owners, these small owners (especially those whose lands are located in *latifundia* territories) have historically emphasized the condition of people who work by their hands and in some cases they positioned themselves on the side of the 'poor', the workers, in opposition to the 'rich', the latifundists (Acosta-Naranjo, 2002). Unlike the large owners, these small owners do not hold power in the local social structure, even though they are well integrated in the local culture and social life in the villages. Their social status remains fuzzy, somewhere between workers and the *bourgeoisie:* A social class of their own. These peasants, *cum* smallholders, have been called the 'awkward class' as intellectuals and political activists find it difficult to classify them (Shanin, 1972; Sevilla Guzmán and Pérez Yruela, 1976).

Today, the landownership pattern of silvopastoral farms is complex and dynamic. Landowners are the primary actors in the management of silvopastoral systems, who take the ultimate and main decisions that structure land use and farm-based activities. Most of these landowners are individuals who have inherited the land from their families, which have owned the land for many generations. However, there are increasingly also divergent profiles, which deserve particular analytical attention, as potential sources of innovation and change. We need to know who these landowners are and what drives their behaviour and management decisions in order to understand them and design strategies for improvement. This chapter explores the most common social structures, both traditional and modern, behind property ownership in silvopastoral systems in Spain and Portugal, focusing on their role in governance and how different discourses influence their decisions.

1.2 The landownership background

The Christian reconquest of the areas currently covered by dehesa/montado is crucial to explain the property rights as they exist today (Guzmán Alvarez, 2016; Grove and Rackham, 2001), with the Tagus River acting as a frontier for the differences between ownership structures. The reconquest and repopulation of the land north of the Tagus River occurred over a relatively long period during the Middle Ages. Here, the strategy was to facilitate ownership transfer to the peasants, promoting occupation while defending simultaneously the conquered territory. South of the Tagus River, the faster conquest of large-scale and unoccupied areas forced the Crown to offer control over large areas to the nobility and military orders. Consequently, larger properties became predominant in the farm structure and this structure influenced the land use practices (Acosta-Naranjo, 2002; De la Montaña Conchiña, 1997).

Later, in the modern age, the industrial and commercial bourgeoisie joined these dynamics, adopted the noble's mindset and started buying land whenever the Crown suffered a financial crisis. In Portugal, landownership of large estates was built on the principles defined when the administrative, juridical and economic structure of the country was reformed after the decline of the *Ancienne Regime,* at the end of the civil war in 1834. The large estates which had belonged to the religious orders, the royal house, the main aristocratic families, as well as the commons belonging to the local community, were all divided into smaller units and sold to new owners: Non-aristocratic and mostly originating from the commercial bourgeoisie, as well as an agricultural elite who had the capacity to buy land. These new wealthy landowners have become the new social and economic rural elite, a minority in the region's population. After the parliamentary monarchy, in 1910 a parliamentary republic was installed followed by a corporate state in 1926. These changes did not affect the land ownership rules established after 1834, which were based on full private property rights with the aim of getting the land into productive use and maximizing production. The result was an unbalanced land and wealth distribution with a small group of landowners, often absent and a large landless and poor population, resident in large villages or the estates themselves. Up to the 1970s, a farm of 1000 hectares could support a small village of more than 50 farm workers (Box 3.1, Chapter 3)

In Spain, the key period influencing the structure of dehesa-properties is the second third of the 19th century, when large tracts of public, common and church land were privatized under the Madoz Disentailment Act (Beltrán Tapia, 2018). Only some common land avoided mandatory confiscation as some commons were exempted for social reasons. The first forest engineers also excluded some forests, arguing that their sale was not beneficial for the government (Casals, 1996). So a few areas remained as common land and do so until now, as shown in Chapter 8. The rest was confiscated and sold to private individuals. The biggest communal and church estates were sold in large lots to wealthy people from the rising urban bourgeoisie. Only in a few cases, where neighbours joined and pooled resources, they were able to re-buy their auctioned land (Sánchez Marroyo, 2015; Alagona et al., 2013; Bernal, 1988; Guzmán Alvarez, 2016; Sánchez Marroyo, 1988). The great properties of the nobility became a commodity available to members of the upper social classes, who acquired full property rights. Recent historiographical debates contest the traditional view of the elites of the time as being the beneficiaries of a process managed by the State (González de Molina, 2014), with peasants and local councils keeping control over part of the land and the elites being the beneficiaries of disentitled Church properties. According to this interpretation, although the process did not foster land redistribution or a more balanced access to land, it did offer opportunities for the consolidation and even the expansion of the small farm units, a certain kind of peasantization, according to Van der Ploeg's concept (González de Molina, 2014; Bernstein et al., 2018). In our opinion, different processes occurred in different areas, and in general, in the dehesa in Andalusia and Extremadura, the disentitlement process intensified the latifundio to different degrees.

In southern Spain and Portugal, large landowners became a distinguished elite, holding much power in state structures and a monopoly over labour recruitment that served to secure their control over land (Giner and Sevilla Guzmán, 1977; Romero Salvadó, 1999). This structure was challenged in the 1930s during the Second Republic but this process was truncated by the start of the Civil War. Once in power, Franco's dictatorship reinstated landowners at the top of the social pyramid in southern rural Spanish territories, including those of dehesa (Pérez Yruela and Sevilla Guzmán, 1981; Martínez-Alier, 1968).

Although migration to the cities started earlier, the '60s and '70s saw a large-scale exodus from rural to urban centres in both countries. Manpower in rural areas became scarce and traditional owners faced a dilemma: To sell up or improve performance. Some responded by investing in infrastructure, fencing and introducing machinery that could allow the farm work to be done by fewer workers. Others sold their land to new upper-middle-class professionals, industrialists, or constructors wishing to consolidate their social status (Gómez, 1991). Some of the latter focused on making the estates profitable and shifted to becoming 'real farmers' while others used the land for leisure, leaving agriculture to the workers settled on or around the property. Even recently, these large scale purchases of land have raised questions about land grabbing in some areas of Extremadura and Andalucía, leading to an unproductive landlord-based ownership. Finally, a small number of tenants and farmers were able to buy part or all of the land they managed, often when landowners got into financial difficulties.

This situation led to the current dehesa and montado ownership which is quite similar in both countries. First of all, there are traditional owners, with large estates that have been managed by the same family for generations (of all dimensions, from 100 to 1000 ha and sometimes more); some of them are co-owned, usually by several relatives, with a manager in charge of the farming issues. There are also large estates (>300 ha) owned by companies or individuals for social status or recreational purposes and managed locally by one or more employees. Finally, there are smaller size properties (100–200 ha) managed by former tenants or their descendants who make their livelihood out of farming. A common pattern, both in Portugal and Spain is for a landowner to also manage others' land, for example properties from other families who do not manage or work their inherited land. It is also very common that cork-dehesa/montado owners despite not having agriculture as their main activity, exploit the cork and rent out the other system's outputs, such as acorns, pastures, etc.

The profile of the ownership and the management models has profoundly changed in recent years, with more professionalized and trained people in charge of the estates who have to deal with increasing challenges in maintaining the profitability and sustainability of the estates (Fragoso et al., 2011; Pinto-Correia and Azeda, 2017).

As for the socio-political dimension, the process of modernization during the final stage of Francoism and Salazarism meant that industry and services came to supersede the importance of agriculture. This, in turn, resulted in the loss of the economic importance of land ownership and the consequent decline in elite position of big landowners. The arrival of democracy coincided with the end of the agrarian bourgeoisie's relevance as part of the ruling class which was diluted by the political and economic framework of modernity. Once rural-urban differences became diluted and more interactions between the two worlds emerged, their role as mediators between the urban world and rural communities was also over, making the previous pattern of clientism obsolete (Acosta-Naranjo, 2008). In parallel, private capital from the agricultural sector was channelled towards the more profitable service sectors and other profitable urban sectors, while new capital from the urban economic elite acquired large properties. This pattern was especially marked after 1986 when Spain and Portugal entered the European Economic Community (EEC) and began to receive agricultural subsidies, although it abated after 2008 when the financial crisis kicked in.

The new democratic context led to a certain social and political discredit of the estate landowners, who were seen as linked to the *latifundia* and a symbol of oppression. These landowners lost their political relevance and were kept outside government circles for a long time, especially in Spain. Even with right-wing governments, this class never regained the influence of previous times. This change has been termed as 'the fall of the gods', both in Spain and in Portugal. Common Agricultural Policy (CAP) subsidies and the subsequent decoupling of aid became an important source income for all landowners, particularly for those with large properties (Leco and Pérez, 2019; Fragoso et al., 2011) who saw their yearly income increase and stabilize, supporting a regained social status.

The small landowners of dehesa had different relevance in different territories. Despite their ambiguous position, they aligned themselves with conservative parties, scared of pressure from the labour movement and looking for protection from strong authorities. Thus, they secured their property rights, although the demands of Agrarian Reform were never directed towards their farms. Consequently, at least in Spain they become active supporters of dictatorial regimes and even a cultural model for Franco's agrarian fascism

(Sevilla Guzmán, 1979). However, policies intended to protect these land-owners ended when Franco's regime shifted to a focus on industrialization. The ongoing process of migration and urbanization ended up leaving these smallholders as relatively autonomous economic households. Yet they were also obliged to participate in modern agriculture systems and markets, start-ing a process of proletarianization that ultimately led to land abandonment (González Rodríguez and Gómez Benito, 2002; Guzmán Casado, González de Molina and Sevilla Guzmán, 2000; Alonso, Arribas and Ortí, 1991). The CAP subsidies, especially those for *dehesa* livestock ended up being one of their main income sources, although most of the aid went to the large scale owners.

1.3 Landowners today

In the second half of the 20th century and prevailing until today, different groups of landowners can be identified in the montado and dehesa.

Large *latifundia* families: The wealthiest group, with very large properties (estates) often more than one, in different locations; they do not live on the farm but in a city or even the metropole. They are urbanites and have close family and friendship relations with the urban high social classes. They also have other income sources besides farming. They spend their holidays at the farm but otherwise manage agriculture through a manager who is on an everyday basis in the farm. The relations with the farm workers are distant. Two different groups can be distinguished:

1 Traditional owners: These are the local elite in Portugal but not so much in Spain, as only a few live in the villages. They live in the local village or municipal town and own farmland concentrated in one or two municipalities. Their properties are generally smaller and they mostly live off the farm income. They do not work themselves in the farm but they manage it directly. They have family and friendship relations with the rural bourgeoisie. Until 1974 in Portugal and 1979 in Spain, they also dominated the professional organizations and the municipal council.

2 Farmers: This group differs from the previous ones, in that they gener-ally work on their farm, or have worked on the (or another) farm before. They have maybe farm managers when they were younger who bought a farm later in their life. They also often lease land from other owners to enlarge their farm unit. In Portugal, they may have good relationships with the farmworkers but identify themselves as part of the local elite. In Spain this group can be found especially among the small and medium size dehesa-properties, most historically relevant in Castilla (Castilla-La Mancha and Castilla y Leon) but also with a presence in Andalusia and Extremadura. They tend to identify themselves with the local elite less so than in Portugal. They work hard and aim for the farm to be inherited by their successors, but this is often very problematic.

After the establishment of democracy other groups emerged:

3 Rural business landowners: In most cases, these are people from the region who have gained money from, say, buying and selling livestock and livestock fodder and have invested in large estates. They may switch to becoming full-time farmers or continue with their business and keep the farm as a secondary activity. Depending on the option, they may work or not on the farm themselves. The main difference in relation to the previous groups is that they do not have roots in farming itself.

4 Urban business landowners: In most cases, those are people from outside the region. They have gained money in the building industry or another sector. They invest in farmland as a status and as an extra business opportunity. They do not move onto the farm and only visit it occasionally, maybe using it to entertain family and friends or as a holiday house. They do not work in the farm and have a manager to look after the daily management. Some of these landowners have initiated processes of innovation and started a new business based on high-quality food production or agro-tourism.

5 Leisure owners: These are the most recent types who have an urban background and are highly paid (liberal) professionals. They buy large estates as an investment in environmental amenities and often also an investment that secures their social status. It should be remembered that in the first years of the new democracies, these large silvopastoral estates were seen as a kind of symbol of the dictatorship and social oppression. After time, there was a kind of Doppler effect (Acosta-Naranjo, 2008) and wealth was no longer seen negatively, in both countries. With the rise of environmental awareness, montado and dehesa started being seen as unique ecosystems. Campos (2013) refers to this as the 'environmental rent' of the landowner who enjoys the benefits of the amenity values of the montado and dehesa that come with ownership. Some of those new landowners are from other countries, mainly northern Europeans. They may be hunters or just use the estate as a summer and weekend house. They do not move onto the farm and maintain their presence in the urban sphere – even though some maybe spend large periods at the farm. They pay a land manager to look after everyday management and often part of the farm work.

6 Stewardship owners: Increasingly in recent years, NGOs, foundations and diverse interest groups are buying silvopastoral estates to protect the public goods that these systems support. This is a very recent and still very limited phenomenon but considering the multiple values of the montado and dehesa and the challenges addressed by new society demands, this type of landowner needs to be acknowledged and could have an influence in the future.

1.4 The discourses that frame landowners today

An analysis of the dominant discourses of landowners is one way to go beyond the surface and obtain a clear understanding of what characterizes the complex motivations of landowners today. A discourse can be defined as an entity of signs that attribute meaning to particular objects, subjects and statements (Pinto-Correia et al., 2019). Discourses guide conduct and shape likely outcomes, the dominance or discourse is more likely to result in certain outcomes. Boundaries of discourses may be difficult to define but their core can be identified and described (Pinto-Correia et al., 2019). When analyzing landowners' position in relation to the farm they manage, it is useful to explore their underlying discourses because these define the courses of action that are seen as legitimate or not. Therefore, discourse analysis can elicit the tacit normative foundation and the taken for granted routines on which the farming system is based. It is an important analytical tool that makes it possible to identify what underlies certain types of management decisions.

As an example, we have performed a discourse analysis on the montado in Alentejo, exploring the argumentations that emerge when farmers describe and explain their individual conduct (Pinto-Correia et al., 2019; Pinto-Correia and Azeda, 2017). Our intention is not to judge particular regimes of knowledge and truth, rather analyze how particular discourses give meaning to the decisions made by various actors. Interestingly land use management is not governed by just one discourse: Our analysis revealed three co-existing discourses that each justify different modes of farming.

The analysis performed is described in detail in Pinto-Correia et al. (2019), and is also grounded on Pinto-Correia and Azeda (2017). Mainly it was based on a mix-methods approach along 10 years of research in central Alentejo, with a review of the scientific and grey literature, media analysis, interviews and participatory observation in workshops and meetings where landowners were present.

The analysis from Spain is similar, although we lack as much evidence to support its conclusions. We compared the discourse analysis that we carried in central Alentejo with those elicited from a study case in the dehesa of Spanish Extremadura (Acosta-Naranjo, 2008) and completed the analysis with key insights extracted from personal interviews with two key women landowners, thought to be in the vanguard of sustainable practices in the sector in south-western Spain. They are active, modern and empowered dehesa professionals, strongly committed to the survival and profitability of silvopastoral systems in Spain. These two examples are described in Boxes 1.1 and 1.2. We believe the situation of

---

TEXT BOX 1.1

Ana Rengifo: stewarding a private state in a National Park

Pedro M. Herrera, Fundación Entretantos

Key points: Modern landowners are aware of their role as land stewards and nature conservation actors, and demand fair levels of support to enhance this aspect of their work and balance the lack of profitability of their ranches.

The family of Ana Rengifo has held their dehesa estate, *Finca El Guijo*, since the Mendizábal Disentailment Act in 1836. The owners of this dehesa estate have followed successive discourses that have followed a path from heritage, to modern farming and then to land stewardship. The later frames their current perspective, especially since the estate was included in the Monfragüe National Park, classified since 2007. Starting from a traditionally managed and owned estate, the number of heirs progressively increased, and the owners decided to transform and update the ownership basis. A private corporation was created, that currently has 110 family stockholders (i.e. owners), an executive board and a managing board. The daily management is in charge of an on-site manager. The manager's position has been commissioned to individuals belonging to the same local family, since 1863. Owners have no right to revenue from the estate but are entitled to use the estate for leisure (shared under a tight schedule), game and firewood. The corporate statutes highlight the social and environmental principles of the dehesa, including local employment and high natural values. The dehesa demands constant investment, and the main revenues (cork, game, pasture) often do not cover the expenses. The cork harvest usually goes to pay off debts.

Ana's diagnosis of dehesa ownership is quite pessimistic, not only because of the lack of profit but also for political and social reasons. The relationship between landowners and the government is tense and landowners often complain of a lack of respect for property rights, aggressive taxation and excessive bureaucracy. Overruling is also perceived as a constraint, especially arbitrary limits on traditional activities under weakly founded nature protection rules. Government hostility is felt as a constraint for improving the condition of the dehesa estates.

On the other hand, Ana embraces the land stewardship discourse with enthusiasm, highlighting the role of landowners. The example of *Finca El Guijo* works in partnership with the University of Extremadura in research and biodiversity conservation activities collaborates in several research and conservation projects (Prodehesa-montado, RUFAN, ARTEMISAN) and pilot projects on Nature Conservation Banks, leads that other local landowners might follow.

landowners in Spain and Portugal may be quite similar, due to the parallels in the past history of land ownership and the shared global societal pressures and expectations, that affect both sides of the border in the Iberian Peninsula.

The discourses, derived from the argumentation that emerges when farmers describe and explain their individual conduct (Pinto-Correia and Azeda, 2017; Pinto-Correia et al., 2019), allowed to us identify three main types of landowners:

TEXT BOX 1.2

Maria Pía Sánchez: a hard choice between well-being and land

Pedro M. Herrera, Fundación Entretantos

Key points: Fear of losing ownership is a real constrain felt by landowners when considering investment, innovation or improving the production of dehesa estates. This prevents new opportunities for the flexible and sustainable use of resources.

The case of María Pía Sánchez, current president of FEDEHESA (National Federation of owners and stakeholders defending the dehesa) member of GER (Network of Women Livestock Farmers) is quite different from Ana Rengifo, although both women share similar values. Pia is the proud owner of a dehesa estate, *Finca La Rinconada*, inherited from her family. The heritage discourse is quite strongly present, as the family has maintained the integrity and value of the estate, transmitting it as a whole to a single heir. Pía abandoned her business career to run the estate, driven by a deep emotional link with her land, and investing all her savings to modernize the farm. Currently, when Pía talks about her estate she seems to be placed within the modern agriculture discourse, although there is a clear concern with the difficulty to draw actual profit from this production model that is compatible with preserving all land values.

A critical issue for Pía and most dehesa landowners is that they share the same fear that erratic political movements could weaken their tenure or even lead to loss of ownership. They also share the same conflict-ridden relationship with the

government (seeking fairer taxation, less bureaucracy, more flexibility and better recognition of their role as providers of ecosystem services). Pía is proposing some new possibilities to improve the management of dehesa, such as developing flexible nested rights associated with the use of different resources. The use of land and resources (pastures, acorns, cork, wood, mushrooms, honey, leisure, game, etc.) could be shared between owners and producers. This could help the dehesa become multifunctional again, diversifying, sharing risks and adopting sustainable production models to improve the farm's economic viability and the public goods they provide. This path demands a profound shift in governance, with new rules and new institutions managing all this complexity.

Pía is also driven by the heritage discourse and wants to keep her family heritage. In the meantime, none of her sons or daughters has shown any interest in taking over the farm, so eventually, Pía may be forced to sell the whole Finca because, in her own words, splitting these estates is lethal for them.

The first group aligns itself to the "Heritage Farming discourse", which has the longest historical trajectory. This discourse is closely related to highly stratified and stable social structures in the region and with the large estate ownership by well-established and wealthy families at the top of the regional social structure. The properties have been passed on from their ancestors, enlarged with marriages among close families, and will be passed on to the next generations. The land is only sold in cases of financial distress. The properties are not divided on succession. This is a highly conservative and closed social group which, despite the modernization of social rules and the revolutionary period in the 1970s, has kept its internal structures and functions almost unchanged, as well as keeping ownership of the land. This land ownership structure, traditional extensive farming practices, the silvopastoral landscapes, as well as the very large white farm buildings, are still seen as the agricultural backbone of the Alentejo region – despite the modern intensive land uses that have emerged in the recent decades (wine, intensive olive, horticulture) that have a much higher financial turnover and visibility. In this discourse, a very important goal of the landowner is to keep the montado healthy, so as to pass it on to the coming generations. Thereby intensification of production and overgrazing in response to new economic contexts and support schemes are viewed negatively, or with a considerable degree of reservation. This group is also unlikely to make investments in the montado, for any type of innovation, be it structure, products, marketing, or the organization of labour. They are simply not on the agenda, not because innovation and investment are negatively perceived: They remain simply unseen, or their potential benefits unappreciated.

In Spain we can see both aspects. While a sector of landowners greatly values emotional links with the land, farm, family and the historical depth of their inheritance, a larger part of them are also engaged in market dynamics and have sold, divided, leased, split, or transferred their properties to new owners, in this losing their emotional umbilical cord to the land.

The second type is embedded in the discourse of *Modern Productive Farming*. This is a more recent discourse, linked with the trends for farming specialization and intensification, which have by and large driven European agriculture change since the middle of the 20th century. It is linked with a strong belief in farm technology as a way of improving the performance of production factors, the farmer as an entrepreneur and the farm business as an income-driven activity. It is also widespread among landowners – often the younger generation – from the large families in the regions, together with those with origins in the business world. A key argument in this discourse is the need to increase profitability in the montado, reducing costs of production and increasing the productivity and the competitiveness of montado farms in an international context. A key step to achieve this has been the reduction of basic-skilled labour, and the intensification of production: Specialization and concentration in one type of livestock, increased production through higher grazing intensity, the introduction of irrigated areas for intensive fodder production and increased dependency on external fodder. This discourse of specialization requires higher technical and professional skill levels yet employs fewer people. While the need for financial and economic sustainability as well as resilience is frequently employed to underpin this second discourse, it does seem to be more focused on short-term (within this generation) profit gains and competitiveness within the current policy and financial framework. This discourse is closely linked to the evolution of CAP measures and related payment schemes and the globalization of food markets.

This discourse can also be found in Spain among large landowners. Families wanting to keep their estates who are under pressure to modernize and intensify their production, adopt new production models and legal status. On the other hand, new landowners who bought estates for leisure or status ended up realizing that owning such an estate without a good business plan was ruinous, so they were forced to sell or intensify. This discourse is also supported by the increasing influence of corporation ownership, the introduction of new production models and innovation for quality production.

The third type can be termed the *Land Stewardship discourse*, the most recent and independent discourse. This is still mostly evident at national and international policy scales, though is slowly trickling down to regional to local policy institutions and is recognized by the new landowners, whom we term leisure landowners. However, it is a discourse that has, as yet, scarcely permeated into regional farmers' mindsets, although it may become more prevalent in the coming years. In this discourse nature conservation and the preservation of natural resources take priority. In terms of production, it is aligned with the concern for vitality and renewal and tree cover. Some landowners see this discourse as potentially providing competitive advantages through tools such as ecological and organic certification (which results in higher prices for the products in the market) and agro-tourism.

In Spain the discourse of *Land Stewardship* has also been adopted by some 'leisure landowners'. These new landowners may not feel strong economic pressure to make farms profitable or the urge to intensify production. They adopt the stewardship discourse as a key argument for their actions, although it can, occasionally, lead to a lack of management and consequently deterioration by abandonment. Conversely, several dehesa owners, both professional and leisure owners, are very aware of the environmental value of their estates and are willing to invest and improve their management systems through sustainability and adaptation criteria. The recent episodes of water scarcity have been an unmistakable warning. Some owners have adopted a land stewardship discourse as a result of confrontations with the environmentalist positions of some institutions and social movements. They assimilate this discourse to publicly justify their position: "If this ecosystem exists, it is because we have maintained it". Within this category, we can find two gradients among two classes: Those motivated by ecological responsibility; and those who want to reach a specific sector of consumers willing to pay for healthier, more environmental-friendly products. Finally, a new actor has arisen as landowner, in Portugal and Spain: Environmental NGOs buying land or developing stewardship leases and agreements mainly focused on nature conservation. They are clearly driven by this discourse

These discourses are not necessarily mutually exclusive. Some landowners may be fully integrated into one discourse and feel confident they are performing the best possible management. But at the same time a landowner or landowner family may simultaneously adhere to aspects of different discourses. Even though landowners are only very rarely explicitly aware of their positioning within one or another discourse, the contradictions among the different discourses can lead to underlying tensions in strategies and management options. Understanding these discourses helps us to grasp what drives landowners to the decisions they take, which otherwise may be difficult to understand fully.

1.5 Conclusion

In this chapter we have shown the particular characteristics of landowners of montado and dehesa. Most of them belong to a specific category, very large and wealthy landowners. Nevertheless, their current situation is singular, due to the extensive nature of the silvopastoral systems and by the historical roots of farm structures in Spain and in Portugal, which is closely related to the socio-political background of both countries. Their profiles are becoming more diversified, driven by the collision between modern global trends and the remaining historical background, which is still reflected in the landownership of today. These complex historical and contemporary profiles can help us to understand the present and future management possibilities for the montado and dehesa.

References

Acosta-Naranjo, R. 2002. *Los Entramados de la Diversidad. Antropología Social de la dehesa*. Badajoz: Diputación de Badajoz.

Acosta-Naranjo, R. 2008. Dehesa de la sobremodernidad. La cadencia y el vértigo. Badajoz: Diputación de Badajoz.

Alagona, P. S., A. Linares, P. Campos and L. Huntsinger. 2013. "History and Recent Trends", Pp. 25–58, in *Mediterranean Oak Woodland Working Landscapes. Dehesas of Spain and Ranchlands of California,* edited by Huntsinger Campos, Oviedo Starrs, Díaz Standiford, and Montero. Dordrecht: Springer.

Alonso, L. E., J. M. Arribas and A. Ortí. 1991. "Evolución y perspectivas de la agricultura familiar: de «propietarios muy pobres) a agricultores empresarios". *Política y Sociedad* 8: 35–69.

Barroso, F. and T. Pinto-Correia. 2014. "Land managers' heterogeneity in Mediterranean landscapes – consistencies and contradictions between attitudes and behaviors". *Journal of Landscape Ecology* 7(1): 45–74.

Beltrán Tapia, F. J. 2018. *En Torno al Comunal en España: Una agenda de investigación llena de retos y promesas.* Sociedad de Estudios de Historia Agraria, Documentos de trabajo.

Bernal, A. M. 1988. *Economía e Historia de los Latifundios.* Madrid: Espasa-Calpe.

Bernstein, H., H. Friedmann, J. D. Van der Ploeg, T. Shanin and B. White. 2018. "Forum: Fifty years of debate on peasantries, 1966–2016". *The Journal of Peasant Studies* 45(4): 689–714. doi: 10.1080/03066150.2018.1439932

Campos, P. 2013. "Renta ambiental del monte". *Actas del VI Congreso Forestal Español* ("Montes: Servicios y desarrollo rural"), 10-14 de junio de 2013, Vitoria-Gasteiz.

Casals, V. 1996. *Los Ingenieros de Montes en la Espana˜ Contemporánea, 1848–1936.* Ediciones del Serbal, Barcelona.

Cutileiro, J. 1971. *A Portuguese Rural Society.* 1st ed. Oxford, UK: Clarendon Press.

De la Montaña Conchiña, J. L. 1997. "Reflexiones en torno a la repoblación y formación de la sociedad feudal extremeña (siglo XIII-XIV)". Norba: *Revista de Historia* 14: 83–101.

Fragoso, R., C. Marques, M. R. Lucas, M. B. Martins and R. Jorge. 2011. "The economic effects of common agricultural policy on Mediterranean montado/dehesa ecosystem". *Journal of Policy Modeling* 33: 311–327.

Giner, S. and E. Sevilla Guzmán. 1977. "The latifundio as a local mode of class domination: the Spanish case". *Iberian Studies* 6(2): 47–58.

Gómez J. M. (Coord.). 1991. *El Libro de las Dehesa Salmantinas.* Ed Junta de Castilla y León. Valladolid.

González de Molina, M. 2014. "La tierra y la cuestión agraria entre 1812 y 1931: latifundismo versus campesinización", in González de Molina (Coord.) *La Cuestión Agraria en la Historia de Andalucía. Nuevas perspectivas,* Sevilla: Fundación Pública Andaluza Centro de Estudios Andaluces, Consejería de la Presidencia, Junta de Andalucía, 21–60.

González Rodríguez J. J. and C. Gómez Benito. 2002. *Agricultura y Sociedad en el Cambio de Siglo.* Madrid: McGraw-Hill.

Grove A. T. and O. Rackham. 2001. *The Nature of Mediterranean Europe. An ecological history.* New Haven and London. Yale University Press.

Guimarães, H., N. Guiomar, D. Surova, S. Godinho, T. Pinto-Correia, A. Sandberg, F. Ravera and M. Varanda. 2018. "Structuring wicked problems in transdisciplinary research using the Social-Ecological Systems framework: an application to the montado system, Alentejo, Portugal". *Journal of Cleaner Production* 191: 417–428. doi: 10.1016/jclepro.2018.04.200

Guzmán Alvarez, J. R. 2016. "The image of a tamed landscape: Dehesa through history in Spain". *Culture & History Digital Journal* 5(1): e003. doi: https://dx.doi.org/10.3989/chdj.2016.003

Guzmán Casado, G. I., M. González de Molina and E. Sevilla Guzmán. 2000. *Introducción a la Agroecología como Desarrollo Rural Sostenible*. Madrid: Mundi-Prensa.

Herrera, P. M. 2016. "El uso pastoral como alternativa de gestión de los hábitats vinculados a los rebollares ibéricos". *Revista Pastos* N° 46(2): 6–23. Diciembre 2016. http://polired.upm.es/index.php/pastos/article/view/3615

Jiménez Blanco, J. I. 2002. "El monte: una atalaya de la Historia". *Historia Agraria* 26: 141–190.

Leco, F. and A. Pérez. 2019. "Desajustes territoriales en la distribución del Pago Básico de la PAC en España". *Cuadernos Geográficos* 58(3): 57–82.

MAPA. Ministerio de Agricultura, Pesca y Alimentación. 2008. Diagnóstico de las dehesas Ibéricas Mediterráneas. Tragsatec.

Martínez-Alier, J. 1968. *La Estabilidad del Latifundismo*. París: Ruedo Ibérico.

Pérez Yruela M. and E. Sevilla Guzmán. 1981. "La dimensión política en la Reforma Agraria: reflexiones en torno al caso andaluz". *Revista de Sociología* 16: 53–91.

Picão, J. S. 1983. Através Dos Campos. Usos e Costumes Agrícolo-Alentejanos. 1st ed. Lisbon: Dom Quixote.

Pinto-Correia, T. and A. Fonseca. 2009. "Historical Perspective of montadomontado: The Example of Évora." Pp. 49–54 in Cork Oak Woodlands on the Edge: Ecology, Adaptative Management, and Restoration, edited by J. Aronson, J. Santos Pereira and J. Pausas. Island Press.

Pinto-Correia, T. and C. Azeda. 2017. "Public Policies Creating Tensions in montado Management Models: Insights from Farmers' Representations". *Land Use Policy* 64: 76–82.

Pinto-Correia, T., J. Muñoz-Rojas, M. Hvarregaard Thorsøe and E. Bjørnshave Noe. 2019. "Governance Discourses Reflecting Tensions in a Multifunctional Land Use System in Decay; Tradition Versus Modernity in the Portuguese montado". *Sustainability* 11(12): 3363.

Pulido F. and A. Picardo (Coord.). 2010. *Libro Verde de la Dehesa. Consejería de Medio Ambiente*, Junta de Castilla y León. https://www.researchgate.net/publication/229812274_Libro_Verde_de_la_dehesa

Romero Salvadó, F. J. 1999. Spain 1914-1918. *Between War and Revolution*. London: Routledge.

Sánchez Marroyo, F. 1988. "La revolución burguesa en Extremadura: acotaciones a un tema polémico". *Alcántara: Revista del Seminario de Estudios Cacereños* 13-14: 63–90.

Sánchez Marroyo, F. 2015. "Consolidación, disolución y estructura patrimonial de las fortunas nobiliarias en la segunda mitad del siglo XIX". *Actas del XII Congreso de la Asociación de Historia Contemporánea* ("Pensar con la Historia desde el siglo XXI"), Diciembre 2014, Madrid, pp. 5383–5399.

Seita-Coelho, I. 1996. "Transferência de propriedade no concelho de Cuba." Pp. 521–29 in O Vôo do Arado, edited by J. Pais de Brito. Lisbon: Museu Nacional de Etnologia, Ministério da Cultura.

Sevilla Guzmán, E. 1979. *La Evolución del Campesinado en España: Elementos para una sociología política del campesinado*. Barcelona: Península.

Sevilla Guzmán, E. and M. Pérez Yruela. 1976. "Para una definición sociológica del campesinado". *Agricultura y Sociedad* 1: 15–39.

Shanin, T. 1972. *The Awkward Class: Political sociology of peasantry in a developing society, Russia 1910–1925*. Oxford: Clarendon Press.

2 Decisions by farmers and land managers at the farm level

José Muñoz-Rojas and Teresa Pinto-Correia

2.1 Introduction: farmers and land managers

Farmers are generally considered as the key actors making decisions on farm management and the most important human agents of change in farming systems (Pinto-Correia and Azeda 2017; Primdahl 1999; Primdahl et al. 2010). This is due to their unique responsibility in implementing decisions at the farm/plot scales on a daily to seasonal basis (Nuthall 2012; Raymond et al. 2016). In contrast to landowners, who merely hold the legal property rights, farmers actively undertake management decisions with direct consequences over the outputs of the farm, and over changes or continuities in land-use and land-cover. Another key difference between farmers and landowners is that the latter hold exclusive responsibilities that are linked to their legal property rights, which entail decisions that can frequently be considered strategic, resulting in actions whose effects usually manifest themselves over longer timescales. This said, in most places in the world, the role of farmers and landowners usually coincides, with one person or legal entity simultaneously acting as owner and manager. However, in the montado and dehesa, these two roles are often separate as land ownership is still tightly connected with family heritage and landowners are often absent from the farm, as described in Chapter 1 (see also Bussoni et al. 2019). This provides the farmers' role in these areas a specific relevance.

In the Iberian silvopastoral systems, farmers can fall into several categories with different combinations of land ownership and management responsibilities. There are those who own their own farm and hold the rights linked to land ownership; close family members of the landowner who have agreed to manage, at times along with their own land, land which is owned by one or more family members; farmers who are simultaneously tenants, leasing the farm or holding the right to manage the farm through an informal agreement; and lastly, there may also be specialized and technical employees of the landowner, or a consulting firm, that undertake key management decisions on a daily and seasonal basis. This last case happens mostly when the landowner is absent from the farm and does not visit it so often – although, in reality, few absent landowners transfer the management decisions to an employee or

DOI: 10.4324/9781003028437-3

a firm. Thus, multiple possible combinations exist, although there is no data on the relative weight of each of these at regional or national levels. Heritage and symbolic values definitively play a key role (Pinto-Correia et al. 2019), as they also do in shaping the typologies of landowners (see Chapter 1), and while the silvopastoral farms are frequently kept in the family, they may be managed by someone outside the family.

Alongside farmers, there are other land managers in Iberian silvopastoral systems, who singly manage specific farm resources, such as forestry products, or part of the pastures for their own livestock production, or hunting, or even biodiversity – as described in Box 2.1. Previous research has shown that there can

---

TEXT BOX 2.1

Dehesa management and production models are determined by biophysical and socio-cultural factors

Gerardo Moreno, University of Extremadura

Key points: Management decisions are based on the potentialities and limitations of natural resources and these are also shaped by socio-cultural factors, often external to the individual farms/farmers, such as the governance context or markets.

Story – The coproduction of goods and services (ecosystem services) is based on feedback processes in which a social system actively shapes and modifies an ecosystem through farm management. In return, the ecosystem provides the physical framework and limits or increases the range of management options based on the ecosystem structure and the ecological processes underlying it. The combination results in a wealth of farmer types, dehesa structures and productive orientations.

Traditionally characterized by multifunctional low-intensity management, owners have in recent decades, shifted toward activities requiring less labour (e.g. cattle breeding instead of sheep breeding and meat breeds instead of milk breeds) and reducing the diversity of their production activities. These management decisions have, to some extent, been influenced by different socio-cultural and biophysical factors, such as land tenure and property size, resulting in progressively contrasting models of dehesa farms.

Based on 42 surveys and 16 quantitative biophysical and socio-cultural indicators, Torralba et al. (2018) applied a hierarchical cluster analysis and identified four main categories of dehesa, as defined by their characteristics and management (Figure 2.1).

A Large and heterogeneous: This category includes large properties, that were also mostly privately owned (as opposed to public, rented, or commonly owned), with a high diversity of land uses. These farms are associated with a high number of products, hunting intensity, housing facilities and proportion of stone walls and low levels of livestock and cereal production and external mineral and capital inputs.

B Small and homogenous: These properties tended to be smaller and more homogeneous in terms of land cover than those in the other clusters; with

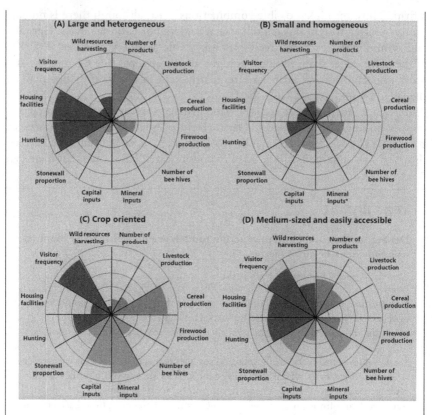

Figure 2.1 Flower diagrams illustrating the quantification of each management indicator by petal length. Each flower represents one dehesa category identified in the hierarchical cluster analysis. Different shades of grey in the
petals refer to the ecosystem service category with which the management indicator is associated (dark grey – cultural, light grey – regulating,
medium grey – provisioning).

Adapted from Torralba et al. (2018)

    a smaller range of products, less wild resource harvesting, lower hunting
intensity, fewer housing facilities and higher levels of capital inputs.

C Crop oriented: This category, less common, is clearly differentiated by specialization in crop production, with high levels of cereal production, use of
mineral and capital inputs and visitor frequency.

D Medium-sized with easy public access: These properties are defined by public access (either through public paths or being part or all common land),
rather than by traits related to property size or the diversity of vegetation
cover. These farms show high values for the number of products, wild
resource harvesting, the number of beehives, proportion of stone walls,
hunting intensity, housing facilities and visitor frequency and low values for
cereal production and mineral inputs.

be more than one land manager simultaneously making decisions about a single plot of land (Barroso 2013; Barroso and Pinto-Correia 2014). This possible combination of different individual interests and related management decisions, on the same land, contributes to the hybridity of Mediterranean agriculture (Ortiz-Miranda et al. 2013), an issue that is clearly identified in Chapter 8.

Decisions by farmers and other land managers are fundamental for the sustainability of silvopastoral systems. Once one acknowledges the farming system as a complex and uncertain playing field for farmers and land managers pursuing multiple, and not necessarily synergistic, goals, one should also recognize that disentangling motivations and decisions by farmers and land managers is a key step to tackling the many challenges encountered in the multifunctional dehesa and montado of Iberia. This is also the case in other similar silvopastoral systems across the Mediterranean. In this chapter the specific focus is on the management conditions, constraints and opportunities, with the aim of identifying what lies behind the complexities inherent in the wide range of decisions and strategies made by farmers and land managers in the Portuguese montado and Spanish dehesa.

The chapter begins by justifying the rationale behind the focus on decision-making at the farm-level. It continues by introducing the methods and approaches applied to analyze the complexity of farmers' and land managers' decision-making. Results obtained from the analysis of a regional case study in Alentejo, Portugal, are then presented and discussed, and pave the way to draw out the main lessons, presented in the conclusions. The diversity of land managers involved with individual elements of the overall land-use system are addressed in a separate section – since such actors need to adapt to the business model followed by farmers, and manage only a part of the farm resources, always in alignment with the farmer.

It should be stressed that the focus of the chapter is specifically on Alentejo, and thus that the results and lessons discussed cannot be directly considered as universally valid for other regions and silvopastoral systems across the Mediterranean. Nonetheless, given the extent and importance of the montado system in the Portuguese region of Alentejo, it is reasonable for the contents of this chapter to be considered as illustrative of the key challenges, threats and opportunities currently faced by farmers and land managers operating in other complex silvopastoral systems across the Mediterranean region.

2.2 Unravelling decision making at the farm-system level and beyond

Given the increasingly complex institutional context under which farmers currently operate (Pinto-Correia et al. 2019; Schermer et al. 2016), management decisions and strategies are far less foreseeable that the classic economic models, based on rationality and linearity, generally indicate.

Farmers exhibit complex, multiple and sometimes contradictory attitudes and behaviours (Barroso and Pinto-Correia 2014; Pinto-Correia and Azeda 2017). This complex combination of attitudes and behaviours, thoughts and actions can drive tensions and conflicts that underpin decisions, frequently resulting in unexpected and ineffective management strategies and outcomes. Furthermore, in the Mediterranean context where land managers and farmers are extremely heterogeneous, the institutional frameworks in which these key actors are embedded plays a key role in guiding their decisions (Barroso and Pinto-Correia 2014; Pinto-Correia et al. 2017).

In response to these challenges, institutional agricultural economics focuses on how inter-connected actors and networks across diverse levels of governance mutually interact (Ploeg and Marsden 2008). According to institutional economic theories, these actors follow complex behavioural patterns where external and internal drivers act as incentives or disincentives for decisions (Darnhofer et al. 2016; Schermer et al. 2016).

Institutional economic approaches emphasize the multiple roles of institutions both formal and informal, private or public, in driving decision making and strategies adopted by economic actors (Beckmann and Padmanabhan 2009; Ostrom 1998) in this case by farmers. In contrast with neo-classic economics, which largely views economic actors as isolated or individual agents operating strictly in their own self-interest and following logical reasoning in response to market and policy incentives, this alternative approach emphasizes the role of context (Ostrom 1998). This context considers, especially, networks and institutions, as key drivers of economic decision-making. According to this framework (Saccomandi 1998; Williamson 2000), transaction costs and thus mutual relationships among individual decision-makers are enacted through institutional arrangements and are key drivers of the efficiency of decisions and strategies. Thus, the farmer needs to be considered as the central piece in a wider farming system. Under such a conceptualization of the farming system, the farm (the economic unit of reference) and household (social unit) are highly dependent on external and internal conditions that materialize at different scales, from the farm plot to the global (Figure 2.2).

Institutional arrangements refer to the materialization of institutions, taking the form of formal government organizational structures as well as norms and beliefs, which insidiously rule human interactions (FAO 2018). Hence, as far as any farming system is concerned, institutional arrangements incorporate the networks of relationships that lead to the production and delivery of multiple goods and services, either public or private. Institutional arrangements result from the combination of market arrangements and public requirements and incentives, with market arrangements taking the form of horizontal cooperation and/or vertical

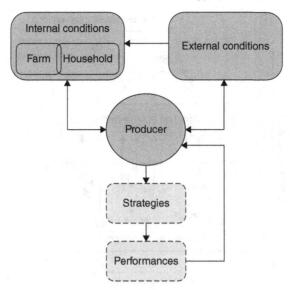

Figure 2.2 Diagram showing the contextual positioning of producers (including farmers and landowners) and how conditions, both internal (i.e. farm dimension,
bio-physical, agronomic and financial characteristics) and external (i.e. local
culture, global markets, policies, climate and social-ecological dynamics),
influence the farm management strategies undertaken, and ultimately its
financial and business performance and the welfare of the farm household.
This is of key relevance in a largely family-oriented farming system, such as
the montado (Muñoz-Rojas et al. 2019; Pinto-Correia et al. 2019).

coordination between actors and institutions (see Figure 2.3). In parallel, the public sector regulates the production and delivery of food and
other services and encourages some practices by subsidizing them (see also
Figure 2.3).

Furthermore, the extent to which costs and risks are shared in both vertical
and horizontal arrangements depends to a great extent on the distribution
of power within these arrangements. The institutional economic approach
recognizes that power dynamics can play a key role within both horizontal
arrangements, such as cooperatives and producer organizations, as well as
vertical structures, such as supply chains.

To identify and characterize the institutional arrangements influencing
farmers' decision-making in the montado, a survey was carried out in 2017
with a sample of montado farmers and land managers across Alentejo, focusing on institutional and other related supply chain arrangements.

A total number of 159 farmers were surveyed, who were identified
through snowball sampling, in collaboration with professional organizations
at the regional level, such as associations of producers of local and regional
breeds of cattle and cork. The farmers surveyed were spatially scattered across

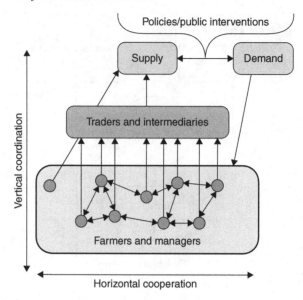

Figure 2.3 Synthesis of the actors, networks and relationships defining the three types of institutional arrangements considered in the analysis of montado farmers and land managers; horizontal cooperation among farmers, vertical coordination between actors and institutions along the supply chain, and policy interventions designed and implemented to correct market failures and mitigate the effects of key power imbalances.

the whole region so that the sample covers the territorial heterogeneity and related socio-economic, ecological and remoteness conditions, which influence farm strategies and performances. The sample is not representative of the full range of social and economic characteristics of farmers (and of their land) that can be encountered in the Alentejo region but covers a wide range of situations under which montado farmers make decisions.

Interviews were carried out face-to-face, in a location suggested by the farmer: their farm, a local association, a *café,* or a slaughterhouse. The interview guide included questions on: business-related, geographic and agronomic characteristics, as well as the dimensions of the farm itself, its geographical context, sales channels, characteristics of the sales agreements, business and farm sustainability, drivers and strategies for farming, policy drivers and the farmers' cooperation and coordination preferences and patterns (Muñoz-Rojas et al. 2019; Pinto-Correia et al. 2019).

2.3 The role of farmers and land managers in influencing the financial sustainability of montado

This study and the empirical evidence it acquired showed the benefits of adopting an institutional-economic approach in terms of unravelling some hidden aspects of decision-making at the farm-system level.

In this sense, the first key point to highlight is that farmers operating in the montado in Alentejo are context-dependent and that territorial aspects matter strongly, confirming original assumptions gained through years of direct experience working with farmers in the field, but which had remained empirically unconfirmed (Muñoz-Rojas et al. 2019). Many of the statements and assumptions arising from the scientific, statistic, legal and even media-based sources of relevance about the montado point to the multiple threats arising from the process of rapid landscape simplification, leading towards ongoing degradation (Godinho et al. 2016). This is recognized by a great number of farmers in the region and creates a great deal of uncertainty about the future of a system that they unequivocally consider worth preserving. Furthermore, these farmers are also largely aware of the relevance of human intervention for the maintenance and sustainability of their farming systems, something that has had already been indicated in previous studies (Bugalho et al. 2011).

Those outside the farming community generally consider financial and policy conditions as the key drivers of farm-based decision-making (see Primdahl 1999; Primdahl 2010; Pinto-Correia et al. 2019). Nonetheless, recently acquired evidence confirms the relevance of socio-cultural aspects, which despite being well documented in the scientific literature, are rarely considered in policy formulation and enactment (Pinto-Correia et al. 2010).

As described in further detail in Chapters 1 (Landowners) and 3 (Local Users and Professions) both the socio-cultural influence of largely conservative social norms still underpinning much of the farming strategies by the elite farming families in Alentejo, and the political connections of different farming models, still play a large role. The presence of these conditions, reaffirmed by many of the farmers surveyed, suggest some limitations for turning the system more sustainable, including the widespread rejection by many farmers and land managers of innovation and the poor level of horizontal cooperation (Barroso, 2013; Guimarães et al. 2019). This is demonstrated by the decline and in some cases, the fall of cooperatives and other associations through which market power was traditionally gained. In addition, these factors stand out as a key barrier for the entrance of new and young farmers who might bring fresh ideas and perspectives (Eistrup et al. 2019). Nonetheless, these are far from the only relevant conditions that affect farmers' strategies. The low quality and outreach of extension and other advisory services, and the lack of alignment and coordination of policies and plans that individually address the various components of the system (Pinto-Correia and Azeda 2017), also contribute to a lack of innovativeness among producers. This should be considered as a key barrier to be overcome in the design and implementation of novel governance frameworks and mechanisms that could help drive the system towards increased future standards of sustainability and resilience.

We identified several institutional arrangements currently adopted by farmers operating in the montado.

A Uniform individual arrangements, characterized by formal agreements established among individual farmers and also including other market and policy actors, generally established in advance. A frequent example of this is individual cattle sales agreements between a single farmer and a supermarket chain on a yearly basis, whereby a secure and yet inflexible price for a certain commodity, such as meat, is fixed *ex-ante*.

B Segmented individual arrangements characterized by formal agreements established among multiple farmers and also including other market and policy actors, set in advance. Other market actors include exporters, intermediaries and traders, who hold most of the bargaining power in this type of transaction, but who also provide certain stability and security to farmers, reducing their costs and workload beyond the scope of the farm.

C Pure market arrangements, characterized by informal agreements that are not legally enforceable, typically in the form of informal verbal agreements. These are more frequent than one might expect, especially in times where public monitoring and control are prevalent, and they raise concerns in both the public and private sectors in relation to quality and traceability.

D Segmented collective arrangements, mirroring the segmented individual arrangements described in cluster B, with the relationship along vertical value chains being quite strong and coordinated, and requiring exclusivity from farmers along with standards for differentiation

E Uniform collective arrangements: these mirrors the uniform individual arrangements, with the main difference being that one of the parties is a collective organization, such as a cooperative, and the agreement rules are part of the rules of being a member of the cooperative. This is frequently enacted through farmers' associations, normally linked to specific breeds or commodities (such as cork or a POD breed of cattle) who negotiate and deal with the stronger market agents, such as exporters, on behalf of a group of farmers, who are linked through a common locality or commodity.

Among these types of arrangements, in Alentejo, the individual arrangements clearly outnumber the others, and this is reflected in the predominant narrative found in the literature and less formal social discourses (Muñoz-Rojas et al. 2019; Pinto-Correia et al. 2019) that characterise many montado farmers as individualists. This finding also helps to explain the many problems encountered in establishing horizontal cooperation and, especially, in improving vertical coordination along the supply chain of the various montado-related commodities, especially involving the two most widespread ones, livestock and cork (Bugalho et al. 2011). Despite these limitations, a relatively wide diversity is found in the strategies followed by farmers to cope with a rising market, policy and environmental challenges, including farm abandonment, land concentration, intensification and credit expansion.

Our survey also revealed an assumption among many farmers that public funding will continue to be widely available in the future, an attitude that potentially hampers their capacity to adopt sustainable alternative strategies including, especially, those related to increased horizontal cooperation and improved value-chain vertical cooperation.

Another key factor that we detected relates to the gap between the aspirations and motivations expressed by farmers, their actual strategies and the various solutions currently in place. Thus, while many montado farmers think of themselves as land stewards (Muñoz-Rojas et al. 2019; Pinto-Correia et al. 2019) and are therefore motivated by a sense of duty to secure that their successors can inherit it (see Chapter 1 for further details on Land Ownership), they face difficulties in finding strategies that secure this in practice. Too few of them have the financial capacity or, at times, even the confidence and personal will to invest any further, specialize their production, externalize their activities or insure against potential losses. These are all strategies that are generally considered fundamental for the survival of any modern farming system across the Mediterranean (Ortiz Miranda et al. 2013).

Nonetheless, in a system such as the montado, some of these findings may prove beneficial for the maintenance of the system. It is important at this stage to remember that this is a complex system whereby high nature (Almeida et al. 2013; Ferraz de Oliveira et al. 2016), cultural (Pinto-Correia et al. 2010, 2016) and landscape (Pinto-Correia et al. 2010) values linked to diversity and heterogeneity ought to be considered as intrinsic and relevant to securing financial sustainability (Muñoz-Rojas et al. 2019; Pinto-Correia et al. 2019). As a matter of fact, montado farmers in Alentejo highly value objectives such as maintaining biodiversity, landscape integrity and related multi-functionality or following socio-cultural traditions.

Another factor strongly influencing the future financial sustainability of these farming systems is related to the strong dependence of farmers on public subsidies. In montado farms in marginal locations with the lowest levels of productivity, and poor accessibility and connectivity, and thus missing key advantages of access to market nodes, public subsidies often make up a large proportion of their net income, and rarely less than half. This creates problems related to the low level of policy and market power exercised by many montado farmers, who are largely dependent on either the public funding bodies granting funds (basically through CAP) or on bigger market operators, such as supermarket chains. It is these external actors, either public or private, who ultimately decide what business model (Pinto-Correia and Azeda 2017; Muñoz-Rojas et al. 2019) they should pursue – with direct consequences on the montado's survival.

Nonetheless, there are some individual exceptions: pioneer farmers who seek to implement alternative strategies, circumventing the restrictions created by centralized policies or by key market agents. Such decisions are made at great financial risk for these pioneering farmers and, of course, also ultimately for the landowner, and too often only those with very large scales or strong personalities are capable of adopting truly innovative and risk-based approaches to management. This limitation in governance capacity adds to the current debate on the decline of the montado systems of Alentejo, a decline that is often linked to the processes of a land management unit and land property concentration (Godinho et al. 2016).

One factor that might help tackle the multiple limitations encountered in governance models enacted by farmers (Pinto-Correia et al. 2019) is the high proportion of these actors that receive technical support and professional assistance. Given the nearly complete absence of public extension services, this advice is largely provided, with some regularity, by private advisory services. Another factor that may help counterbalance the poor levels of horizontal cooperation among farmers is strong and culturally determined set of social support ties and networks that persist between many farmers. Such ties are mainly related to family, and contrasts with the poorer levels of cooperation expected of, or existing with, colleagues and neighbours. The last factor that needs to be highlighted is the ageing of montado farmers, although their relatively high levels of education and motivations might soften the inherent problems with this (Eistrup et al. 2019).

To summarize the key lessons learned, it is clear that farmers in the montado of central Alentejo are far from homogeneous in their motivations, aspirations and decisions.[1] Furthermore, they are far from acting as rational economic agents, with cultural and historical traditions weighing strongly upon the factors driving their decisions. As a result of all of this, multiple gaps can be detected in institutional arrangements, including in horizontal cooperation and vertical coordination. These factors need to be added to the lack of a uniform and widely agreed vision for where the system should head,[2] resulting in the problems encountered in defining strategies for the future that satisfy the aspirations of the multiple types of farmers, while at the same type securing the future financial sustainability and self-sufficiency of this farming system

2.4 Other land managers

Much of the existing literature on land management in the montado and dehesa focuses on the traditional roles of farmers: those who take the definitive and decisions on land use (Bussoni et al. 2019; Kay et al. 2019; Pinto-Correia and Azeda 2017). Yet alongside these farmers, who are involved in the management of the farm system as a whole, other land managers may exist, including individuals who manage single components of the overall farming system, or the production of one specific commodity in the farm but not the farm production system as a whole. They may be in charge of forestry products, or specific grazing areas, or hunting or even biodiversity. Previous work has shown there can be more than one land manager simultaneously making decisions about the same plot of land (Barroso 2013; Barroso and Pinto-Correia 2014).

To the best of our knowledge, this results in a rather unique complexity. The combination of different individual interests and their management, in the same unit of land, contributes to the hybridity of Mediterranean agriculture (Ortiz-Miranda et al. 2013). In one single farm or estate, there may be different individuals who take decisions, and all need to be aligned with each other. This largely happens without conflicts, based on a strongly regionally embedded customary form of management.

The landowner – or the farmer when the landowner is absent – is the one who decides which component in the farm is managed by whom. He or she is also the one to whom the other managers are linked. In this context, one-to-one relationships are frequently formed, although the management structure may result in a quite complex decision network.

These other land managers do not have rights to the full use of the land, but to part of this use, on a semi-permanent basis – usually for a period of many years. Their right to use the land and/or resources in the farm may be defined through a formal contract or, more frequently, an informal agreement. Such land managers are often people who own livestock but not enough land to graze his/her animals, as described in Box 2.2. In such cases, this livestock

---

TEXT BOX 2.2

A livestock producer as a land manager in a montado farm, Alcácer do Sal, Alentejo

Teresa Pinto-Correia, MED, University of Évora.

Key points: The combination, on the same farm and for the same production system, of management by the landowner and farmer along with at least another land manager.

The *Serra dos Mendes* farm covers 111 hectares of thin soils of the hilly northern part of the municipality of Alcácer do Sal, in Coastal Alentejo. The farm used to be a larger estate of more than 500 hectares and was successively divided since the 1970s through a process of farm succession and the need to repay bank loans incurred by the last owner of a traditional land-owning family. In the late 1990s, the farm was bought by new owners, who settled there to produce Iberian pork (Figure 2.4). The montado, mainly cork oak, was already in decay and the strategy adopted was to protect and renew the tree cover and create a business model with low-intensity impact over tree regeneration, after the shrub was cleaned. Due to the relatively small size of the farm and the fluctuations in the market price for Iberian pork, this strategy did not generate an adequate income and was restricted to part-time production. The landowner returned to his part-time job outside the

Figure 2.4 Iberian pigs grazing under the tree cover in the montado.
Photo by José Muñoz-Rojas. Taken in Barrancos (Alentejo), Portugal in 2020.

farm. In the meantime, a young entrepreneur in the village was increasing his sheep production but had limited access to land. This young producer does not belong to a landowner's family, but rather to a farm worker's family. An informal agreement was achieved: pig raising was maintained in a limited area of the montado, managed by the landowner. The remaining montado areas, as well as the plots with open pastures were exploited by the sheep producer, who also grazed and took care of a small number of sheep belonging to the landowner. This producer became the land manager of the largest part in the farm, although subject to the agreement by the landowner, who also maintains his management activities. In this way, the landowner secures a diversity of income sources (pigs, sheep, cork and his other job). And also, not less importantly, while having another job, he also has secured the daily management of a large part of the farm by the younger producer, who keeps the shrub under control, maintains the fencing and addresses the daily operations of the farm.

producer may enter into an agreement with a farmer from another property, to graze the most distant plots in the farm, or to use the grazing only seasonally, while the farmers are compensated by the livestock producer who agrees to undertake responsibilities such as fence maintenance, or shrub control. Similar arrangements in the Spanish dehesas in Extremadura and Andalucía, where pig herd managers pay landowners for leased land based on the seasonal gains in their herds' weight.

In this way, in the most complex land management agreements, one may encounter on a single farm a situation whereby:

- One or more individuals are the landowners;
- There may also be a farmer who looks after the everyday management of the main farm system;

- A second land manager may exist who has the right to part of the grazing area using it for feeding his own livestock;
- Another land manager may also be engaged who has the right to manage hunting (e.g. an association). providing land management interventions to support game diversity,
- There may also be someone who keeps beehives on the farm and has the right of access to the farm for part or all the year but with no land management rights;
- Lastly, other actors may have access to the farm to pick some wild plants over a limited period of the year.

Of course, this is only an illustrative example, and such a multiplicity of actors is not the norm. Their joint actions in the one single farm is based on a clear hierarchy, not necessarily legally written down but acknowledged by all, and depending on the coordination, and main decisions, centralized in the landowner or farmer.

To further add to such complexity, one could also consider as land managers those individuals who might make use of in-farm resources but who bear no responsibilities of land-use management, since they are only redrawing/collecting resources, in many cases with access to the farm only for limited periods in time. These other actors and their roles are described separately, in more in detail, in Chapter 3.

2.5 Conclusions

Up to now, the motivations of montado farmers have been addressed mainly by identifying the discourses (Pinto-Correia et al. 2019) and motivations (Muñoz-Rojas et al. 2019) underpinning their resulting strategies. The analysis of other relevant factors pertaining to institutional theory, such as conditions and institutional arrangements has been largely neglected. To fill this void, the authors of this chapter have applied institutional economics approaches to help enrich this picture, providing a broader understanding of the key conditions that influence farmers' strategies and thus the performances (current and future) of the farming system as a whole. Our findings indicate a complex and diverse picture where key gaps exist in farmers' capacity for mutual cooperation and alignment, and for the ideation and implementation of more innovative strategies to enhance the sustainability of the system. The capacity for those strategies is mainly hampered by socio-cultural factors, but also policy and administrative factors. Evidence shows how farmers' reliance on, and expectations about, public subsidies is in sharp contrast with their fierce individualism and close affinity with the heritage values of the system which is seen as changing too fast and not in positive ways.

Some potential pathways could help these systems move in more sustainable directions and help overcome the multiple barriers towards adopting governance models capable of underpinning the future sustainability of the sector. These pathways include:

- Fostering more receptive attitudes among farmers (especially related to trust) and structures for horizontal cooperation;
- Providing improved support for moving towards more efficient vertical coordination strategies along the supply chain, and;
- Facilitating a shift of mentality of farmers to move from beneficiaries of public funds towards land stewards who are well-positioned for providing multiple services to society.

Whether these changes become implemented ultimately depends upon farmers themselves.

Lastly, it remains to be seen whether the necessary governance changes, that are made even more urgent by climate change, environmental and landscape degradation and demographic trends, will be made possible through generational renewal, institutional reform, cultural shifts or more efficient incentives or even, what seems like the optimal, and yet least likely option, a combination of all of the these. These possibilities are specifically addressed in Section C of this book.

Acknowledgements

The authors would like to acknowledge the European Commission's HORIZON 2020 programme which through the SUFISA project (n. 635577, 2015–2019) funded the research tasks, analyses and findings underpinning the arguments in this chapter. They would also like to thank María Rivera Méndez (MED-Universidade de Évora) for her invaluable support with the design and improvement of the graphic figures in this chapter.

Notes

1. For a more complete picture of the results of the survey the reader can consult the following sources and reports: https://www.sufisa.eu/wp-content/uploads/2019/02/Summary-Beef-Portugal_def.pdf, and https://www.sufisa.eu/wp-content/uploads/2018/09/D_2.2-Portugal-National-Report.pdf.
2. The following synthesis report details the scenarios constructed for this and other farming systems across Europe,: https://www.sufisa.eu/wp-content/uploads/2019/07/Deliverable-4.2.pdf

References

Almeida, M., Guerra, C., & T. Pinto-Correia. 2013. "Unfolding relations between land cover and farm management: High nature value assessment in complex silvopastoral systems". *Geogr. Tidsskr. J. Geogr* 113, 97–108.

Barroso, F. 2013. "How can land managers and their multi-stakeholder network at the farm level influence the multifunctional transitions pathways?" *Spanish Journal of Rural Development* IV(4):35–48.

Barroso, F., & T. Pinto-Correia. 2014. "Land managers' heterogeneity in Mediterranean landscapes - consistencies and contradictions between attitudes and behaviors". *Journal of Landscape Ecology* 7(1): 45-74.

Beckman, V., & M. Padmanabhan. 2009. *"Institutions and sustainability: Political economy of agriculture and the environment – essays in honour of Konrad Hagedorn 2009th Edition"*. Springer-Verlag, 2009.

Bugalho, M.N., Caldeira, M.C., Pereira, J.S., Aronson, J.A., & J. Pausas. 2011. "Mediterranean oak savannas require human use to sustain biodiversity and ecosystem services". *Frontiers in Ecology and the Environment* 5: 278–286.

Bussoni, A., Alvarez, J., Cubbage, F., Ferreira, G., & V. Picasso. 2019. "Diverse strategies for integration of forestry and livestock production". *Agroforestry Systems* 93(1):333–344.

Darnhofer, I. 2020. "Farming from a process-relational perspective: Making openings for change visible." *Sociologia Ruralis* 60(2): 505-528.

Darnhofer, I, Lamine, C., Strauss, A., & M. Navarrete. 2016. "The resilience of family farms: Towards a relational approach". *Journal of Rural Studies* 44:111–122.

Eistrup, M.s, Sanches, A.R., Muñoz-Rojas, J., & T. Pinto Correia. 2019. "A 'young farmer problem'? Opportunities and constraints for generational renewal in farm management: an example from southern Europe". *Land* 8(4):1–13.

FAO 2018. *"Institutional Capacity Assessment. An Approach for National Adaptation Planning in the Agriculture Sectors"*. *Briefing Note* [Available at: http://www.fao.org/3/I8900EN/i8900en.pdf] FAO, Roma

Ferraz-de-Oliveira I, Azeda C, & T. Pinto-Correia. 2016. "Management of montado and *dehesa* for high nature value: An interdisciplinary pathway". *Agroforestry Systems*90:1–6.

Godinho, S., Guiomar, N., Machado, R., Santos, P., Sá-Sousa, P., Fernandes, J.P., Neves, N., & T. Pinto-Correia. 2016. "Assessment of environment, land management, and spatial variables on recent changes in montado land cover in southern Portugal". *Agroforest Systems* 90:177–192.

Guimarães, M.H., Esgalhado, C., Ferraz-de-Oliveira, I., T. Pinto-Correia. 2019. "When does innovation become custom? A case study of the montado, Southern Portugal". *Open Agriculture* 4(1):144–158.

Kay, S., Graves, A., Palma, J.H.N., Moreno, G., Roces-Díaz, J.V., Aviron, S., Chouvardas, D., Crous-Duran, J., Ferreiro-Domínguez, N., García de Jalón, S., Măcicăşan, V., Mosquera-Losada, M.R., Pantera, A., Santiago-Freijanes, J.J., Szerencsits, E., Torralba, M., Burgess, P.J., & F. Herzog. 2019. "Agroforestry Is paying off – economic evaluation of ecosystem services in European landscapes with and without agroforestry systems". *Ecosystem Services* 36:100896.

Muñoz-Rojas, J., Pinto-Correia, T., Hvarregaard Thorsoe, M., & E. Noe. 2019. "The Portuguese montado: A complex system under tension between different land use *management paradigms"*. *In Silvicultures – Management and Conservation*, edited by F. Allende Álvarez; G. Gómez-Mediavilla, N. López-Estébanez, 146–164. United Kingdom: IntechOpen.

Nuthall, P.L. 2012. "The intuitive world of farmers – the case of grazing management systems and experts". *Agricultural Systems* 107:65–73.

Ortiz-Miranda, D., A. Moragues-Faus, & E. Arnalte-Alegra. 2013. *"Agriculture in Mediterranean Europe. Between old and new paradigms"*. 1st ed. Emerald. Bingley, UK.

Ostrom, E. 1998. "A behavioural approach to the rational choice theory of collective action". Presidential Address, American Political Science Association, 1997. *The American Political Science Review* 92(1) :1–22

Pinto-Correia, T., Almeida, M., & C. Gonzalez. 2017. "Transition from production to lifestyle farming: New management arrangements in Portuguese small farms". *International Journal of Biodiversity Science, Ecosystem Services and Management* 13(2):136–46.

Pinto-Correia, T., & C. Azeda. 2017. "Public policies creating tensions in montado management models: Insights from farmers' representations". *Land Use Policy* 64.

Pinto-Correia T., Barroso F., & H Menezes. 2010. "The changing role of farming in a peripheric south European area – the challenge of the landscape amenities demand". In: Wiggering H., Ende HP., Knierim A., Pintar M. (eds) *"Innovations in European Rural Landscapes"*. Springer, Berlin, Heidelberg.

Pinto-Correia, T., Muñoz-Rojas, J., Hvarregaard Thorsøe, M., & E. B. Noe. 2019. "Governance discourses reflecting tensions in a multifunctional land use system in decay; tradition versus modernity in the Portuguese montado". *Sustainability* 11(12):3363.

Ploeg, J. D. van der, & T. Marsden. 2008. *"Unfolding Webs. The dynamics of regional rural development"*. 1st ed. Assen, NL: van Gorcum.

Primdahl, J. 1999. "Agricultural Landscapes as places of production and for living in owner's versus producer's decision making and the implications for planning". *Landscape and Urban Planning* 46:143–150.

Primdahl, J., Kristensen, L., Busk, A., & H. Vejre. 2010. "Functional and structural changes of agricultural landscapes : how changes are conceived by local farmers in two Danish rural communities". *Landscape Research* 35(6):633–653.

Raymond, C. M., Bieling, C., Fagerholm, N., Martin-Lopez, B., & T. Plieninger. 2016. "The farmer as a landscape steward: Comparing local understandings of landscape stewardship, landscape values, and land management actions". *Ambio* 45(2).

Saccomandi, V. 1998. *"Agricultural Market Economics. A neo-institutional analysis of the exchange, circulation and distribution of agricultural products"*. Van Gorcum-*Assen. Netherlands. ISBN 9023229398.

Schermer, M., Darnhofer, I., Daugstad, K., Gabillet, M., Lavorel, S., & M. Steinbacher. 2016. *"Institutional impacts on the resilience of mountain grasslands : an analysis based on three European case studies"*. *Land Use Policy* 52:382–391.

Torralba, M., Oteros-Rozas, E., Moreno, G., & T. Plieninger. 2018. "Exploring the role of management in the coproduction of ecosystem services from Spanish wooded rangelands". *Rangeland Ecology & Management* 71(5):549–559.

Williamson, O. 2000. "The 'new institutional economics: Taking stock, looking ahead". *Journal of Economic Literature* 38(3):595–613.

3 Professions and local users

Rufino Acosta-Naranjo, Teresa Pinto-Correia and Laura Amores-Lemus

3.1 Introduction

Dehesa and montado farms are multi-functional complexes and extensive agro-ecological systems that require the presence of many different workers/ professions, on a daily, seasonal or sporadic basis, alongside the local population which use the farms for the provision of diverse resources and services. In this chapter, we differentiate between professionals, the various workers on and around the farms, and users: other people who enjoy or make use of the dehesa and montado for diverse purposes. The reason for dedicating a chapter of this book to these groups is because a complete map of all the actors involved is needed to plan and implement governance mechanisms that take all stake-holders' interest into account. Identifying all of the systems' components is an important first step to evaluating and enhancing the systems' resilience.

Although there are numerous economic studies on the dehesa and the montado, they usually are focused on the landowners as decision-makers. Research on the other social groups present in this agroecosystem, and especially on conflicts and work conditions are far less common. Thus, there is a dearth of details about the people who work and use the montado and dehesa: they seem to be invisible, something we aim to remedy in this chapter.

The traditional dehesa and montado contained a wide variety of employ-ees, permanent or seasonal, who managed a great diversity of tasks (Sánchez Gómez, 1993; Picão, 1983; Acosta-Naranjo, 2002). They are listed in Table 3.1. But this plethora of specialized occupations has almost completely disappeared. The farms and the silvopastoral system have undergone pro-found transformations and adaptations to social, commercial, technological and public policy changes that have led to the loss of many activities, the outsourcing of others, and specialization in livestock or forestry, especially for the production of cork.

At the same time, some of those who use and work, and therefore know this system are still there, not as employees but as users: hunters, gatherers, tourists and others. To engage in processes of governance, it is important to understand which types of users are to be found in the montado and dehesa along with their interests and their skills.

DOI: 10.4324/9781003028437-4

Table 3.1 The variety of employees, permanent and seasonal, working in a dehesa or montado estate until the middle of the 20th century, and how their function is secured nowadays (SVWM: single versatile worker-manager)

Permanent workers	Present status
Housekeeper	Disappeared. Exclusive in very few farms. SVWM
Encargado (SP) / *encarregado (PT)*(manager)	Assumed by SVWM
	Still exists in farms where a landowner is absent
Guard	Disappeared, except in hunting farms. SVWM assumes the charge
Bicheiro (PT) Controller of predators	Disappeared
Worker in charge of mules or oxen	Disappeared
Pavero (turkey keeper)	Disappeared. Very unusual occupation
Swineherd	Assumed by SVWM and maintained at big farms with important pig herds
Goatherd	Almost disappeared. Only maintained in very few big farms
Cowboy	Assumed by SVWM and maintained at big farms with important cattle herds
Shepherd	Assumed by SVWM and maintained in some large farms
Shepherd's assistants, Swineherd assistant Goatherder assistant Cowboy assistant	Disappeared
Ploughman	Disappeared. Assumed by SVWM or specialized worker on machinery managing in a few cultivated dehesas
Aperador (SP) (responsible for cultivation work)	Disappeared. Assumed by SVWM
Accountant	Disappeared. Assumed by SVWM or manager. Outsourced to administration/ consultancy companies
Rapa (SP) (boy who ran errands)	Disappeared
Yunteros (SP)/ *seareiros* (PT) (landless cultivators who rented cultivation areas at the dehesa)	Disappeared

Temporary workers	Present status
Coalmakers	Highly reduced by mechanization
Woman coal gatherers	Disappeared
Sachadores(SP)/*Desmatadores* (PT) (workers weeding)	Disappeared
Guards of acorns	Disappeared
Cork harvester	Maintained
Pruner	In Spain, highly reduced by mechanization and because of the delay of pruning time. In Portugal, maintained
Castrator	Assumed by veterinarian

Table 3.1 (Continued)

Temporary workers	Present status
Carnero, in charge of male sheep outside the covering period	Disappeared
Temporil, shepherds' assistant during *paridera* (farrowing time)	Disappeared
Gordero (SP)/ Porqueiro (PT), swineherd during *montanera* (acorn pig feeding time)	Disappeared
Tosquiador (PT) Shearer	Maintained but enormous reduction by mechanization
Acorn picker	Almost disappeared – is returning with the retro-innovation in acorn use for human consumption
Arrancadora (women for legume harvest)	Disappeared
Sowers	Disappeared. Assumed by SVWM or specialized worker on machinery managing in a few cultivated dehesas
Reaper	Disappeared. Assumed by SVWM or specialized worker on machinery managing in a few cultivated dehesas

3.1.1 Permanent workers

Even just 50 years ago the workers on some single estates, along with their families, would have enough to constitute a small village. On a farm of 1000 ha, there could have been up to 50 workers (see Box 3.1). These workers had multiple professions, as described in Boxes 3.1 and 3.2. Some of these tasks could be performed by external, temporary workers – but when the estate was large, it would employ many full-time workers.

Today, the *single versatile worker-manager* (SVWM) is a very common figure. Specialized silvopastoral farms generally require only one multi-purpose worker who performs a set of tasks that were previously assigned to a group of specialized employees in the former multi-purpose farms: general manager, livestock manager; cowboy; shepherd; goatherd; swineherd, tractor driver; housekeeper; fence keeper; tree pruning specialist; guard or handler of temporary workers. In the present context, there is no longer a need for a worker to take care of animals or longer accompany them, as fencing is used to keep the livestock within designated plots (Acosta-Naranjo, 2008; Caballero et al., 2009).

It is remarkable that, despite the high unemployment rates in the regions where dehesa is prominent, such as Andalusia where it reaches 20% (Junta de Andalucía, 2017), farm owners can have problems finding a reliable SVWM (and almost impossible to find one who accepts to live on the farm). This is a problem, despite salaries being quite reasonable, exceeding usually 1200€/a month on average in 2020 (when the legal minimum salary in Spain is 950€/month). The same is reported in Portugal, in the region of Alentejo. Apart from the SVWM, it is rare to find farm units with other permanent

TEXT BOX 3.1

Herdade das Parchanas, Alentejo region: the multiple and specialized professions in a silvopastoral estate in the 1970s

Isabel Manoel is a daughter in law of the landowners; she knows well the estate since the 1970s and was interviewed by Teresa Pinto Correia March 2020

Key points: The multiplicity of professions found in a silvopastoral farm until the 1970s and the number of workers who spent their lives on such a large estate

Herdade das Parchanas is a large estate of 1350 ha, 1170 ha of cork oak, with some plots mixed with pine trees. Otherwise, there is an olive grove, a vineyard and rice fields. Today the estate is still in the hands of the same family as in the 1970s, but continues with very few workers. Only 50 years ago, there were more than 60 workers living in the estate, with their families – a whole village. There were various professions, some very specialized: a farm manager, a bookkeeper, a driver, a mechanic, a carpenter, a mason, a forestry guard and two wildlife controllers, a school teacher and even a social assistant to teach women about their housekeeping and children's education. The pig keeper, the cow keeper and the two sheep shepherds, were also specialized and often helped by their wives or children. And then there were undifferentiated workers, with their families. The work done followed the seasons, and the same man or woman could do different tasks throughout the year: In October and November they would plough 300 ha with oxen or a tractor, and then sow oat and rye, all by hand, with 10 men in a

Figure 3.1 Cork harvester.

row; In December and January, there was the olive harvesting; and during the winter and until May they would clean the shrub, with the men cutting the shrub and the women putting it together and burning it. In May and June, the women harvested the cereals while the men transported them and managed the thresher. In July and August, the women whitewashed all the buildings. In September some helped in the harvest from the vineyard and then with the winemaking. Some workers were specialized in forest work: pruning the trees, collecting the wood, harvesting the cork (Figure 3.1) and harvesting pine cones. In years when much cork was harvested other workers helped with the cork. Some women worked 'in the house', for the landowners: cleaning, cooking and baking. The owners lived in Lisbon, and the landlord visited the farm once a week to check the whole farm management with the manager; he would take products and food from the farm to the family house in Lisbon. Very little was work externally contracted, except with the rice culture, calving sows and pruning the vineyard.

TEXT BOX 3.2

Sierra Prieta; 800 ha estates, all holm oaks

Rufino Acosta-Naranjo, of the University of Seville, collected data on this estate during his PhD research in 1990s. In February 2020 he collected more up-to-date and accurate data on the employees' profile through interviews with Manuel Garrote Rosa, who started as a work team member, and later became work-er-manager of the farm until his retirement.

Key points: The multiplicity of uses and tasks in the 1960s and the large number of employees who worked and lived on the estate and the subsequent reduction of manpower and specializations at the farm.

In the 1960s, the farm was part of the properties of a family of large landowners, who had three other dehesa properties nearby and who lived in Llerena, a town 20 kms away. There was a manager (the son-in-law of the owner) for the four large estates. There was a permanent work team of eight workers who travelled through these four farms doing seasonal tasks such as shrub cleaning, weeding, harvesting, pruning and building and repairing fences. The farm was divided into different spaces that were used on a rotational basis. In first and second years, a plot was cultivated with wheat grown for human consumption, together with other cereal and legume species for the animals. There were seasonal tasks of sowing, weeding and harvesting. For the following three years the plot was used as pasture land, and was cultivated again in the sixth year. Holm oaks were pruned every five years.

There were around 22 permanent workers in Sierra Prieta exclusively working and living on the farm, with their families. In addition, there were occasional employees in pruning, weeding, mowing, acorn gathering and assistants to those in charge of animals at farrowing time and the *montanera* (the time for fattening pigs on acorns). This is the list of the jobs:Housekeeper and his wife, taking care of the *cortijo* (farmhouse) and the poultry. Guard. *Aperador*, responsible for crop related tasks, especially ploughing. Five *mozos de mulas,* workers in charge of the mules, and another man in charge of the oxen, all used for cultivation tasks. Three herds of sheep, each with a head shepherd and his assistant. Four herds of pigs, each with a head swineherd and his assistant. A cowboy and his assistant.

Today, this estate and a neighbouring property (180 ha) are managed directly as a unique estate by the current owners, two brothers living in Madrid, grandchildren of the former owner and sons of the former manager, with two permanent employees for all the farm work. Crops have been abandoned, and production is exclusively livestock breeding with hay being cut for the animals. Decades can pass between each oak pruning and cultivation. Pruning (Figure 3.2) is done by

Figure 3.2 Prunner.

external contractors who take firewood in exchange, to sell or transform it into charcoal. The farm is a hunting reserve rented to hunters from a nearby village. The two permanent employees live in nearby towns. One of them is the worker-manager and the other is a multipurpose employee; they divide up the tasks of taking care of animals, although with a certain specialization. They are also in charge of the cultivation when needed, and of the maintenance of the farm. Some people are temporarily employed, especially for tasks related to animal welfare.

employees, except in the very large estates or farm units with more intensive livestock raising. When there is more than one employee, there is a certain specialization: the secondary worker is dedicated to some specific tasks, for example, a certain livestock species, especially those requiring more labour, such as sheep or pigs. In Spain, the farms in which there are more permanent employees are usually those where the intensive fattening of Iberian pigs has been developed. This situation is less common in Portugal. Goats require more labour, but they are rarely found in large silvopastoral farms today. In the farms producing cattle, the manpower demand is lower, except in the large farms that raise fighting bulls (Vargas, Huntsinger and Starrs, 2013), in which there are generally more than two permanent employees.

Dehesa/ montado farms may need more permanent workers when they invest in exploiting hunting commercially, which are only the larger farms, with more than 1000 ha of hunting area. Nevertheless, few farms have staff dedicated exclusively to the care of the game and the management of hunting. The hunting employee acts as a guardian and in some cases feeds the hunted species, such as partridges or deer.

In many cases, the second employee is a relative (son, son-in-law or nephew) of the worker-manager. Only in very exceptional cases, and in very large estates, do these secondary employees live on the farm. In recent years a new group of people has started working in the dehesa and montado, immigrant workers, who may live on the farm. There is still no statistical data or case-studies data focusing on these workers. The wife of the worker-manager living on the estate is usually the only one woman working at the dehesa, caring for the *cortijo/monte*, the house. An important thing to take into account is the gender dimension (see Chapter 5), as with the reduction of the number and diversity of workers, women's work seems to have vanished from this agroecosystem; the SVWM are, to the best of our knowledge, always men. Women's work in the montado and dehesa is now in more specialized professions as veterinarians, agronomy consultants, forest agents, accountants and so.

3.2 Temporary workers

The number of temporary employees has dropped dramatically since 1960s/70s, as shown in the text boxes and Table 3.1. The reduction of employment in the farms resulted in a high unemployment rate in the villages, especially in Andalusia and Extremadura and also Alentejo, the most latifundist areas, and led to continuous emigration to the cities and industry. Usually, seasonal employees only work in the dehesa or montado for a few days in a year.

Today, extensive livestock farming focused on meat production rarely requires the hiring of temporary workers, except for very specific tasks or to cover the permanent worker's holidays. Specialized staff is hired at specific times: ironwork, castration, shearing or sanitation of animals, sometimes during the *montanera* (the acorn fattening time); pruning, clearing shrubs or cork extraction; or harvesting the few crops that are still cultivated. This happens in very similar ways both in Spain and in Portugal. These specialized workers are skilled individuals, generally men, from the region who live in the region, who have learned some of these techniques from their elders.

One occasional occupation that remains is that of shearer, since sheep shearing is an essential activity and requires a certain experience. The sheep population registers variations along time (ex. for the Alentejo region, 1989: 1 505 314; 1998: 1 883 000; 2008: 1 761 000; 2018: 1 361 000; data from INE-National Agricultural Statistics) while the number of shearers is constantly decreasing, due to the mechanization of the profession – which has reduced the time required to shear a sheep by 80%. The task is carried out by teams who work at most one month a year. These are temporary workers, belonging to the group of skilled workers mentioned above. They do this task in the sheep shearing period, and have other work/employment in the rest of the year. Sometimes the shearers are small livestock owners, who have the necessary equipment, razors and electric generators. Often the work teams are made up of several relatives. Not all the villages or towns have such specialized workers, so those doing this work usually work across a regional range. Frequently, the same team performs the same task on a farm, every year. However, in Spain there are some work-teams, consisting of 4–5 shearers, sometimes of foreign workers that move between different regions. They are generally paid for every sheep sheared. The associations of sheep breeders also form and organize groups of shearers to provide services to their members, trying to achieve more profitability for the wool produced.

The case of cork harvesters (Figure 3.1) is the most unique of all temporary workers in the dehesa or montado, for various reasons. The extraction is still carried out systematically and very little has been mechanized, except for transport. In the extraction year, the cork offers significant revenue with hardly any investment from the owners. The periodicity of nine years is maintained as defined by law and related to the survival of the trees. Stripping is a delicate operation that requires great skill. Because of this strong specialization, the salaries to cork strippers paid are the highest and the most stable in the dehesa / montado for temporary workers, between 80–90 €/day both in Spain and in Portugal (Coca and Quintero Morón, 2018; Oviedo et al., 2013). The difference with other temporary work is significant: in Alentejo, these other tasks, such as pruning, is paid at 50–60 €/day (data from fieldwork in Alentejo 2020 by the authors).

Cork collectors are organized into informal groups, usually represented by a contact called *manageiro/manijero*, who recruits and manages several axes or pairs (couples) of debarkers per tree. Currently, groups vary from 12 to 20 axes (6 to 10 pairs of extractors), but there are some groups with 40 axes. The work teams also include other auxiliary professions: cork collectors, a tree painter, loaders and stackers (Coca and Quintero Morón, 2018; Acosta-Naranjo, 2008). Working conditions have improved, and many teams now travel from their villages to the farm by car, sometimes more than 100 km away, without needing to stay in the place where they harvest, as they did before, often in uncomfortable conditions.

In Alentejo the individuals skilled in tree pruning also know about cork harvesting, and may also be engaged in sheep shearing – they are knowledgeable in three or four different tasks related to the silvopastoral system, that is performed sequentially along the year. They learn from more experienced workers, as described in Box 3.3. This succession of different specialized

---

TEXT BOX 3.3

Cork harvesting – a skill transmitted from worker to worker. The case in the Alentejo region, South of Portugal

Teresa Pinto-Correia, MED, University of Évora, Portugal,

Key points: The specificity of informal training in such a crucial part of cork production – a living tradition that secures quality

Cork harvesting requires very particular skills, in order to safeguard the trees' balance and to obtain regular cork plates. The cork needs to be extracted in a very particular way: with the use of a specific cork axe. The tree bark to which the cork plate will be detached from the tree with a controlled cut, that does not touch the tree's internal film in order to protect the next generation of cork. This requires a specific degree of humidity, not too much and not too little, in order that the cork can be separated from the tree with a simple movement. If the tree is too dry, the cork will crack; if the tree is too humid, the cork will not release itself - this depends much on the season and the weather. An experienced cork harvester knows if the conditions are not favourable and stops if the cork is going to crack.

How does one learn to be a good cork harvester? In Portugal, there is neither school nor formal training. The transmission of knowledge, skills and sensitivity from older workers to newer ones, is done by practice, during the cork harvesting. It is a fundamental component of the montado's productivity, and remains an informally learned skill, as it has been for many generations. The cork is always harvested by teams of harvesters, normally organized by a master. How does the master ensure he hires a quality team of cork harvesters? Only by word of mouth information – asking around and collecting informal information. The cork harvest pays almost twice the daily rate than other tasks in these silvopastoral systems and attracts many people – normally skilled workers who have other regular jobs during the year and take a couple of weeks of holiday to engage in cork harvesting. It is a seasonal occupation that is done with pride, dominates the conversations around cafés and market places during the harvest and provides one more example of the informality and hybrid characteristics of Mediterranean farming systems.

tasks throughout the year, performed by the same person, reflects what happened with the permanent workers in a large estate (see Box 3.1). In Portugal, these skilled independent workers do not have contracts, but work on a daily basis, as freelancers often with informal contracts; they are dependent on informal recommendations about the quality of their work, transmitted between landowners.

Alongside diseases and pests, excessive pruning and extraction is an important issue in the management of cork trees especially when the owners do not do the pruning directly themselves but sell the cork on the tree, so buyers try to extract the maximum cork. The same happens when the cork harvesters are not paid per day, but according to the amount of cork they extract. Workers generally receive a payment for the cork they extract, which varies from year to year although as noted before it is one of the highest-paid jobs in the agricultural and forestry sector and the highest in the dehesa/montado. In Spain daily rates are more common, whereas in Portugal workers are paid by the cork that they extract.

Other forestry operations were also traditionally done by groups of professionals who worked only in the periods or seasons suitable for the respective operations: pruners, charcoal makers, shrub cleaners and unskilled auxiliary workers. These professionals still exist today, but in smaller numbers, working individually through informal contacts or organized in informal groups. There are also companies in the market that work in these areas, as we will describe later.

The pruning of the trees has decreased greatly, due to the high cost of labour. It has been spaced evenly for more than 20 years, which has meant a reduction in employment. In a large number of areas, these temporary workers are not paid in cash but more frequently are given the firewood in exchange for pruning. Therefore, it is mostly people or companies who are involved in the sale of firewood or the manufacture of charcoal who carry out these tasks (McAdam et al., 2009). This involves the risk of abusive pruning, which is now (unlike before) performed with chainsaws. The pruners are mostly organized in informal groups, usually of between 2 to 6 workers, represented by a *manageiro/manijero*. The groups are smaller than they used to be as many of the pruning and other cleaning operations are now mechanized. These groups may join up with the cork extractor groups if they have the relevant experience.

A group with particular characteristics, with an activity that has kept a similar pattern from tradition until today, are the charcoal makers. These are local people, normally one responsible and a few employees (often family members), who buy wood from pruning and process it in traditional charcoal piles, which can sometimes be seen alongside the road, often close to villages. Another group collects and commercializes the wood resulting from the pruning of the oaks.

In Portugal, the firewood merchant supplies charcoal manufacturers, who make different types of charcoal in technically advanced ovens located in permanent facilities, rather than on farms. In Spain, the pruners-charcoal makers can be people from the villages that cut the firewood in their locality, those from other nearby villages and, in some cases, from towns with a strong charcoal making activity, who may move from state to state, sometimes up to hundreds kilometres away, living on estates where they are working for a few weeks. The pruners-charcoal makers usually work for most of the year, although they stop work in the summer as the burning of firewood is restricted due to the risk of fire. They sometimes complement their incomes with occasional farm work. Family networks play an important role in this activity, with working gangs consisting of people from the same domestic group or related groups in logging-coal making work. They are usually self-employed, and own their means of production, such as tools, machinery (for example, tractors), and/or have land and infrastructure for the manufacture of coal (Acosta-Naranjo, 2008).

Large farms with their own machinery and one or more employees dedicated to its management will usually perform the tasks of clearing, preparing or maintaining forest tracks, firewalls, etc. But in most cases, when clearings are carried out sporadically, this is done by a company or a self-employed worker with the necessary equipment and knowledge. These contractors will often hire seasonal employees for these tasks. Reforestation is carried out by forestry companies that have their own permanent workers and hire temporary workers when necessary. The owners are usually people with skills in forestry tasks, some of whom even have Bachelor's degrees from University, who created small companies, contracting their teams of (almost inevitably) men from their villages or towns as permanent and temporary workers. It is attractive for farm owners to contract them to carry out these seasonal tasks as they manage the whole process from start to finish.

Finally, hunting also provides sources of employment, with some companies dedicated to breeding certain game species (such as red-legged partridge) and others dedicated to the organization of hunting expeditions for both small (red-legged partridge, wild rabbit, hare) or big game (deer, wild boar), through contracts with the owners (MADRP/PDR, 2020). Some farms provide driven hunts, which can involve hiring Houndsmen and beaters (Macaulay, Starrs and Carranza, 2013), usually from nearby villages. Large game hunts may also employ a butcher to prepare the carcass of a deer or boar, and if the meat (large or small game) is to be sold then it needs checking by a vet.

Veterinarians also provide skilled services to the farms. Their work includes not only the prevention and treatment of diseases but also the castration of pigs and serologically testing the herds. Sometimes the farms hire them directly, at others they are hired by sanitary farmers' associations. Some are employees of companies that provide these services and others self-employed. Some of them (as well as agronomists) are employed by organizations such as the Regulatory Councils of Protected Designations of Origin or Geographical Indications, such as those for ham, lamb, *retinto* beef and cheese, which have supervisory and control functions, or of the administrations themselves.

In administrative control agencies, we can find forestry agents, environmental agents and the nature protection service of the police, which are involved in infractions or crimes related especially to pruning, hunting or the use of banned chemicals, but who also track populations of certain species of birds or mammals. Other employees of administrations include forest firefighters, and crews dedicated to the extinction of fires in spring and summer that perform clearing work on public farms during the rest of the year.

3.3 Other users

The bushy and herbaceous diversity of vegetation that characterizes the dehesa/montado, carries many blossoms at certain seasons (Moreno and Pulido, 2009), and offers fine opportunities bees to forage and the production of different honey types. This has led to the spreading of apicultural interest in the dehesa/montado, an activity that now has a broad territorial range. In some villages there are several beekeepers who set up their hives in the surrounding villages, counties and provinces, often on silvopastoral estates, taking advantage of their access to multiple agroecosystems. Very few farm owners are beekeepers; more usually they make deals with hive owners who settle their beehives in the properties during a certain period, which does not interfere with any other farm activities. This brings evident benefits to the landowner, as bees are important pollinators for the natural pasture flora. There is usually no formal contract for this use of the land resources, nor any payment in money; although usually the beekeeper will give some honey to the landowner each year.

Following the modernization that took place following the 1960s in Portugal and Spain, the number of hunters has increased exponentially. Rural dwellers have joined urban hunters in practicing hunting as leisure, with the improvement of living standards, increase in free time and increased mobility. In Portugal alone, there are 250, 000 registered hunters today.

Portugal and Spain have similar categories of hunting management regimes which are legally classified as either a) touristic, b) associative, c) municipal and d) national. In both countries, associative hunting areas are very important. Hunters create civil associations and manage hunting through dedicated deals with farm owners. Those associations have played a crucial role regarding the animation of rural communities, involving local people in hunting practices and management activities. The social dimension of the hunting activities also provides a link between the local communities and the hunting groups.

The hunting in touristic hunting areas is mostly rented to outsiders, urban people from outside the area. Usually, those who rent the hunting resource, for a whole season, are informal groups of friends. Less frequently, the owner will arrange driven hunts, directly or through other people or companies, and receives a payment for each stationary position or each trophy.

Collecting edible, medicinal and aromatic plants and mushroom gathering is another important activity. Traditionally these activities have not been exploited by landowners, and today this is still rare. As explained in Chapter 8, customary rights allow other people to enter private properties in order to gather plants and mushrooms. Recent legislation makes it possible for the owner to claim the right to this resource and forbid collection by others on his/her land – although this needs specific permission from the public authorities, and only a few landowners make use of this possibility. Most often in the extensively used silvopastoral areas, access is easy and almost impossible to control by landowners. There are gatherers who pick mushrooms and wild plants for self-consumption. But it is also common that mushroom gathering is done for sale, both informal and formal. This is more significant in the regions where the soils and climate are more favourable to the production of mushrooms. In Portugal, there are some rural families who make a significant proportion of their income from mushroom picking and selling. This is more usually hidden in the informal economy and therefore with the precise (or approximate) values not known (Pinto-Correia, Barroso and Menezes, 2010). Although there is a potential market, it hasn't been developed, especially because it is a highly informal economy and provides additional (tax-free) income for collectors of mushrooms (and asparagus). In Andalusia (Spain) it has been estimated that the average annual value of mushrooms collected could be worth more than 43 Million euros (Martínez-Peña et al., 2015). In some villages, there are specific events for selling mushrooms and more recently mycological societies have started to proliferate, in both villages and cities.

Asparagus gathering is also very important, although this is more of a leisure activity, done for self-consumption or to make gifts. Only in a few cases is such gathering commercialized: only three small companies exist in Extremadura. Yet, asparagus gathering is a very important social phenomenon in Extremadura and Andalusia, with many villagers (dwellers or vacationers), and a growing number of urban collectors participating.

The most frequent visitors to the dehesa and montado are probably local people who visit the countryside for aesthetic appreciation, recreational activities or other experiences that provide an escape from their daily routines or to link with their surrounding territory (Surová et al., 2018).

Hiking is an activity born in the cities, originally practiced by urbanites, but now adopted by rural people too. Itineraries for visitors have been marked, for fostering rural tourism, or as a leisure activity for local people. There is a growing number of hiking routes, organized by rural tourism companies or by hikers themselves across the montado and dehesa regions. They take advantage of the routes marked by local institutions or organizations, or by themselves, and publicize them through various media and social networks such as Wikiloc, which has minimized the need for meeting or interacting with local people. The sociological profile of hikers is quite broad. In this connection, there are also some estates that are sites of festivals, as *romerías*, or sometimes farms are crossed by pilgrim's ritual paths, as described in Chapter 8.

Birdwatching is another activity that has grown in recent years (Bugalho, Pinto-Correia and Pulido, 2018) started by foreign visitors and nationals, but now also enjoyed by locals. Extremadura in Spain is the most important region in the Peninsula for birdwatching, receiving the most birdwatchers. There are a few companies in the dehesa/montado that offer services for bird watching or nature photography, providing guides and hides.

Environmentalists have a direct, but mainly indirect, presence in the dehesa/montado estates. For environmental NGOs the dehesa/montado is the main interest point concerning rural land use in the Southwestern part of the Iberian Peninsula. Many of these NGOs have run campaigns focused on these systems. These NGOs are mostly based in cities, although they have quite a few members in dehesa/montado villages and rural areas. They have few direct interactions with farms, except for some collaboration initiatives with some particular landowners, such as territory stewardship agreements.

Finally, we want to point out the constant presence of researchers in the dehesa/montado (Acosta-Naranjo and Rodríguez-Franco, 2016). Being such complex systems, many different disciplines work in montado/dehesa, observing different phenomena. Their motivations are mostly to better understand the system, its conditions and the driving forces, in order to develop the knowledge basis to support more sustainable management practices (Guimarães et al., 2018).

3.4 Professions and users: a transversal approach for governance

Our goal is to identify some relevant features of the sociological dimension of the dehesa/montado that considers not only the interest of the landowners but, also other stakeholders. Obviously, the landowners play a central role in the agroecosystem, but the dehesa/montado would no longer function without the labour and knowledge of working people, who despite a vast reduction in their numbers in recent decades, still numerically far outweigh the landowners. The dehesa/montado is immersed in a specific sociological universe and without acknowledging the particularities of this universe, it is impossible to explain the past and present or to design governance strategies for the future. We are aware of the problem of making generalizations on this issue, as the dehesa/montado is diverse, geographically, agronomically, socially and culturally, but it is evident that history and the current situation have created specific sociological features that need to be considered.

A first outstanding matter is the footprint of latifundism which can be said to be the consequence of the long-established structure of land holdings on local society and culture. Indeed, as noted in Chapter 1, although not all dehesa/montado areas are latifundist (and even in the most latifundist areas, not all the farms are latifundios), this land ownership system has marked the history and the society of very large regions. It is historically a system of social domination too, which has created specific social mechanisms in Alentejo,

Extremadura and Andalusia and in some areas of Castilla-La Mancha which, together, make up the vast majority of dehesa/montado spaces. This resulted in not-integrated agrarian societies, strongly socially polarized; with the social universe divided between 'us' and 'them; the poor and the rich (González de Molina, 2014; Naredo and González de Molina, 2004; Acosta-Naranjo, 2002; Giner and Sevilla Guzmán, 1977; Cutileiro, 1971; Martínez Alier, 1968). Land ownership rights were one of the key triggers of confrontation, resulting in revolutionary movements in Spain and Portugal, and in (finally) failed agrarian reform processes. This domination was exacerbated since both countries were subject to dictatorships until the 1970's.

With the arrival of democracy, the large landowners lost their political power at the local and national level and the same thing happened to their economically dominant role after the agriculture crisis and the emergence of agroindustry which further impacted their political influence at the state level. Many workers migrated to the cities but those who didn't migrate saw an improvement in their living standards. However, the social and economic distance between workers and large landowners remains enormous, and the marks of history represent a handicap to the two groups having a common interest. Dehesa/montado as idiosyncratic cultural landscapes have symbolic values that could be shared by all those related to the areas, although the legacy of the latifundist dehesa/montado estates has left a historical mark rooted in centuries of oppression. We have to take into account that most landowners do not live in the local villages, so they are not part of the local moral community, a group with common interests and projects. Sometimes their only link with local society is the existence of a worker/manager (whom we called the SVWM earlier in this chapter).

An added difficulty is the weakness of civil society and the lack of social capital in Spain and Portugal until recent times (Alberich Nistal, 2007), especially in low-density rural areas. In Southwestern Iberia, there are associations and various collective entities, but these are usually restricted to leisure, sporting, religious or sociability activities. Except for the largest landowners, some of whom belong to land owners' unions, this lack of associative tradition represents an additional difficulty for governance, as there are no intermediation structures for public administration. There are hiking associations, religious and festival brotherhoods who perform rituals at dehesa/montado farms but organizations such as environmental ones are weak in the rural areas.

Trade unions are more political than civil society organizations, and they have had an uneven presence in the rural areas of Spain and Portugal. Montado areas within Portuguese territory have a historical record of social clashes and fighting. In some agro towns in the Portuguese montado (e.g. small towns where most economic activity is linked to agriculture), the elite lived in large houses in the noble urban centres, and most workers and their families in small houses on the outskirts of town or in villages. This visible difference fostered the growth of the trade unions and communist political parties, which have been very important, with the later still in power today in some municipalities (since the post-1974 revolution). One of the main goals of the 1974 Revolution was the agrarian reform of the Alentejo latifundist estates. Conversely, in Spain, although there was a strong worker's movement in Andalusia, this was mainly not in the mountainous regions and didn't prosper in the dehesa areas. This can be partly explained by the structure of settlements (the lack of agro towns) and because the work processes related to husbandry meant that work teams were not common. The same is true of Extremadura (Acosta-Naranjo, 2002).

When democracy arrived in Portugal (1974) and Spain (1977) the introduction of unemployment benefits and their related political patronage also helped de-escalate class conflict (Ortí, 1984; Sánchez López, 1986). In both cases, the dehesa/montado workers' groups were not very united in defending their own interests. In the case of day workers, only cork workers have sometimes undertaken collective action to get better labour conditions and salaries (Coca and Quintero Morón, 2018). We have to bear in mind that they are a special case: cork workers gather in work teams and have a specialized and crucial role in an activity that is highly profitable for the landowners.

Beekeepers are a relatively well-organized group, in both Spain and Portugal. They are self-employed and own their own means of production, but they also need to be organized in order to commercialize their products. This leads them towards collective action, particularly when negotiating with other economic stakeholders and with administrations. This applies to some extent to coal-makers, but only when there is a sufficient density of them in one village.

Hunters are perhaps one of the main stakeholders for the governance of the montado and dehesa, both in the local communities and at a higher political and societal level. Hunters have strong links to the estates, managing the hunting resources by diverse agreements with the owners. Hunting gives them a territorial cognition of these spaces and a close bond with them which allows for the recreation of local knowledge. In addition, they belong to local, regional and national organizations and have experience and a background in negotiating with administrations.

3.5 Conclusion

From a sociological point of view, it is pertinent to consider the different links that users and professions have with the silvopastoral systems of Iberia, the territories of which vary considerably. We can see a gradient of strong to weak bounds, from permanent employee to temporary employee, liberal professional, charcoal-maker, beekeeper, pruner and so, down to hikers or tourists, who have just occasional links with the landscape and the places. Latifundism, and livestock enclosure in present times, represent a physical separation of the majority of the rural population, the non-landowners, from the dehesa/montado. Working in the field is the main link of rural people with the agroecosystem. But the work at farms and estates has decreased enormously, and very little manpower is currently employed. When we add in extensification and mechanization, we can see that this formally active bond has now become very weak.

The same is true for another crucial element: knowledge. The knowledge that workers and users have, and pass on to the coming generations, on the agroecosystem and its functioning is closely linked to the intensity of their relationship to the estates and their territories. As the active relationships have decreased enormously in most cases, people's knowledge of the agroecosystem has decreased and faded away over time. We can find a scale from the intense, practical, broad and generic knowledge of SVWMs to that other quite fragmentary and specific of, for instance, the beekeepers, shearers or hikers.

In the light of this situation, the lack of inhabitants on the estates, the decrease of work teams and the enclosures, have further reduced the possibilities for knowledge transmission. In this sense, hunters and gatherers have gained a tacit claim to the strongest bond with the agroecosystem and being cognitive masters of it, as discussed in Chapter 8 on rights of use.

On the other hand, expert systems play an increasing role in dehesa/montado management. The knowledge of farms, livestock, pastures, trees and crops is more expertized, with veterinaries and agronomic engineers managing more aspects of production and taking decisions on more matters. The same applies to the administration's staff, especially in the forestry and livestock sectors. Environmentalists also hold this kind of expert sectoral knowledge; if they are native and living in the villages, this may be hybrid wisdom. The most paradigmatic case of expert and theoretical knowledge, with a sector perspective and in-depth capacity, is of course that of the researchers.

In terms of governance, it is important to find communication mechanisms that can convey multiple knowledges and interests from experts to laypersons and vice versa, enabling knowledge dialogue and citizen science (Delgado and Rist, 2016; Dankel, Vaage and van der Sluijs, 2017). Progress towards integration is high on the agenda, as we can see in Chapters 13 and 14. We need to take into account the partiality of knowledge and the plurality of interests of all the people and groups who work with or use silvopastoral systems or who have an interest in them. Promoting resilience needs to be a collective task that is weakened by exclusion.

References

Acosta-Naranjo, R. 2002. *Los entramados de la diversidad. Antropología Social de la dehesa*. Badajoz: Diputación de Badajoz.

Acosta-Naranjo, R. 2008. *Dehesas de la sobremodernidad. La cadencia y el vértigo*. Badajoz: Diputación de Badajoz.

Acosta-Naranjo, R. and R. Rodríguez-Franco. 2016. "Persistence and obscurity: scientific and political uses of the contemporary dehesa", in *World Congress Silvopastoral Systems, Silvopastoral Systems in a changing world: functions, management and people*. 27–30th September. Portugal: Évora.

Alberich Nistal, T. 2007. "Contradicciones y evolución de movimientos sociales en España". *Documentación social* 145: 183–210.

Bugalho, M., T. Pinto-Correia and F. Pulido. 2018. "Human use of natural capital generates cultural and other ecosystem services in montado and dehesa oak woodlands" in M. L. Paracchini, P. C. Zingari and C. Blasi (eds.): *Re-connecting Natural and Cultural Capital. Contributions from Science and Policy.* Brussels: Joint Research Commission. European Commission, 115–124.

Caballero, R., F. Fernández-González, R. Perez Badia, G. Molle, P. P. Roggero, S. Bagella, P. D'Ottavio, V. P. Papanastasis, G. Fotiadis, A. Sidiropoulou and I. Ispikoudis. 2009. "Grazing systems and biodiversity in Mediterranean areas: Spain, Italy and Greece". *Pastos* 39(1): 9–153.

Coca, A. and V. Quintero Morón. 2018. "Otro mundo es posible, o el movimiento (ambiental) de los corcheros y arrieros en Andalucía", in Cortés Vázquez and Beltran (coords.): *Repensar la conservación. Naturaleza, mercado y sociedad civil.* Barcelona: Universitat de Barcelona Edicions, 179–196.

Cutileiro, J. 1971. *Ricos e pobres no Alentejo: uma Sociedade Rural Portuguesa.* Lisboa: Lisboa Livraria Sá da Costa.

Dankel, D. J., N. S. Vaage and J. P. van der Sluijs. 2017. "Post-normal science in practice". *Futures* 91: 1–4.

Delgado, F. and S. Rist. 2016. *Ciencias, Diálogo de Saberes y Transdisciplinariedad. Aportes teórico metodológicos para la sustentabilidad alimentaria y del desarrollo.* Cochabamba: Agruco.

Giner, S. and E. Sevilla Guzmán. 1977. "The latifundio as a local mode of class domination: the Spanish case". *Iberian Studies* 6(2): 47–58.

González de Molina, M. 2014. "La tierra y la cuestión agraria entre 1812 y 1931: latifundismo versus campesinización", in González de Molina (coord.): *La Cuestión Agraria en la Historia de Andalucía. Nuevas perspectivas*, Sevilla: Fundación Pública Andaluza Centro de Estudios Andaluces, Consejería de la Presidencia, JUNTA DE ANDALUCÍA, 21–60.

Guimarães, H., N. Guiomar, D. Surova, S. Godinho, T. Pinto-Correia, A. Sandberg, F. Ravera and M. Varanda. 2018. "Structuring wicked problems in transdisciplinary research using the social-ecological systems framework: An application to the montado system, Alentejo, Portugal". *Journal of Cleaner Production* 191: 417–428. DOI:10.1016/jclepro.2018.04.200

Instituto Nacional de Estatística. 2016. *Inquérito à Estrutura das Explorações Agrícolas 2016*, https://ine.pt

de Andalucía, Junta. 2017. *Plan Director de las Dehesas de Andalucía. Consejería de agricultura, pesca y desarrollo rural.* Consejería de Medio Ambiente y Ordenación del Territorio.

Macaulay, L. T., P. F. Starrs and J. Carranza. 2013. "Hunting in managed oak woodlands: Contrast among similarities", in Campos, Huntsinger, Oviedo, Starrs, Díaz, Standiford and Montero (eds.): *Mediterranean Oak Woodland Working Landscapes. Dehesas of Spain and ranchlands of California.* Dordrecht: Springer, 311–350.

MADRP/PDR2020 - Programa de Desenvolvimento Rural (2014-2020) www.pdr-2020.pt

Martínez-Alier, J. 1968. *La estabilidad del latifundismo.* París: Ruedo Ibérico.

Martínez-Peña, F., J. Aldea, P. De Frutos and P. Campos. 2015. "Renta ambiental de la recolección pública de setas silvestres en los sistemas forestales de Andalucía", in P. Campos and M. Díaz (eds.): *Biodiversidad, usos del agua forestal y recolección de setas silvestres en los sistemas forestales de Andalucía. Memorias científicas de RECAMAN.* Volumen 2. Madrid: CSIC.

McAdam, J. H., P. J. Burgess, A. R. Graves, A. Rigueiro-Rodríguez and M. R. Mosquera-Losada. 2009. "Classifications and functions of agroforestry systems in Europe", in A. Rigueiro-Rodríguez, J. McAdam, Mosquera-Losada (eds.): *Agroforestry in Europe: Current status and future prospects*, Springer Science: 21–41.

Moreno, G. and F. J. Pulido. 2009. "The functioning, management and persistence of dehesas", in A. Rigueiro-Rodríguez, J. McAdam, Mosquera-Losada (eds.): *Agroforestry in Europe: Current status and future prospects*, Springer Science: 127–160.

Naredo, J. M. and M. González de Molina. 2004. "Reforma agraria y desarrollo económico en la Andalucía del siglo XX", in González de Molina and J. Parejo Barranco (eds.): *La historia de Andalucía a debate. Vol. II: El campo Andaluz*. Barcelona: Anthropos, 88–16.

Ortí, A. 1984. "Crisis del modelo neocapitalista y reproducción del proletariado rural", in E. Sevilla (ed.): *Sobre Agricultores y Campesinos*. Madrid: MAPA, 167–250.

Oviedo, J. L., P. Ovando, L. Forero, L. Huntsinger, A. Álvarez, B. Mesa and P. Campos. 2013. "The private economy of dehesas and ranches: Case studies", in Campos, Huntsinger, Oviedo, Starrs, Díaz, Standiford and Montero (eds.): *Mediterranean Oak Woodland Working Landscapes. Dehesas of Spain and ranchlands of California*. Dordrecht: Springer, 389–424.

Picão, J. S. 1983. *Através dos Campos: usos e costumes agricolo-alentejanos (concelho de Elvas)*, Lisboa: Publicaçoes Dom Quixote. Publicaçao original 1903.

Pinto-Correia, T., F. Barroso and H. Menezes. 2010. "The changing role of farming in a peripheric South European area: The challenge of the landscape amenities demand", in H. Wiggering, H.P. Ende, A. Knierim and M. Pintar (eds.): *Innovations in European Rural Landscapes*. Berlin-Heidelberg: Springer, 53–76.

Sánchez Gómez, L. A. 1993. *Las dehesas de Sayago. Explotación, trabajo y estructura social*. Valladolid: Castilla Ediciones.

Sánchez López, A. J. 1986. "La eventualidad, rasgo básico del trabajo en una economía subordinada: el caso andaluz". *Sociología del trabajo* 3-4: 97–128.

Surová, D., F. Ravera, N. Guiomar, R. N. Sastre and T. Pinto-Correia. 2018. "Contributions of the Iberian silvopastoral landscapes to the well-being of contemporary society". *Rangeland Ecology and Management* 71: 560–570. DOI 10.1016/j.rama.2017.12.005

Vargas, J. D., L. Huntsinger and P. F. Starrs. 2013. "Raising Livestock in Oak Woodlands", in Campos, Huntsinger, Oviedo, Starrs, Díaz, Standiford and Montero (eds.): *Mediterranean Oak Woodland Working Landscapes. Dehesas of Spain and Ranchlands of California*. Dordrecht: Springer, 273–310.

4 The role of large companies in the cork exploitation of dehesas and montados

Francisco M. Parejo-Moruno,
José Francisco Rangel-Preciado, Amélia Branco,
Antonio M. Linares-Luján and Esteban Cruz-Hidalgo

4.1 Introduction

Processing companies that buy the product of the montados and dehesas are another key actor group in the management of these silvopastoral systems. They are usually in direct contact with landowners and managers and generally their primary link with the market, thus they have a large impact on the economy of these systems. They are not only involved in marketing cork and meat, but also acorns, mushrooms and aromatic and medicinal plants. There are different types of companies, from very small ones attuned to niche markets, such as those processing acorns, to the very large ones processing meat or cork acorns (see Text Box 4.1 later in this chapter). Cork companies are a key actor as cork is a highly valued product but one whose value is strictly market-driven (i.e. not supported by the EU or state subsidies). This chapter explores their role. It also briefly examines the role of meat processing companies, especially those dedicated to processing pork meat from Iberian pig into sausages and the well-known *pata negra* ham.

Since the end of the 19th century, when the cork factory system was established, and especially after the emergence of the modern company, the global cork business has had a dual production structure, characterized, on the one hand, by many SMEs and micro-enterprises (definitions according to European Commission (2003)) that are technically poor, labour-intensive and with an explicit artisanal tradition; and on the other hand, by a few medium and large companies (more than 250 workers and a turnover of more than 50 million Euros). This second group is much more highly capitalized (APCOR, 2016; Sala, 1998) and based more on the division of labour and strategies for process integration and product diversification. These large companies have exercised an undisputed leadership in the cork trade throughout their history, competing and imposing their market power on the small ones, thus conditioning the technical and productive evolution of cork processing. Small businesses have often acted as satellites of the larger firms, cooperating with them in the diversified configuration of their supplies, resulting in relationships of dependence that have persisted over time (Branco and Parejo, 2011; Sala, 1998).

DOI: 10.4324/9781003028437-5

The role of large and medium-sized firms in the industrialization of cork and its contemporary internationalization has been extensively studied in recent years. However, despite their contribution to our knowledge of raw cork markets, most of these analyses have overlooked the crucial role of these private companies in exploiting dehesas and montados, and provide very little information on the strategies they adopt for ensuring an adequate supply of cork (in terms of quality and quantity). The objective of this chapter is to try to fill this gap and explore the role of these companies in the markets for raw and processed cork, especially from the demand side, as they cannot really be thought of as suppliers. We also suggest measures that might be adopted by public and private agencies to improve the functioning of the cork market, and the efficiency of cork exploitation in the dehesas and montados.

The analysis of the influence of large companies on the governance of silvopastoral systems in the regions of the Iberian southwest, and in particular on the supply of raw materials, is best undertaken from a historical perspective, as a long-term approach allows us to observe changes in the business and social behaviour of these large companies and their ability to adapt to changes in the environment and regulatory framework.

In methodological terms, the contents of this chapter build on a review of the existing literature, which is triangulated with primary data obtained from some interviews conducted with different informants and the results of previous research carried out by the research team. Those interviewed are private agents who work (or worked) at different points in the cork value chain. The interviews were anonymized and the qualitative information that emerged from them is useful, as it illustrates the relative and absolute importance of large companies in the productive and commercial exploitation of Iberian dehesas and montados, as well as suggesting ways to rebuild the networks of public and private agents that have been created around these silvopastoral systems that would make them more sustainable.

4.2 Large companies in the Iberian cork market: agents and the cork purchasing process

There is considerable opacity surrounding the exact figures for the cork market in the Iberian Peninsula. A report from the Autoridade da Concorrência (2012, 56) indicates that over the period 2000–2010, the market was worth close to an average of EUR 390 million per year, of which two-thirds would have been harvested and processed in Portugal and the remainder in Spanish territory. This amount would correspond to an average of 200,000 tons of cork a year although the amount of cork extracted oscillates hugely from year to year. This is partly due to the opaque functioning of the market, and partly due to cork's non-perishable nature, which allows it to be stored when prices are too low to make it worth processing/selling. Other factors which affect the cork harvest include adverse weather conditions (in dry years it is not possible to obtain the entire expected harvest) and legal and technical

Table 4.1 Raw cork. Different raw materials for different branches of the industry

Cork type	Features	Industrial application
Good quality cork and calibre		+ Natural cork stoppers
Cork	If good quality and calibre	+ Natural cork discs
(reduced calibre)	Corks for trituration and	+ Agglomerated cork stoppers
Cork pieces	granulation	+ Microgranulated stoppers
Virgin cork (first		(technical stoppers)
debarking) and *refuge*		+ White agglomerate
Other poor quality corks		manufactures (pure,
Cork manufacturing waste		composite)
Falca	Corks for trituration and	+ Black agglomerate
Burnt cork	granulation	manufactures

Source: Own elaboration.

restrictions concerning the minimum cycle for cork extraction (9 years in the Iberian southwest).

As a raw material, cork is highly heterogeneous and its quality determines the industrial applications for which it can be used. This is a fundamental feature of the market for cork. On average only 20% of the cork extracted each year has the quality and calibre necessary for the manufacture of natural cork stoppers (Santiago, 2016), the most important product in the cork industry. The remaining 80% is only suitable to produce other products, mainly cork discs or agglomerated cork products, depending on the calibre of the cork, the size of the plank, or its qualitative condition (Table 4.1). Finally, the *falca* and burnt cork are also used in the agglomerate industry, in this case, for the manufacture of black agglomerated cork products which are mostly used for insulation and coating in construction.

These different market segments give rise to a divergent and oscillating trajectory in the prices of the raw material that are paid from one year to the next. However, little is known about this trajectory, which depends on supply and demand factors and on the functioning of the market, which is opaque and in which agents act with asymmetrical information. It is also shaped by speculative behaviour that is possible because of the non-perishable characteristics of cork and with the different bargaining power exercised by large and small companies.

The buying and selling of natural cork are the first linkages in the value chain relating to the production of raw material and the production of cork manufacturers. It involves, or may involve, in addition to the forester or the owner of cork, who act as the suppliers, different agents located at different stages of the production chain, with different roles (see Figure 4.1). These agents act as buyers of cork raw materials, competing in the cork market but also using part of the cork supply network built by large companies for getting the cork needed for manufacturing. In this sense, the main channel used by large companies to source cork is direct contact with the owners, through

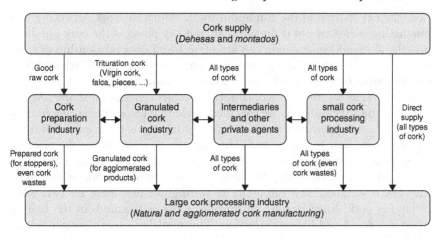

Figure 4.1 Cork supply network for large cork processing companies in Iberian cork market.

Source: Own elaboration.

their purchasing managers. However, an essential part of the cork that these large companies transform reaches them through other agents, mainly the cork preparation industry (good cork for making cork stoppers), but also private intermediaries who work in the cork market on commission. Smaller competing companies also cooperate with the large companies in meeting its supply, guaranteeing themselves a source of income through such collaboration, as shown in Figure 4.1.

According to data available for the Portuguese market (we have not found similar data for the Spanish, French or Italian markets), approximately 70% of the cork traded in the Iberian dehesas and montados is directly traded by the processing industry (mostly by large companies), with the preparation industry and intermediaries (small firms) responsible for around 30% of the remaining purchases (Autoridade da Concorrência, 2012, 43). Large industrial companies make relatively little use of intermediaries for the supply of cork, which are is increasingly acquiring less prepared cork from the cork preparing firms, as some of the large firms have perfect vertical backward integration in order to directly acquire the cork they need in the dehesas/montados (Rangel, Tejeda and Parejo, 2016).

The process of buying cork in dehesas and montados is regulated through the process map of the large companies' quality management system. This process informs the purchase managers of the company of the detailed instructions to be followed and the documentation to be collected at each stage of the process. Usually, the negotiation for buying cork is made by the regional purchasing manager of the company, although it is possible to find intermediaries buying cork for large companies and even small industries who acquire part of the supply network of the large companies. In any case, the negotiation of the conditions under which the cork debarking and purchasing is carried out is undoubtedly the key phase of the cork purchase process; although, such conditions are often based on a relationship of trust between the owner and the purchasing company (or its agent), which is the result of years or decades of business and contractual relations.

In general terms, the contractual relationship is concise and simple. At the outset, technical conditions are specified, including the exclusive purchase of cork with a minimum of 9 years ageing, which, in general, is extracted from the tree from June to September. The contract also determines the price to be paid for a Castilian quintal of 46 kg, as well as the assumption of the costs associated with the debarking by the seller or owner of the cork, at least until the cork is weighed. It is important to distinguish that there are two ways of buying cork: 'in the tree' and 'in-pile'. Cork is acquired 'in the tree' is when it is purchased before extraction. This modality poses a greater risk to the buyer; hence it is normal for the buyer to require discounts or bounties associated with such uncertainty; it also requires greater expertise, since the purchase is made without knowing exactly the qualitative composition of the cut, which makes the subsequent profitability of the cork acquired more complex and uncertain. In this sense, it is essential to have historical information of the farm concerned, data that the large cork demanding firms usually have available in their information systems and which help reinforce their negotiating position with the owners. When the cork is purchased 'in the tree' it is common for the buyer to accept the expenses of the debarking. This type of contract is common when cork oak management is not the main activity of the owner's business (which is very common) or when the owner has financial problems, both situations that increase the buyer's capacity to affect the price. The seller, however, manages to reduce risks related to the cork extraction, so it is usual to hire 'in the tree' when dealing with this kind of owner.

The 'in-pile' purchasing system offers more certainties for the buyer, as the buyer can qualitatively inspect the cork to be acquired and can offer a more suitable price. In this case, it is usual that the costs of extracting the cork are assumed by the seller, giving them the advantage of controlling the process and avoiding any damage to the trees that might impair their future performance. Chapter 3 describes the profession related to cork stripping. In any event, in both systems, it is normal that the party that bears the costs of the extraction subcontracts the services of third parties specialized in 'the draw', as this is an activity that neither owners nor buyers are capable of vertical integrating into their operations.

4.3 Asymmetric information and purchasing power: size matters

The role of large companies in exploiting the cork of the dehesas and montados is in their condition of cork buyers, not as suppliers. This is because the supply in this market is highly fragmented, with no large estates or leading firms with the market power to influence prices. In contrast, demand is highly concentrated, among a small number of large firms, which have become even more concentrated over the past two decades through fusions and absorptions. This market structure conditions the commercial relations established in the exploitation of cork in the dehesas and montados, which are characterized by two factors. First, there is an obvious information asymmetry in favour of the large cork buying firms, compared to the smaller competitors and also against the suppliers (foresters or cork producers). Secondly, the market power that these large companies wield because of their financial capacity and their dominant market position, enables them to condition the market situation at any given time (Parejo, 2010; Sala, 2003).

What is the reason for the fragmentation of supply in the cork supply market? In principle, this is due to the existence of many small-scale producers of cork, which is paradoxical when talking about these cork oak forests being located in *latifundia*. It can be explained by the dynamics of cork production: as debarking is done each nine years, even if the cork oak area can be important, the total production potential is relatively low. It is also related to the difficulties in getting a concentration of producers. More than 95% of the cork extracted in the Iberian Peninsula comes from cork oak producers that can be classified as small-scale producers in what concerns cork. The group of the twenty largest producers barely supplies the remaining 5%, and none of them has even 1% of the Iberian *subericola* production (Autoridade da Concorrência, 2012, p. 98). This fact, which is often favoured by the wealth fragmentation resulting from the transmission *mortis causa*, has consequences in terms of productive efficiency since it makes the unit cost of extraction of cork higher than if producers were larger. The rentability of the exploitation of cork oaks is based on the price of cork, which is highly variable and therefore undermines the profitability of the exploitations. This consequently affects the supply of cork in the medium term, as it provides a disincentive to preventing the stagnation or decline in the cork oak area. This decline is, according to Zapata (2002), one of the main problems that the sector has faced in recent decades.

Further obstacles are provided by the regulatory frameworks governing cork extraction activities, in particular, the regulations restricting the land use of dehesas and montados and those that specify the actions that can be carried to maintain cork oaks (cutting, pruning or uprooting). Such restrictions discourage the entry of potential investors into the sector, thus limiting the growth of the cork oak area. In addition, there are at least three types of barriers to the entry of new producers into the sector, financial, geographical and regulatory barriers. The normative ones refer to the legal framework governing the dehesa and montado, and have already been outlined above. The geographical ones are not easily controllable, as they are related to the soil conditions of the land where cork oak forests could be developed. The financial ones are undoubtedly the most important because becoming a cork producer involves mobilizing substantial resources and making large investments that only have a very long-term return.

Considering the inefficiencies of the raw material market and the highly fragmented supply, it should be noted that the difficulty of making cork exploitations profitable ends up affecting the bargaining power of producers. It also increases the transaction costs for buyers, as they are forced to deal with numerous different sellers to obtain the cork they need. One way to reduce these costs is through intermediaries (commercial agents), a resource that the large companies use (albeit very occasionally), especially in regions where supply fragmentation is more pronounced (Faísca, 2019; Parejo, Faísca and Rangel, 2013). Another strategy followed by large buying firms to min-imize the effects of this dispersed supply is supply planning, in the sense of geographically adding purchases. This strategy requires powerful infor-mation systems containing quantitative and qualitative historical production records, something that is only available to large industrial companies. This again increases information asymmetries as small-scale suppliers do not have this information about the total supply of raw materials coming onto the market each year.

In short, the relative fragmentation of cork production affects the land-owners' bargaining power in relation to large firms buying cork when agree-ing on the transaction conditions. A significant proportion of the owners are small individual entrepreneurs, responsible for small farms, usually not organized through associations of owners (although there are some examples of cooperatives of owners) for whom the exploitation of cork is frequently a complementary activity to their agricultural or livestock work. In this sense, they consider cork to be a premium that comes every nine years; so it does not matter so much to them if the price obtained is not optimum and could have been better. There are some cork producer organizations, both in Spain and Portugal, but their membership is quite small and they do not have much influence. It is no coincidence, therefore, that cork producers often lack a network of contacts in the sector and adequate information, as well as the specific technical training that would allow them to properly assess the potential value of their cork harvest, all aspects that are decisive in their negotiations with large cork buyers.

The demand side is substantially different from the supply side. On this side, SMEs and micro-enterprises co-exist with medium and large compa-nies, competing for the purchase of cork under conditions of inequality. The purchase of raw materials requires a high financial capacity, in addition to a capacity for forecasting and anticipation, while the cork acquired in a cam-paign will probably not be transformed to produce manufactured items until the following year. In this sense, the size of the buyer is relevant, because this dimension is decisive for the company's ability to finance its supply and to lower the permissible margin in its purchasing decisions. For large companies in the sector strong market fluctuations can lead to an imbalance between purchases and sales forecasts which can harm the liquidity of the company or lead them to sell a stock at lower prices than intended. In short, the success of large companies in the cork market is based on their financial capacity and the implementation of a purchasing strategy based on an accurate forecast of the future behaviour of the cork manufacturing market.

Related to the above, the financial strength and strong bargaining power of large companies exaggerate the imbalance of the relationships between buyers and suppliers. This often leads to relationships of dependence between small producers and intermediaries and the large cork demanders firms. This becomes clear when examining the role of the Amorim Group, the main buyer of cork in the Iberian market (Branco et al., 2017). This dependence is not homogeneous, but is more intense in some regions than compared to others, something that is not only due to the presence of other large buying companies but to the average size of land ownership in each region.

In recent years demand for cork has become concentrated within a few large business groups. This trend deepened after the international financial crisis of 2008, which led to the bankruptcy of some of the main companies in the sector in both Spain and Portugal (including Grupo Suberus, Fernando Oliveras Cortica and Empresa Industrial de Paços de Brando). By 2010 more than 70% of the cork traded in Portugal went through seven business groups: Corticeira Amorim, Abel Costa Tavares, Fernando Oliveira Cortiças, Grupo Bourrasé (Socori), Relvas, Alvaro Coelho and Piedade. Although we do not have precise figures for the Spanish market, similar conclusions can be drawn for the whole Iberian market, since these same companies acquire, more than half of the cork traded in the peninsula (Autoridade da Concorrência, 2012, p. 102). The concentration is even higher when one takes into account that Corticeira Amorim has acquired some of these seven companies in the last decade.

Corticeira Amorim's power in the Iberian cork market is recognized as a fact by the actors in the market, who recognize the ability of this group to shape the market and affect the prices paid for cork in each harvesting season (Branco et al., 2017; Branco and Lopes, 2018; Branco and Parejo, 2011).

As mentioned before, financial capacity is one of the factors on which bargaining power in the buying and selling process is based on, this is exercised not only against suppliers but also against demanders with less business and financial capacity. The supply of cork requires a significant investment that does not produce an immediate return for the high average period of economic and financial maturation of the business. In the case of the Amorim Group, its greater financial capacity gives it an advantage over its competitors, for at least three reasons: first, it can pay higher prices for higher quality and calibre corks, giving it a competitive advantage over its competitors that cannot offer comparable prices; second, it can better position itself against sellers, insofar as it can offer advances for the cork purchased, or fast payment for them, thus allowing it to finance the process of selling to the sellers; and third, it can handle the first cork buying operations of each season under better conditions than the competition, as its financial capacity is a buffer that cushions it against errors arising from market uncertainty at the very beginning of the season. Only companies that are financially solvent can buy cork at times of high uncertainty, which offers them access to the best quality cork and enables them to set market price trends.

Another factor that gives bargaining power to large companies is tradition. The trust and personal and professional ties that are built up over time reinforce an agent's reputation in the cork market. Many forest owners maintain their sales commitments with the same agents for a long time, based on trust from past accumulated experience. The trust relationship offers a mutually beneficial arrangement for both parties in terms of prices and purchase conditions.

It follows that the exercise of power by large undertakings stems, in part, from the existence of asymmetric information on the market. The Amorim Group has information on the raw material market that is hardly available for landowners or other competitors. Apart from the historical records of farms and operations that the group has systematized in its databases, it has a wide network of purchasing agents (and other intermediaries) in the producing regions that have complete information about the farms that will be debarked in the following season, as well as the quantities and qualities of cork that is expected to be extracted. This same network provides the group with information on the transactions that are being carried out on the market and, therefore, on the prices paid and quantities purchased by the competition, information that is decisive for the strategic direction of the group in terms of cork supply. In short, this network of agents and intermediaries creates an asymmetric information system in favour of the group on the raw material market in the broad sense, because through it manages to capture information on quantities and transactional qualities, prices paid and agents that are operating; It is also an informal channel that the group can to use to transmit information to the market, which ends up affecting its development.

Nowadays, various programs led by public institutions serve to correct, in part, the existing information asymmetry among the agents participating in the cork market. Specifically, we want to draw attention to the Suberoteca project being developed by the ICMC (Cicytex) of the Junta de Extremadura in Mérida (Spain), and the advisory, evaluation and quality report program of the Junta de Andalucía (in addition to the Andalusian Suberoteca located in Alcalá de los Gazules), both aimed at improving the information to cork oak forest owners, but whose information is also useful for small industrial firms.

4.4 Conclusion: lessons learned and recommendations for better governance

This chapter has highlighted the structural inefficiencies in the exploitation of Iberian cork as a result of imperfections of the cork market, characteristics that have remained constant over time. It is an opaque market, in which agents have asymmetrical information, and there are significant differences in companies' trading capacity, due to their different business and financial capacities.

What, if anything, can public and private institutions do to correct these imperfections? In our view, institutions working in this sector must promote higher informational symmetry at two levels: first, between suppliers and buyers, and; second, between the agents from different companies with differing business and financial capacities. One possible intervention would be for public administrations to publish detailed and updated statistics on the production and marketing of cork, specifying the raw materials' qualities and prices by region. It is true that public institutions, such as the Instituto de Conservação da Natureza e das Florestas in Portugal and the Instituto del Corcho, la Madera y el Carbón Natural in Spain, to mention but two of the institutions of relevance to the sector, make a significant effort in this direction. However, it is also true that this has not yet led to the release of a regular source of statistical information that is easily accessible to all actors in the sector. In this sense, the private initiatives of producer associations should be applauded, as their publications and reports do shed light on the Iberian cork market.

A very concrete proposal on how to implement an information system on the cork market is made in Autoridade da Concorrência (2012) and seems to us to be a good start. This report suggests the creation of a database with information relating to the cork market, including the list of owners who will extract their cork each year and the estimated amounts to be extracted (by type of raw material, qualities, prices, etc.). A public institution would manage the database, and the information could be available online, with the data protection and confidentiality specifications required for such sources. The platform could have a pan-Iberian nature, and the notification of information to the database by all producers would be mandatory. It would be continuously updated information with the items being contracted in each campaign so that cork prices received by producers could be published on a monthly basis. The information in the database could be completed each campaign by surveying producers, manufacturers, intermediaries and other actors involved in the cork market, which discusses the various aspects of interest in the cork extraction campaign (quantities extracted, qualities, prices paid, payment methods, etc.). In short, this intervention would provide all actors with basic market information that would substantially reduce uncertainty in decision-making, reducing economic and financial risks especially in smaller actors and ultimately improve the normal and competitive functioning of the market.

Another aspect related to information asymmetry has to do with the procedure used by buying companies to know the quality of the cork of the farms that they intend to buy from. Unlike large firms, which have their historical information system that provides them with such information, smaller buying companies bear transaction costs in monetary and time terms (performing the tests, delays in the performance of the business, costs in the analysis of the quality of the cork being sampled, etc.), since they are forced to carry out samples that, while allowing them to reduce the risk in their buying and adjust the price to the quality of the cork they are going to acquire. These costs could be reduced if the public administration made this information available to agents, which would mean having a qualitative map of the cork of the main producing regions. We think that the Suberoteca project by ICMC (Cicytex, Junta de Extremadura, Spain) goes in the right direction, which is associated with a Cork Quality Sampling Plan, whose purpose is to estimate the quality of commercial cork consignments, both in the tree and in the pile, of private and public owners. In the same vein, there are also several associations of owners of cork oak forests in Portugal that also sample the quality of cork among their associates; APFC, in Coruche, which does a large number of samplings every year is a notable example.

The position of small cork-demanding companies would also improve in relation to the industry's leading large firms if cork transaction lists were created (as it happens for some agricultural products). This type of initiative would dilute the power of large business groups, by having duly ordered lists for the adjudication of the existing cork consignments in the market. Each buyer would acquire the cork, of known quality, in the quantities desired and in the fixed order in each season, not being obliged to meet with the large groups for the contracting of this. This system could also be implemented by creating cork lots or farm lots, with established conditions of purchase, which buyers could be eligible to participate in on an equal footing. The tenders in these lots would not be known during the campaign, in order to avoid strategic movements of large business groups versus small cork buyers, as has historically happened in the sector.

Finally, it will also be helpful to intervene in order to reduce supply fragmentation and improve the efficiency of farms. The easing of some legislation would allow the owners of dehesas and montados to manage their holdings more autonomously and efficiently, for example by giving them greater freedom in decision-making on the crops they wish to grow on their farms; or authorizing them to manage, with minor administrative and legislative obstacles, the renovation of their woodland (to bring down old or sick cork oaks, replace them with new ones, etc.), as well as being able to prune under a less restrictive and controlled regime. All this could be achieved without prejudice to the public obligation to be vigilant with regards to the possible environmental or ecological damage that may arise from these interventions. The fragmented supply could also be alleviated by legislative regulations (in the sense of restricting the cases where fragmentation of property is possible). The concentration of the supply side through associative organizations could also be desirable. In fact, the dehesa owners' association, in general, is already moving in this direction and achieving positive results improving training and information for producers, as we see in the cases of FEDEHESA and UNAC. It is reducing the degree of monopsony that exists in the cork market today. Both interventions can help improve the bargaining power of producers in the face of large business groups that are buyers of cork, which, in the authors' view, needs to be corrected in the medium-long term.

TEXT BOX 4.1

A family-owned company processing products from Alentejano pig, Estremoz, Alentejo

Rui Charneca, MED, University of Évora

Key points – This is a locally well-known company with strong connections with pig farmers that uses new production technologies to improve product quality but whose production basis is clearly based on ancestral/artisanal know-how.

The Alentejano pig breed (the Iberian pig in Spain) is an animal intrinsically linked to the montado/dehesa. This breed was on the edge of extinction in the 1980s. Since then there has been a recovery of the breed and its traditional production systems, partly enhanced by grants from several entities. Nowadays, this breed represents an economic, ecological, social and symbolic added value in the regions where it is produced.

One of the entities responsible for the relaunch of the breed and its products was SEL – Salsicharia Estremocense, S.A., a company established for more than 35 years, which is today a reference point in the national agro-food sector, and an increasingly relevant player in the international panorama. It is an innovative company, certified, according to NP EN ISSO 9001 and IFS FOOD standards. It offers selected products of superior quality, that are widely sought after for a variety of reasons. It has been awarded the status of pioneering SME and has won more than 15 international and national awards.

One important aspect of the company's approach is to establish lasting partnerships with regional pig producers, ensuring confidence, quality of raw materials and animal welfare. It aims to promote and stimulate entrepreneurship in the free-ranging pig sector, and this ties in with its long-term business plan of future expansion.

As a family business, spanning several generations, SEL has tradition and family values as its core, while adopting and adapting new technologies which are an essential tool in adding value to the manufacturing process. In short, it is an excellent example of a supply chain actor that valorizes the montado agro-sylvo-pastoral systems and its animals and products.

References

APCOR (2016). *Cortiça. Estudo de caracterização setorial. Estatísticas e prospectiva.* Lisbon: APCOR.

Autoridade da Concorrência (2012). *Relatório Final Análise do sector e da fileira da cortiça em Portugal.* Lisbon: Autoridade da Concorrência.

Branco, A., & Lopes, J. C. (2018). "Cluster and business performance: Historical evidence from the Portuguese cork industry". *Investigaciones de historia económica*, 14(1), 43–53.

Branco, A., Lopes, J. C., Rangel, J. F., & Parejo, F. M. (2017). "From a Portuguese small firm to world leader in the cork business: The role of the internationalization of Corticeira Amorim to Spain". In C. Perrin (Eds.) *Petites entreprises dans l'histoire industrielle* (pp. 165–183). Paris: L'Harmattan.

Branco, A., & Parejo, F. M. (2011). "Distrito industrial y competitividad en el mercado internacional: la industria corchera de Feira en Portugal". In J. Catalan, J. A. Miranda and R. Ramon-Muñoz (Eds.) *Distritos e Clusters en la Europa del Sur* (pp. 123–142). Madrid: LID Editorial Empresarial.

European Commision (2003). *Commission Recommendation of 6 May 2003 Concerning the Definition of Micro, Small and Medium-Sized Enterprises.* Brussels: European Commission.

Faísca, C. (2019). *El negocio corchero en Alentejo: Explotación forestal, industria y política económica, 1848-1914.* Badajoz: Universidad de Extremadura.

Parejo, F. M. (2010). *El negocio del Corcho en España durante el Siglo XX.* Madrid: Banco de España.

Parejo, F. M., Faísca, C., & Rangel, J. F. (2013). "Los orígenes de las actividades corcheras en Extremadura: El corcho extremeño entre catalanes e ingleses". *Revista de estudios extremeños*, 69, 461–490.

Rangel, J. F., Tejeda, A., & Parejo, F. M. (2016). *Plan Estratégico para la Especialización en la Transformación de Productos Corcheros.* Badajoz: OCICEX.

Sala, P. (1998). "Obrador, indústria i aranzels al districte surer català (1830-1930)". *Recerques*, 37, 109–136.

Sala, P. (2003). *Manufacturas de Corcho S. A. (antiga Miquel & Vinke).* Palafrugell: Museu del Suro.

Santiago, R. (2016). *"El descorche del alcornoque con nuevas tecnologías. La agricultura y la ganadería extremeñas. Informe 2016".* Badajoz: Fundación Caja de Badajoz.

Zapata, S. (2002). "Del suro a la cortiça: el ascenso de Portugal a primera potencia corchera del mundo". *Revista de Historia Industrial*, 22, 109–137.

5 Gender and women in the governance of silvopastoral systems

Elisa Oteros-Rozas, Victoria Quintero-Morón, Federica Ravera, Ignacio García-Pereda and María E. Fernández-Giménez

5.1 Introduction: women as stakeholders in the governance of silvopastoral landscapes

The management of silvopastoral systems connects very different social actors at diverse scales and levels of governance. Understanding women's roles in the governance of silvopastoral systems requires inquiring into the relationships between people who identify with different genders and institutions, such as legislation, markets, governments, associations, rural enterprises, etc. How does women's engagement with these institutions influence their ability to make decisions, negotiate and interact with other actors, thus contributing to governance? The issues that shape the interactions between gender, natural resource management and governance are multiple and complex, including differential socialization processes with regard to nature and care, differential access to resources, work, power, prestige and trust.

Many studies have underlined that gender differentiation permeates decision-making related to natural resource use (see for example Agarwal, 2010; Villamor et al., 2014). In many societies, use and access rights to natural resources, including land, trees, water and animals, are differentiated along gender lines. In most contexts, including forests, women have fewer ownership rights than men (Elias et al., 2017; Willy and Chiuri, 2010), and women's access rights are often mediated by their relationships with men via marriage, divorce, or widowhood. Additionally, under a gendered division of labour, women are frequently excluded from decision-making spaces, community administration and institutional roles that affect the management of natural resources (e.g. Karmebäck et al., 2015; Aregu et al., 2016; Buchanan et al., 2016).

The participation of women in the governance of social-ecological systems is associated with greater efficacy, especially under scenarios of environmental crisis and in promoting more sustainable resource management. For instance, research in forest community governance shows that the inclusion of women's voices contributes to a more democratic debate, negotiation of norms, and greater compliance with these (Agarwal, 2010). Additionally, studies of both forestry and silvopastoralism have shown that women demonstrate greater

DOI: 10.4324/9781003028437-6

capacities for cooperation and for inclusion of more diverse knowledge types, such as gender-specific knowledge, which may support more robust decisions and help to strengthen women's empowerment (Agarwal, 2010; FAO, 2013; Aregu, 2016; Elias et al., 2017). Finally, studies in pastoralism demonstrate that when women are excluded from institutions that define the access and use rules of communal lands, this negatively affects the resilience of the communal pasture (Perez et al., 2015; Po and Hickey, 2018). This occurs because the exclusion of women's knowledge leads to future adaptation options being overlooked with the result that women start to question the legitimacy of informal institutions, as a result of their failure to address women's needs, which weakens the capacity of these institutions to respond to crises (Aregu et al., 2016).

A gender perspective to the governance of natural resource management implies an intersectional approach (Crenshaw, 1989). Intersectionality means observing and accounting for power relationships based on the interactions between different conditions, as intersecting axes, therefore beyond gender. A variety of social factors, such as race, ethnicity and class, intersect with gender to shape both oppressions and privileges that ultimately make particular people vulnerable to social and environmental changes and give them differential response strategies (Kaijser and Kronsell, 2014; Nightingale, 2011). Thus, women's different environmental practices, as well as their access to natural resource decision-making arenas are related not only to their gender, but also to their social class (e.g. large owners, small owners, day labourers, etc.), origin or culture (e.g. Romanian, Maghreb, Andalusian, Gypsy, Pallaresos-Catalan, etc.), age and their historical relationship with the countryside (e.g. rural origins, married into a farming family, newcomers, etc.), among other factors. Women's knowledge, their contact with pastures, forests and animals and sometimes their demands and needs, differ according to the varied embodied, psychological and social experiences associated with multiple intersecting social identities and positions (see for example Aléx et al., 2006).

Although women's roles within silvopastoral systems are being reconfigured by new uses and perspectives in certain regions, the twin processes of abandonment of less productive land, and the industrialization and mechanization of the most productive areas in Iberian rural systems over the past 60 years, have contributed to the progressive invisibility and exclusion of women from forestry and agropastoral management. For instance, women's knowledge and practices were the most affected by the industrialization of agriculture in Spain (Siliprandi and Zuloaga, 2014). Official statistics show that women account for only 24% of farming labour today in Spain (Sabaté, 2018). The division of labour between the productive and reproductive dimensions was deepened by agricultural industrialization, linking men to the agricultural, livestock and forestry tasks - increasingly commodified - and associating women with work related to the domestic sphere, caring for people, animals or the household, or a secondary supporting role for the men's functions. However, the segmentation between productive and reproductive tasks does not take into account the interrelation and interdependence between these two dimensions (Durán, 2012). Though several studies demonstrate this interdependence (see for example García-Ramón et al., 1995), statistics and public discourses continue to perpetuate the idea that women do not play an important role in agricultural or forestry tasks (see above). This highlights the need to deepen our analyses of gendered practices and perceptions in order to understand the differential experiences of women and men in silvopastoral systems.

Additionally, the current institutional architecture of agropastoral and forestry sectors, both formal and informal, perpetuates power relationships and gender inequities related to access to resources and participation in decision-making that can hinder future social, and hence ecological, sustainability in the mid to long-term. For example, the Spanish Law of Shared Ownership (Ley 35/2011) was passed in 2011, largely thanks to the pressure exerted in courts and the streets by women farmers from Galicia organized in the trade union *Sindicato Labrego Galego*. This law provides institutional and economic acknowledgment of women's contributions to farming and defends their rights of access to land and to be paid for agricultural work. Yet as of 2017, only 339 women throughout Spain had requested and obtained the acknowledgement of shared ownership (Senra Rodríguez, 2018). Certain regions such as Catalonia or Andalusia also have local formal laws related to women's farm ownership.

The objective of this chapter is to explore the intersection between women/ gender and governance of the silvopastoral systems of the Iberian Peninsula. We first carry out a systematic review of the academic literature and then present preliminary qualitative results from three case studies in Spain. Although rural studies with a gender or feminist perspective have increased over the past two decades, these works do not relate directly to the dehesas, montado or other silvopastoral systems. Thus, our first aim was to scrutinize the academic literature on gender and governance, identifying key findings and knowledge gaps. The second aim was to share observations from three qualitative ethnographic studies about the role of women in the governance of extensive livestock farming in the Dehesa and other silvopastoral systems of Andalusia and Catalonia (Spain), as well as in cork harvesting in Andalusia (Spain).

5.2 Approach

We carried out a systematic review of the literature and focused on three concepts (keywords in parentheses): 1) silvopastoral systems (silvopastoral, *dehesa, montado*, wood pastures, agroforestry, forestry or other systems of trees such as olive or chestnut combined with pastoralism/livestock/cattle, etc.); 2) women and gender (women, gender, feminism, feminine, family); and 3) governance (institutions, trade unions, cooperatives, associations, tenure/ property/ownership, land rights, access to land).

We also draw on three qualitative case studies from different silvopastoral contexts to analyze women's relationship to governance and management from a gender perspective. The cases we present come from different social-ecological and governance settings and the data were obtained through different research processes. Data from all cases were analyzed using a common analytical lens of gender and women's relationships to resource management and governance. In all cases, we used qualitative methodologies combining participant observation and in-depth life-history interviews of women livestock farmers, shepherds, temporary workers and other rural women involved in the management of silvopastoral systems.

5.3 Women in the governance of dehesas and montado: what the current literature reveals (or not)

The review revealed a surprisingly sparse body of literature related to gender in silvopastoral systems. Within the total of 26 references identified, none focused on the governance of silvopastoral systems from a gender perspective. Research has largely adopted an androcentric approach (Díaz, 2015) to the study of Iberian farming and forestry practices and knowledge. The existing literature either addresses governance, but barely mentioning women, or it focuses on gender relations or the role of women in rural areas where silvopastoral systems are present (entrepreneurship, tourism, cork harvesting, etc.). Accordingly, women's contributions to the socio-economic network of rural areas with silvopastoral systems, despite being a fundamental pillar, have not been sufficiently accounted for, leading to invisibility and persistent asymmetries (Flores and Barroso, 2011). Traditionally, the domestic arena has been associated with women, while the public spaces have been associated with men (Do Carmo, 2007). This division, in Mediterranean cultures, has been reinforced through honour-shame codes. However, regional differences exist, e.g. between the south and the north of Portugal, which have different agrarian tenure structures.

With the incorporation of men into the industrial sector, women not only kept the power they had over domestic issues but also gained central roles in subsistence farming tasks. In parallel, one of the most striking trends is the masculinization of the rural population, particularly in mountain regions (Valle, 2018). In Andalusian mountainous areas, including dehesa, women make up 48% of the rural population but represent only 25% of the population officially employed in the agricultural sector, and within this, half of them are dedicated to management tasks (Valle, 2018). Although in recent decades women have led rural and agrarian economic diversification with activities such as agritourism, food transformation and handcrafts, only 27% of long-term employment positions created by the CAP LEADER II programme were occupied by women (Flores and Barroso, 2011). Women's work also remains invisible in terms of land tenure and access to decision-making spaces around silvopastoral systems (Carretero and Avello, 2011). Young women in particular tend to abandon agricultural tasks and knowledge and are typically replaced by more disadvantaged populations such as day labourers and immigrants.

It is interesting to note also that, due to social structural conditions, women are less prone to individual activities and more likely to engage in cooperation as a way to cope with difficulties (Baylina et al., 2019). For instance, women participated in land occupations during the Portuguese agrarian reform after 1974 (Pires de Almeida, 2018). However, today women face challenges to participation in cooperatives, and even more so to assuming leadership roles, due to their role in care tasks (Baylina et al., 2019; García Pereda, 2011; Pires de Almeida, 2018). As a result, only 9% of the boards of agricultural cooperatives have equal representation of women and men, and 78% are composed entirely of men (Hernández Ortiz et al., 2018). Resilient forms of traditional masculinity, including in young men, persist when criticism from a gender perspective renders visible certain hidden elements of patriarchy (Baylina et al., 2019). However, it seems that some of these barriers might loosen once at least one woman enters the board of the association/cooperative (Carretero and Avello, 2011).

5.4 Case studies from Andalusia and Catalonia

The three cases presented here show some transformations and trends that facilitate the understanding of how women are currently inserting themselves into governance and management of the dehesa and montado. We observe that women are developing and claiming new roles in governance, which should be understood as a long-term process.

Based on ethnographic descriptions of traditional dehesa and montado (Acosta-Naranjo, 2002; Quintero-Morón, 2001; Coca, 2008), women's work and presence in farming had greater significance in the past than today. Women were highly sought-after as day-labourers to harvest legumes and acorns, as they were thought to be more flexible to kneel or crouch down, and were paid lower wages. In the past, small landowners and skilled workers usually lived permanently on the dehesa or montado with their families, including women. Women's main tasks included care of poultry, small orchards or sick animals, harvesting wild asparagus, mushrooms and medicinal plants, etc. However, women carried out many other tasks, ranging from feeding and moving livestock to assisting with sowing and harvesting crops, as well as processing food products. In the 1960s and 1970s, people mostly abandoned farms and migrated to the villages, agricultural work decreased and the availability of employment opportunities in agriculture declined greatly. As a result, women were excluded from salaried work (e.g. cork harvesting and extraction). In addition, many of the tasks performed by women while they lived with their families in the countryside diminished.

Women not only disappeared from the countryside; fewer and fewer live in small rural villages. The masculinization of rural towns, particularly within the active working population, is a long-term process in Spain and Portugal (Camarero and Sampedro, 2008; Fernández Alvarez, 2017). Although the number of women over 65 in the villages is much higher than that of men, in the younger cohorts these rates are reversed. Young women do not find employment in agricultural tasks – or this is very scarce and undervalued and families tend to leave farms to sons and give more formal education to daughters. The limited labour market for these women and barriers to their social and political development further contribute to the masculinization of rural spaces (see case studies).

5.4.1 Andalusian women in livestock farming

Most of the nine interviewed women (identified here with pseudonyms) entered livestock husbandry through family, by inheriting land, livestock and/or occupational identity, or via marriage/partner, although at least two of the Andalusian women started 'from zero'. Most owned or co-owned their enterprises with their spouse, while others were silent and unpaid workers in their partner's family enterprise. The enterprises varied from small subsistence farms and moderate-sized farms specializing in one type of livestock, to diversified farms involving an agri-tourism and/or crop component, to large estates with cattle, sheep and pigs.

Although all the women felt empowered to some degree and most enjoyed their work, several of the women interviewed perceived sexism when selling or negotiating the prices of their products. Male partners, fathers and brothers, or the merchant with whom they were dealing, preferred that women not do this task. As the managers of the household economy who know the costs of living, women bargain hard and do not sell for less than what they consider a fair price. Some interviewees said that male buyers and family members don't trust their capabilities. Other women, such as Maria, the sole owner of a cattle and Iberian pig operation, and Carla, owner of a small organic sheep, goat and Iberian pig farm, reported they never felt discriminated against as women. Maria attributed this to the fact that since childhood she has always been around other livestock producers in the community, with her father.

Women in our sample participated in both traditional livestock organizations and new initiatives and networks that influence governance. In the Sierra Norte of Sevilla, it is not rare for women to be official owners of the operation, and in recent years, both Maria and Carla have been elected as president of their local livestock farmers' association. Maria recognizes that being part of collective organizations is critical to access relevant information, including training opportunities, as well as other local events, that they would not otherwise know about.

Participation in these organizations is not without its challenges for women. Susana, part of a family that raises sheep, goats and organic almonds, remembered that when she officially stepped into the business, she also entered the board of the cooperative. She recalled being the first woman in its history, and how things have changed since then. "At first, my colleagues accepted this and voted for me 'as a flower vase.' 'What a nice girl, let's have her here so that at least we enjoy the view.'" Well, the girl struggled to do her homework. We would gather at night, from 21 or 22 until we would finish all bullet points in the agenda, at 1, 2 or 3 at night. [...] For instance, I introduced a percentage that each association had to dedicate to training, which is so essential and they had never done it, so I brought the idea of training [...]. With patience and my 'left hand' and the ideas written down, when they accepted me, after 9 years ... I decided to leave voluntarily, and they begged me not to leave" (Susana, livestock farmer, female, 51 y, 2019).

Susana also recalls her early times in an agrarian trade union. She remembers how that space was the opportunity for many women to leave their homes and the oppression of their partners. Ana, a small-scale sole operator who grazes a few sheep on her retired partner's farm in the Serranía de Ronda (Málaga), is also a strong defender of cooperatives. She argues that the economic unsustainability of operations in Spain is related to people not trusting cooperatives. In contrast, she belongs to three cooperatives, to which she sells her lambs, cereals and olives, all her production except that which she keeps for self-consumption.

In addition to participating in more traditional organizations like the stockgrowers' associations and trade unions, women also participate in collectives advocating for extensive livestock and the dehesas (Box 5.1) and

---

TEXT BOX 5.1

Back to the roots to foster governance changes

Elisa Oteros-Rozas, Elisa.oterosrozas@gmail.com

Key points – After half a life in the city, coming back to her family land, Pia claims that a dehesa in Extremadura, setting up a farm and gathering people together for advocacy has fulfilled her life.

Pia was born in a family dehesa in La Rinconada, Extremadura. Her childhood memories are associated with life in the countryside, the children of the shepherds, with whom she shared the games, the women washing by the well, the sounds of the bells at sunset and the smoke that came out of the chimneys of the farmhouse. With a soul full of human and natural landscapes, she moved with her father to the city in order to study law at university. She studied, travelled, worked and lived always waiting for the moment to return. For 26 years she was director of a bank office, which allowed her to raise her three children. In 2004, she accepted the proposal to head the list in Badajoz for the elections to the Spanish Congress. After one term in office, she realized the political arena was not attractive so she returned to her position in banking.

However, Pia explains how the image of the rural, the light of the dehesa and the sound of the bells haunted her until, in 2012, she returned and began to fulfil

Figure 5.1 Pia in her dehesa.

her dreams (Figure 5.1). She bought a flock of sheep, rebuilt the house, arranged the warehouses, planted thousands of trees, removed stones, cleaned brush, pruned oaks, made fences, built ponds and she "found the immense solitude of the dehesas today: the individualism of their owners, the abandonment of towns". Since then, she is has applied her political vocation to gather people together and create collective projects, embodying the paradigm of women new leadership in dehesa governance. She created the Association of Dehesa Managers and co-founded Fedehesa, she joined the Platform for Extensive Livestock and Pastoralism and promoted Ganaderas en Red. Pia proudly states: "I became a peasant for love, conviction and intelligence and now my life is completely full of achievable and utopian projects, of wonderful people, of dawn lights, of dreamy sunsets. I sit at the top of the mountain to contemplate the stars, I write, I read, I learn, I share. I have become again what GeR's motto states: a woman with "the land in the soul, the wind in the hair and the cattle in the heart".

form new institutions to support each other in pastoralism and to transform agricultural systems more broadly. Most of the interviewees belong to Ganaderas en Red (GeR), a network of some 180 women pastoralists from throughout Spain. GeR is an online community with daily virtual communication on pastures and herding, veterinary and livestock husbandry, but also on the challenges and barriers encountered by women in livestock farming and the strategies they develop to overcome them.

Founded in 2016 and already awarded several prizes, GeR has become a leading example of women's empowerment and social communication on rural and farming issues, with thousands of followers on social networks and important impacts on social media. Susana recalled the day she was invited to a seminar on pastoralism and was one of the very few women in attendance. She attributed the absence of women partly to family and partner barriers, but also a lack of women's empowerment. Together with the other women present that day, Susana helped organize the first meeting of the group that became GeR. GeR members are becoming leaders in the sector thanks to mutual support and empowerment. Participants report that GeR and other networks support them in improving direct marketing of their products, and accessing information that men get from hanging out at the bar. Ana said that being connected to other women and having a network in which to ask questions, learn about subsidies and other opportunities for public support is very valuable to her. She explained how when her children were younger, she didn't have time to participate in community initiatives, and that her partner controls and restricts her movements. She expressed how being part of an online collective such as GeR has changed her life so that she no longer feels isolated and alone. Several other interviewees also reported that it was not until they joined GeR that they made friends with other women pastoralists and formed a personal and professional support network.

Participants like Maria, who is active in several associations and networks, recognized, however, the trade-off between being active in collective spaces, and having sufficient energy and time to organize, monitor and plan her own business. Several interviewees also pointed out the multiple labour burdens women bear, including active roles in organizations, livestock management, family caregiving obligations and frequently an off-farm job for the family's economic sustainability. In addition to trade-offs in their personal lives, women's increasingly visible roles in governance may entail other costs. Carla argued that when men witness women's empowerment, they assume defensive positions. Susana and Ana perceived a recent regression in girls' empowerment and an increase in sexism and micro-aggressions in response to greater women's empowerment. However, Carla remained convinced that women have more opportunities than men and this is the time for women, pointing to tourism and value-added food processing as engines for rural women's empowerment and development. Entrepreneurship, she believes, is in women's hands.

5.4.2 *Women pastoralists, new peasantries and fire prevention in Catalonia*

Catalonia hosts a growing community of young women shepherds and pastoralists, many of whom are part of the 'new peasantry' who grew up in cities, not on farms. Most of them started to work in the extensive livestock sector after being trained at the School of Shepherds or agrarian and veterinarian schools. They participate in the women pastoralist regional network of *ramaderes*. In contrast to older women from multi-generational rural agricultural households, the young women we interviewed were highly educated and motivated by a degrowth philosophy and agroecological principles, as well as a desire for a rural lifestyle. Cloé represents some of these experiences: she came to the countryside when she was pregnant, to run away from an enclosed city apartment and look for the "magic of producing my own food". Montse adds that she wanted to "live with animals and nature" and she recently started her own small but profitable project. Their political engagement through their choice of extensive agroecological livestock management is also highlighted by the slogan of their social network, 'Without shepherdesses, there will be no revolution'. All the women interviewed in Catalonia demonstrated innovative examples of private-public alliances and were engaged in short value chains, such as producing their own cheese or selling directly to cooperatives, restaurants and local shops. Most of these initiatives emerged as adaptive responses to increasing economic challenges and climate change risks, such as the occurrence of destructive wildfires (see https://visors.icgc.cat/focalbosc/).

In Catalonia; young new peasant women shepherds are breaking barriers of participation and empowerment in innovative silvopastoral management initiatives, especially around the use of livestock to mitigate against wildfire risk. Montse described a project she was involved within the Collserola Natural Park, where shepherds received direct economic payments for removing biomass by grazing the forest margins. Today she is involved in a similar project with goats in the Garraf Natural Park. Another interviewee is part of the *Ramats de Foc* alliance between public and private agents, through which shepherds are supported to graze the forests. As land access is a major challenge for young pastoralists and women, this project has mutual benefits for pastoralists and other stakeholders. In addition, the Artisanal Butchers Guild of Girona adds value to the products of participating pastoralists, through a label that certifies that products originate from herds that reduce wildfire risk.

Although proud of their involvement in these innovative institutions, the women also shared critiques. Montse remarked that the government administration capitalized on her image as a young women shepherd through social media, but did not initially fund the project sufficiently. Only recently, after several years, have local governments promoted the project enthusiastically. Similarly, Joana reports that, despite the recognized public benefits of the *Ramats de Foc* project, collaboration among actors is weak, consumers have not responded to certification, and shepherds and their work remain targets of discrimination. Additionally, they suggest that these pilot experiences are limited and the role of silvopastoral management in reducing fire risk remains largely invisible to society.

5.4.3 Collective organization around cork harvesting in Andalusia

Dehesas in Andalusia have followed the trend of increasingly excluding women, as described above. Although forestry activities have traditionally been associated with men, in the past women were present at tasks such as *carboneo* (charcoal production), the extraction of essences or the harvesting of wild plants. Many of these activities were carried out by family groups, in which women were assigned tasks considered as minor or secondary (García Pereda, 2011). Present forms of farm management, mostly aiming at maximizing economic profits, have minimized the presence of women around cork oak lands.

However, in contrast to other areas, women in Andalusia have a visible role in organizing labour for cork harvesters. In 2017 the Association of *Corcheros* (cork harvesters) and *Arrieros* (cork extractors who work with mules) of Andalusia (hereinafter ACOAN) was established in Jimena de la Frontera (Cádiz), bringing together the cork workers of the province of Cádiz and some from Málaga. This social movement has introduced a gender debate and perspective into what had traditionally been very masculine spaces.

The ACOAN movement includes very active women, demands labour rights, salary improvements and better working conditions (Coca and Quintero-Morón, 2019). Examined from a relational perspective within which gender roles are constructed and intertwined (Coca and Quintero-Morón, 2018), this movement has incorporated new and remarkable elements such as increasing the use of digital social networks such as Whatsapp and Facebook to share information and mobilize participants; and certain 'pro-environmentalist' and − tentatively − 'pro-feminist' discourses, which could transform silvopastoral governance in this region.

The appearance of women in cork harvesting is relatively recent, but ACOAN's manifesto (2017) demands the "incorporation of women in specific cork harvesting tasks they are capable of" and in 2018 ACOAN created a women's committee. Our fieldwork, in which we interviewed some active participants, confirmed women's involvement in certain harvesting activities (see Box 5.2). The women rarely work as cork harvesters but are frequently involved in other roles related to management and accounting. Although the proportion of women workers is small, it is no longer rare for women to be hired to weigh the cork (especially on public farms).

Despite these advances, debates on digital social networks question women's capacity to take on the role of harvesters as amateurs or teachers and to physically wield an axe. In this debate, both men and women argue that women lack the strength to manage this tool, although others advise that "if they can be gatherers and carry [loads] like a man, why wouldn't they be able to harvest [with an axe]? Not all women; those that have what it takes" (Mario, corckharvester, man, 40 y, 2018). This viewpoint is also captured in an ACOAN video where a mother demands that her daughter should have the opportunity to be a cork harvester in the future if that is what she wants.

TEXT BOX 5.2

Women working in the Andalusian cork forests

Agustín Coca Pérez, acocper@upo.es

Key points – The importance of an intersectional gaze to make the women of today visible in the context of highly masculinized forest activities in which, in one way or another, women have always been present.

Rossana Acevedo, born in Paraguay, has been living in Cortes de la Frontera (Malaga) for 14 years. She is currently unemployed. She is one of the few women who has worked in cork production in the south of Andalusia. She spent a few years loading this 'brown gold' onto trucks to be transported to Portugal. She was also involved in the process of *cabria* (weighing) and became a *fiel* or trusted member of the crew. "I get up around four thirty in the morning. It's about an hour and a half's journey from the village. We wait for the cork to arrive and we weigh it (…). Then we are the last ones to leave, until extraction comes to a standstill and they stop bringing it. So many hours working in the countryside. You leave your home, you leave it all…when you work in the forests you have to arrive very early and you get back home very late…and then I have to make sure the food is ready and do a few things to keep my family going". She remembers how, for the first few days, when she was loading the trucks, she worked non-stop, day and night: "When I got home and was going to sleep, I would repeat that same mechanical movement I had been doing with my arms all day but in my sleep, and no one could sleep near me because I would be thrashing out around me …."

Maria is employed as an *arriera* (specializing in transportation with mules) in the Finca de Murtas cork plantation in Los Barrios (Cadiz). She is completing her degree in social work and sociology within Andalusia's public university system, continuing her family's trade during the summer months. Inma and Luisa also work in the process of *cabria* (weighing) in the mountains around Alcalá de los Gazules (Cadiz), while in Cala (Huelva) Cristina came from Romania to learn her trade as a *juntaora* (a stacker of cork bark) and *tractorista* (tractor driver), replacing the traditional *arrieros*. Only the job of *maestra corchera*, which involves stripping the trees, wielding an axe, is forbidden to women in patriarchal Mediterranean cultures. Meanwhile, women are gradually taking on complementary non-specialist tasks and also holding positions of responsibility and trust. At the Association of Corcheros and Arrieros of Andalusia (ANOAN), Manuela Barberá states: "Women must be taken into account. Not only so that the *corchero's* wife is looking out for them to make sure they are ok… No! They must be taken fully into account because they work just the same as men! That way, in the world of cork production, men and women would be treated under the same conditions of equality" (Figure 5.2). In Andalusia, *la corcha* (cork bark) takes the feminine gender, all the time it is linked to the tree, the mountains, its territory, its culture, its people. It only becomes masculinized as *el corcho* when it has been mechanically transformed when it is no longer of the earth and becomes separate from its men…and its women.

Figure 5.2 Woman working in crock extraction in Portugal.

These expressions, from digital social network debates, in different interviews and other contexts, might be interpreted as signs that women's groups and pressure are providing new spaces for work, social insertion, discourse changes and therefore also a new legitimacy to become part of collective spaces, such as the assemblies of organizations and other decision-making arenas.

5.5 Lessons learned and insights for future research

The central message outlined in this chapter is that the involvement of women is part of the process of modernization of the political, social, economic and cultural life of the dehesas and montado. Our case studies demonstrate how women are engaging in tasks linked to silvopastoralism, either working in the countryside or organizations and associations, or both. At the same time, women's increased participation is influencing silvopastoral systems. Despite decades of invisibility of women's roles in the management of natural resources in the Iberian Peninsula, reflected in the scarcity of academic literature on women in the governance of silvopastoral systems, women's aspirations are evident in their expanding roles, especially among younger women. Some of these women claim an identity based on individual autonomy and not on subordination to the home or family farm, opting to enter the labour market and decision-making spaces in economic (e.g. cooperatives) and political (e.g. councils) arenas, as our case studies illustrate. Although still few in numbers, and not always politically progressive, women are gaining visibility within the governance of silvopastoral systems.

Social networks, particularly digital ones, and new forms of communication are rapidly changing women's influence within the governance of the dehesa and montado. Such virtual spaces generate new areas of autonomy, mutual aid and empowerment that enhance self-confidence and facilitate participation in public arenas. Debates and controversies that question or transcend the boundaries of traditionally accepted gender roles take place in these networks more easily than in other spheres, hence generating new discourses.

In the current context of rural masculinization, the opening of social and workspaces for women is essential for generational turnover in natural resource management. The struggle for women's empowerment and visibility in rural areas is long-standing and slow, but the ways in which some women, whether local or newcomers, develop strategies to overcome barriers and occupy a diverse role in the different dimensions of silvopastoral systems, remains underexplored (Fernández-Giménez et al., *under review*). Unravelling the ways women are developing new activities related to the processing and elaboration of products (food, handicrafts, etc.) and the processes of landscape heritagization also remains a challenge. It is also urgent to acknowledge and understand the barriers women still face, such as those reported by interviewees in all three case studies, and to identify the changes that might act as leverage, in order to promote them through public policies.

References

Acosta-Naranjo, R. (2002). *Los entramados de la diversidad. Antropología social de la dehesa.* Badajoz: Diputación Provincial de Badajoz.

Agarwal, B. (2010). *Gender and Green Governance: The political economy of women's presence within and beyond community forestry*, New Delhi: Oxford University Press.

Aléx, L., A. Hammarström, A. Norberg, and B. Lundman. (2006). "Balancing within various discourses - The art of being old and living as a Sami woman". *Health Care for Women International* 27(10):873–892.

Aregu, Lemlem, Darnhofer, Ika, Tegegne, Azage, & Hoekstra, Dirk & Wurzinger, Maria. (2016). "The impact of gender-blindness on social-ecological resilience: The case of a communal pasture in the highlands of Ethiopia". *Ambio* 45:287–296. 10.1007/s13280-016-0846-x.

Baylina, M., Villarino, M., Garcia Ramon, M. D., Mosteiro, M. J., Porto, A. M. and Salamaña, I. (2019). "Género e innovación en los nuevos procesos de re-ruralización en España". *Finisterra LIV*(110):75–91.

Buchanan, A., Reed, M. G., and Lidestav, G. (2016). "What's counted as a reindeer herder? Gender and the adaptive capacity of Sami reindeer herding communities in Sweden". *Ambio* 45:352–362.

Carmo, Renato Miguel Do. (2007). "Género e espaço rural: O caso de uma aldeia Alentejana". *Sociologia, Problemas e Praticas* 54(2007): 75–100.

Camarero L., and Sampedro R. (2008), "¿Por qué se van las mujeres? El continuum de movilidad como hipótesis explicativa de la masculinización rural". *Revista Española de Investigación Sociológica* 124:73–105.

Carretero, M. J. and Avello, G., (2011). *La Participación de las Mujeres en las Cooperativas Agrarias. Estudio de diagnóstico y análisis acerca de las barreras para la participación de las mujeres en los órganos de gestión de las cooperativas del sector agroalimentario.* Madrid: Proyecto Integra, Mujeres de las Cooperativas y Liderazgo Empresarial. Fundación Mujeres. https://bit.ly/2xYUeKf

Coca, A. (2008) *Los Camperos: Territorios, usos sociales y percepciones en un. "espacio natural"andaluz.* Sevilla: Fundación Blas Infante.

Coca, A. and Quintero-Morón, V. (2018)" Comprender las dehesas desde metodologías feministas: Revisitando el descorche". *Actas del VII Congresos de Agroecología.* OASAL: Córdoba, 76–83.

Coca, A. and Quintero-Morón, V. (2019) "El movimiento ambientalista de corcheros y arrieros en Andalucía". In Beltrán, O y Cortés, J.A. *Repensar la Conservación. Naturaleza, mercado y sociedad civil* Barcelona, Bellaterra, 179–196.

Crenshaw, K. (1989). "Demarginalizing the intersection of race and sex: A black feminist critique of antidiscrimination doctrine, feminist theory, and antiracist politics". *University of Chicago Legal Forum* 140:139–167.

Díaz, C. (2015). "La perspectiva de género en la investigación social". In: García Ferrando, M. et al., *El Análisis de la Realidad Social: Métodos y técnicas de investigación* (4a edición). Madrid: Alianza Editorial.

Durán, M. A. (2012). *El Trabajo no Remunerado en la Economía Global.* Bilbao: Fundación BBVA.

Elias, M., Stevens Hummel, S., Sijapati Basnett, B., and Piece Colfe, C. J., (2017). "Gender bias affects forests worldwide". *Ethnobiology Letters* 8(1):31–34.

Fernández Álvarez, O. (2017). "Las que sostienen el campo mujeres y trabajos en el medio rural" en Vicente Rabanaque, García Hernandorena, Vizcaíno Estevan (coord.), *Antropologías en transformación: sentidos, compromisos y utopías,*Valencia: FAAEE, págs. 1577–1587

Flores, D, and de la O Barroso, M. (2011). *"La mujer en el turismo rural: Un análisis comparativo de género en el Parque Natural Sierra de Aracena y Picos de Aroche."* Revista de Estudios Sobre Despoblacion y Desarrollo Rural, Ager, no. 10: 39–69.

FAO (2013). *Governing Land for Women and Men. A technical guide to support the achievement of responsible gender-equitable governance land tenure.* Roma: FAO.

García Pereda, I. (2011). *Mujeres Corcheras.* Lisboa: Euronatura.

García-Ramón, M., Cruz, J., Salamaña, I., and Villarino, M. (1995). *Mujer y agricultura en España. Género, trabajo y contexto regional.* Vilassar de Mar (Barcelona): Oikos-Tau.

Hernández Ortiz, M, J., Ruiz Jiménez, C., García Martí, E., and Pedrosa Ortega, C. (2018). "Situación actual de la igualdad de género en los órganos de gobierno de las sociedades cooperativas agroalimentarias". *Revista de Estudios Cooperativos* 129(129): 66–83.

Kaijser, A., and Kronsell, A. (2014)." Climate change through the lens of intersectionality". *Environmental Politics* 23(3): 417–33.

Karmebäck, V. N., Wairore, J. N., Jirström, M. and Nyberg, G. (2015) "Assessing gender roles in a changing landscape: diversified agro-pastoralism in drylands of West Pokot, Kenya". *Pastoralism* 5(21): 1–8.

Nightingale, A. J., (2011). "Bounding difference: The embodied production of gender, caste and space". *Geoforum* 42(2): 153–162.

Perez, C., Jones, E. M., Kristjanson, P., Cramer, L., Thornton, P. K., Förch, W., and Barahona, C. (2015). "How resilient are farming households and communities to a changing climate in Africa? A gender-based perspective". *Global Environmental Change* 34:95–107.

Pires de Almeida, M. A., (2018). "Contributos da história oral para uma abordagem da situação da mulher em meio rural no século XX". *Actas de Congreso Transiciones en la agricultura y la sociedad rural.* II Congreso Internacional. XVI SEHA Rural Report, TranRuralHistory, Santiago de Compostela.

Po, J. Y. T., and Hickey, G. M. (2018). "Local institutions and smallholder women's access to land resources in semi-arid Kenya". *Land Use Policy* 76: 252–263.

Quintero-Morón, V. (2001). *Las matanzas en el Andévalo Occidental. Viejas y nuevas estrategias domésticas.* Huelva: Diputación Provincial de Huelva.

Sabaté, A. (2018). *¿Qué Significa ser Mujer en Zonas Rurales? Mujeres y mundo rural: nuevos y viejos desafíos.* Madrid, Spain: Dossier FUHEM Ecosocial.

Senra Rodríguez, L. (2018). *Un Antes y un Después de la Ley 35/2011. Mujeres y mundo rural: nuevos y viejos desafíos.* Madrid, Spain: Dossier FUHEM Ecosocial.

Siliprandi, E., and Zuloaga, G. P. (2014). *Género, Agroecología y Soberanía Alimentaria. Perspectivas ecofeministas.* Barcelona: Editorial Icaria.

Valle Ramos, C. (2018). "Mujeres agrarias en las comarcas serranas de Andalucía occidental: caracterización sociodemográfica y trabajo". *Acta De Congreso Missing conference Title?*

Villamor, G. B., Noordwijk, M., Djanibekov, U., Chiong-Javier, M. E., and Catacutan, D. (2014): "Gender differences in land-use decisions: shaping multifunctional landscapes?" *Current Opinion in Environmental Sustainability* 6:128–133.

Willy, D. K. and Chiuri, W. (2010). "New common ground in pastoral and settled agricultural communities in Kenya: Renegotiated institutions and the gender implications." *European Journal of Development Research* 22 (5):733–50.

6 Societal views on the silvopastoral systems

Diana Surová and Teresa Pinto-Correia

6.1 Introduction

Over their existence, the silvopastoral systems have been shaped by human activities and constantly adapting to human demands generated by the socio-economic situation at both the local and increasingly global level. Over recent decades European society has been increasingly expecting that rural areas provide amenities. Local cultural values, the production of quality food from a healthy environment, recreation possibilities, attractive landscapes and contact with nature are all highly appreciated by many European citizens. These new societal demands imply the need for new modes of governance that take into account the perceptions and preferences of different groups of users. The growing awareness of the scientific community about the critical linkage between public attitudes and the practical solutions in land management, also leads to an increasing research focus on the socio-cultural perspectives on land-use systems.

The silvopastoral systems of the Mediterranean are characterized by overlapping private and public uses; they have multiple functions and provide a colourful mixture of benefits (see for example, Barroso et al. 2012; Hartel and Plieninger 2014; Gaspar et al. 2016; Surová et al. 2018). However, in a rapidly changing and globalized world, the silvopastoral systems have been changing, and they face significant challenges to their future (Hartel and Plieninger 2014). In this regard, a public voice could help broader acknowledgement of the essential characteristics of these systems that should be maintained, even if changes are needed to adapt to evolving global social, economic and environmental situations (Oteros-Rozas et al. 2014).

Civil society usually does not have a direct influence on land management decisions. But it represents users who often sensitively perceive the results of the land management decisions, especially in areas where people live, visit or are interested in (Guimarães et al. 2018). The way that silvopastoral systems are seen by people, the worldviews and value systems that shape people's appreciation of silvopastoral systems, governance structures and everyday activities can be decisive for the maintenance of these systems, especially when environmental quality and intangible values are at stake (Almeida et al. 2014).

DOI: 10.4324/9781003028437-7

Thus, knowledge about peoples' perceptions and preferences can be a powerful tool in developing proper:

- Contextual policies for silvopastoral systems, as this knowledge, can provide insights about the societal relevance of the landscapes under various bioclimatic, socio-cultural and economic settings;
- Result-based payment schemes, where landowners are compensated for the contribution of their farms to amenities valued by society;
- Newmarket-based solutions with payments for the use of specific facilities;
- Landscape based strategies for silvopastoral dominated areas, and;
- Management strategies for silvopastoral systems at the landscape scale, including ways to reconnect societies to the care of these systems (Hartel and Plieninger 2014).

Most of the existing research on society's perspective on silvopastoral systems has emerged in the last few decades. Despite this recent interest, there is not a substantial body of literature that explores societal views about these systems. However, it has been recognized that knowledge about the socio-cultural perspective is critical if we wish to tackle land management issues and their impact on human well-being successfully (Martín-López et al. 2012).

Studies in Spain and Portugal, including quantitative assessment of public preferences for different land-uses, reveal that, in the regions where they exist, the silvopastoral systems are the most preferred landscape types. García-Llorente et al. (2012) performed a comparative analysis between different typical landscapes in south-east Spain, including the dehesa, which was among the most valued in terms of visual preferences and willingness to pay for its maintenance. It was also considered the landscape with the highest capacity to provide multiple benefits. In Portugal, a study by Surová and Pinto-Correia (2016) highlights montado as the most preferred landscape type, satisfying the expectations of a wide variety of user groups: inhabitants, second-home residents, regular visitors, hunters, tourists, eco-tourists, land managers and as the second-best land cover for new rural inhabitants in the Alentejo region. Another study on the specific expectations of urban residents of the rural Alentejo landscape again reveals the highest preferences for montado (Almeida et al. 2016). Even, when people are asked to compose their favourite Alentejo landscape from different land cover types, they create configurations that mostly include or are similar to landscape-level mosaic of diversified silvopastoral systems (Carvalho-Ribeiro et al. 2013; Surová et al. 2014).

The goal of this chapter is to describe public views about these silvopastoral landscapes through reporting on expressed perceptions and preferences of those engaged with these systems for a variety of goals and activities. The chapter draws on the modest amount of information gathered from the available literature concerning landscape perceptions and preferences in Portugal and Spain, including published empirical studies by the authors. The 'society' in focus are people who live, work or otherwise interacts with the regions where these landscapes can be found. These values are primarily shaped by people's interactions with the landscape, so the chapter also examines the interrelations between society's views, management practices and public policy. It also highlights the main approaches employed to date in getting the public and landowners involved in participating in the assessment of silvopastoral landscapes. Part of the chapter focuses on the challenges and opportunities that societal perceptions of these silvopastoral systems hold for future policy, management and research.

We conducted a review of both qualitative and quantitative research published in scientific journals and books that were accessible through the Web of Science. Keywords used for a search were the combination of the following terms: 'montado' or 'dehesa' or 'silvopastoral and Mediterranean' and 'public or 'society' or 'users' or 'actors or 'stakeholders' or 'farmers'. Based on this search, we provide an overview of the methodological approaches applied to date to assess societal views on the silvopastoral systems. This is followed by a section describing specific perceived and preferred aspects of the Mediterranean silvopastoral systems. To do this we analyzed and classified all the research findings relevant to the topic in one or more categories in two main groups – benefits and challenges. The benefits are the aspects of the silvopastoral systems that people perceive as positive. The challenges are those aspects that people see as the weaknesses related to these systems or their management. Both themes were subdivided into three sections compatible with the three pillars of sustainable development: socio-cultural, economic and environmental.

6.2 Approaches and methods applied to assess societal views on the silvopastoral systems

In general, the approaches that capture societal views on landscapes are based on a holistic vision that (ideally) integrates information about ecology, societal values and governance, covering a diversity of views and different stakeholders (Plieninger et al. 2015; Torralba et al. 2017).

The approaches applied to assess societal views on the silvopastoral systems are predominantly based on frameworks based around the landscape or ecosystem services. The scientific community argues that the knowledge generated by these approaches can be a powerful tool in developing contextual policies for silvopastoral systems (Pinto-Correia et al. 2015; Plieninger et al. 2015). More recently, some studies have applied a societal well-being approach.

The landscape framework focuses mainly on the identification of landscape preferences and values attributed to systems studied. This approach captures peoples' expectations, demands or needs from a territory (Gómez-Limón and de Lucío Fernández 1999; Barroso et al. 2012). Some studies go further and address the linkage of societal expectations and decision-making system at the local level or territorial planning at a regional level (Surová et al. 2011; Carvalho-Ribeiro et al. 2013; Pinto-Correia et al. 2015).

In general, the ecosystem services approach has focused primarily on monetary and biophysical perspectives but, in recent years, an increasing number of studies explore socio-cultural preferences regarding ecosystem services (Martín-López et al. 2012; Pinto-Correia et al. 2015). Recent studies have also considered social complexity, analyzing the distribution of benefits and differing values attributed to, and interests in, ecosystem services (e.g. Oteros-Rozas et al. 2013; Nieto-Romero et al. 2014).

The well-being approach has been applied to reveal how society perceives not only the benefits but also the challenges that these systems currently face. This perspective can provide a holistic outlook on silvopastoral systems, and take into account the complexity of land users' values, priorities, strategies and actions (Fagerholm et al. 2016; Surová et al. 2018). Ideally, an sustainable well-being approach will focus on the interdependency of well-being; e.g. individual, societal as well as natural well-being and thereby illustrate and resolve how different conflicts and tensions impact land-use systems and reflect a holistic and ongoing adaptive process rather than a fixed state.

These approaches can be applied at a local or regional level and use different methods of social research. The advantage of the local-level assessment is that it allows spatially specific discussions between stakeholders and can reveal some information about the land and resources (Garrido et al. 2017). Local-level studies can also provide vital information for larger-scale assessments, as more detailed approaches permit a more in-depth understanding of specific social-ecological systems.

These qualitative socio-cultural methods can enhance quantitative methods and have the following advantages:

- Facilitate an in-depth understanding of the intertwined natural and anthropogenic as well as socio-cultural complexity of silvopastoral systems;
- Provide new ways of understanding the rationale behind social preferences towards silvopastoral systems and highlight societal values. thus fostering cultural landscape conservation in the long term, and;
- Help identify potential trade-offs and synergies among the benefits (services) expected by different stakeholder categories and therefore provide critical evidence-based input for landscape planning, management and stewardship (Garrido et al. 2017).

Table 6.1 provides an overview of the methods and regions of application in the recent research assessing societal perspectives on silvopastoral systems. The studies covered the region of Alentejo in Portugal and Extremadura and Andalusia in Spain. Several studies used qualitative semi-structured interviews (e.g. Pinto-Correia and Azeda 2017) and quantitative questionnaires (e.g. García de Jalón et al. 2018) to stakeholders at different analytical scales, ranging from local, landscape and regional levels. Additionally, a combination of various methods as interviews, questionnaires, workshops, focus groups and spatial analysis were useful in contextualizing the montado as a social-ecological system (Guimarães

Table 6.1 Overview of the methods and regions of application in recent research assessing actors' views on Mediterranean silvopastoral systems.

Methods	Scale of analysis	Region of application	References
Semi-structured interviews	Local and regional	Extremadura	*Garrido et al. 2017* *Torralba et al. 2017*
	Regional	Alentejo	*Surová et al. 2018*
	Local	Alentejo	*Carolino et al. 2011* *Pinto-Correia and Azeda 2017*
Photo-questionnaires	Landscape	Guadarrama mountains (Madrid)	*Gómez-Limón and de Lucío Fernández 1999* *Surová and Pinto-Correia 2008*
		Alentejo	*Pinto-Correia et al. 2011* *Barroso et al. 2012* *Carvalho Ribeiro et al. 2013* *Surová et al. 2014* *Almeida et al. 2014* *Surová and Pinto-Correia 2016*
		Andalusia	*Serrano-Montes et al. 2019*
Questionnaires	Landscape	Andalusía	*García-Llorente et al. 2012*
		Cáceres	*Fagerholm et al. 2016*
		Portugal and Spain	*García de Jalón et al. 2018*
Focus groups	Landscape	Sierra Morena, Castilla la Mancha, Andalusia Extremadura	*Oteros-Rozas et al. 2013* *Gaspar et al. 2016*
Review of historical sources	European	Iberian Peninsula	*Jørgensen and Quelch 2014*
Combination of methods	Farm and landscape	Alentejo	*Pinto-Correia et al. 2019*
	Landscape	Alentejo	*Guimarães et al. 2018*
Methodological literature	Landscape	Alentejo	*Carvalho Ribeiro et al. 2013*
	Landscape	Europe	*Pinto-Correia et al. 2015*
Linkage between landscape preferences and spatial planning	Plot level	Alentejo	*Surová et al. 2011*
Landscape Typology based on societal expectations			
Implementation of public preferences in decision support system			

et al. 2018). Pinto-Correia et al. (2019) used a discourse analysis approach to understand the representation and values attributed to the montado by different stakeholders, mainly landowners. The method used was based on a combination of an analysis of the relevant literature and research results from diverse projects, including a review of media representation of the montado, participatory observations and in-depth interviews with key stakeholders.

Apart from using interviews and questionnaires, participatory methods proved particularly useful when exploring stakeholders' knowledge, preferences, practices, perceptions and values in silvopastoral systems (Guimarães et al. 2018). The focus group technique applied by Gaspar et al. (2016) also revealed itself as a useful tool for the study of silvopastoral systems and the ways they are perceived by society.

Some studies used visual stimuli such as photographs which increased respondents' curiosity and willingness to participate as well as help to overcome possible confusion in the use of expert and non-expert language The precise identification of the information which is intended to be differentiated by the observer is crucial for the quality of the study, so photo editing is often applied (Barroso et al. 2012). Maps and satellite images may bridge knowledge divides and help to achieve a fair distribution of benefits through improved visualization and communication across stakeholder groups. In particular, participatory mapping approaches enable an assessment of the system's complexity, including multiple place-based practices and values, emerging from everyday embodied subjective experience and accumulated knowledge (Fagerholm et al. 2016). Maps were also useful in the spatial representations of landscape expectations by relevant user groups and helped to bridge the gap between landscape preference research and landscape planning, translating local scale preferences into regional-scale planning settings (Carvalho et al. 2013; Almeida et al. 2014). Surová *et al.* (2011) made efforts to integrate public preferences in the decision support system at the plot level.

6.3 The perceived benefits of silvopastoral systems

At national and even at the international level, the montado and dehesa in south-western Iberia have been recognized as some of the few sustainable land-use systems in Europe, as unique high nature value farming systems (Oppermann et al. 2012; Pinto-Correia et al. 2018), and as valuable biocultural heritage (Jørgensen and Quelch 2014). Due to their outstanding universal value, there are moves to get the dehesa and montado recognized as cultural landscapes on UNESCO's World Heritage List. The justifications for this recognition include a combination of many cultural and natural values. The authentic values of these areas include:

- Important cultural values,
- Functional systems adapted to the limitations of the Mediterranean physical environment,

- Ethnological value in terms of agri-food know-how,
- The scenic beauty of the landscape,
- High biodiversity value, and,
- Diverse natural habitat containing tree species from the primitive Mediterranean forests (Silva Pérez and Fernández Salinas 2015).

Regional and local level studies show the particular benefits of these silvopastoral systems in more detail. Figure 6.1 illustrates multiple and diverse values inherent in the silvopastoral systems in terms of the three dimensions of sustainable development. Table 6.2 lists the particular benefits according to the literature review developed for this chapter. This literature review shows that the vast majority of the studies about social perceptions of these

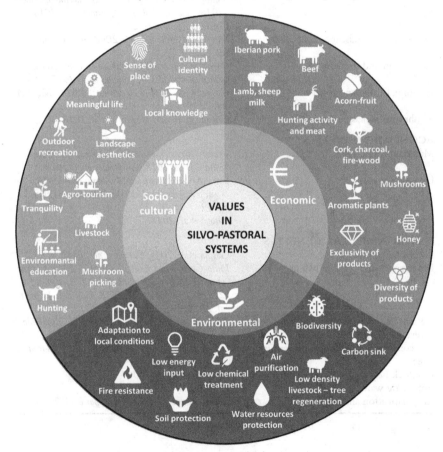

Figure 6.1 Multiple and diverse values are recognized in the silvopastoral systems, according to the three dimensions of sustainable development. This figure illustrates the multiple benefits that society currently receives, and expects, from these systems.

Source: Literature review.

Table 6.2 Benefits from the silvopastoral systems recognized by society

Benefits	Literature references
SOCIO-CULTURAL BENEFITS	
Cultural identity, sense of place	Jørgensen and Quelch 2014; Surová and Pinto-Correia 2016; Gaspar et al. 2016; Torralba et al. 2017; Garrido et al. 2017; Surová et al. 2018; Guimarães et al. 2018; Pinto-Correia et al. 2019
Local traditional knowledge and practices	Garrido et al. 2017; Surová et al. 2018
Meaningful life and spiritual experience	Gaspar et al. 2016; Garrido et al. 2017; Surová et al. 2018; Guimarães et al. 2018
Environmental education	Oteros-Rozas et al. 2013; Garrido et al. 2017
Landscape aesthetic quality	García-Llorente et al. 2012; Barroso et al. 2012; Surová and Pinto-Correia 2016; Fagerholm et al. 2016; Garrido et al. 2017; Surová et al. 2018 García de Jalón et al. 2018
Livestock as a landscape component	Surová and Pinto-Correia 2008; Serrano-Montes et al. 2019
Recreation and tourism	Oteros-Rozas et al. 2013; Gaspar et al. 2016; Garrido et al. 2017; Torralba et al. 2017
Hunting, fishing and tourism	Gaspar et al. 2016, Garrido et al. 2017; Surová et al. 2018
Mushroom and asparagus picking	Garrido et al. 2017; Surová et al. 2018
ECONOMIC BENEFITS	
Material products	Oteros-Rozas et al. 2014; Gaspar et al. 2016; Surová et al. 2018; Guimarães et al. 2018
Diversity of goods	Jørgensen and Quelch 2014; Surová et al. 2018; García de Jalón et al. 2018; Pinto-Correia et al. 2019
Exclusivity of products	Surová et al. 2018
ENVIRONMENTAL BENEFITS	
Local adaptability	Gaspar et al. 2016; Surová et al. 2018
Biodiversity and wildlife habitat	Pinto-Correia and Carvalho Ribeiro 2012; Gaspar et al. 2016 Torralba et al. 2017, Surová et al. 2018; Guimarães et al. 2018; García de Jalón et al. 2018; Pinto-Correia et al. 2019;
Natural hazard regulation, disturbance prevention (fire resistance)	Gaspar et al. 2016, Torralba et al. 2017, Garrido et al. 2017, Surová et al. 2018
Soil conservation - erosion prevention by trees	Gaspar et al. 2016, Garrido et al. 2017, García de Jalón et al. 2018
Livestock helping with tree regeneration	Oteros-Rozas et al. 2014
Improving water quality	Gaspar et al. 2016
Air purification and carbon sink	Oteros-Rozas et al. 2014, Gaspar et al. 2016

silvopastoral systems reveal combinations of benefits that include socio-cultural and environmental ones (e.g. García de Jalón et al. 2018; Surová et al. 2018;). The public also recognizes the economic benefits (e.g. Oteros–Rozas et al. 2014, Surová et al. 2018). It appears that society, in general, is aware of multi-functional role of these silvopastoral systems and desires them to be maintained. When people were asked about these systems, they recognize a package of multiple benefits that are interconnected and cannot exist independently from other existing benefits.

6.3.1 Socio-cultural benefits

Cultural heritage and identity, both the *individual identity* of local people as well as *territorial identity* are among the most appreciated benefits of silvopastoral systems, among different groups of land users (Garrido et al. 2017). Identity is the value most frequently recognized by montado users in the Alentejo (Surová and Pinto-Correia 2016). People with higher formal education levels are most likely to appreciate cultural identity: land managers, whom one might expect to have a unique focus on production aspects, also attach considerable value to having a tie with previous generations and the sense of family heritage in their land (Pinto-Correia et al. 2019). The cultural heritage can be attributed to the long-term existence of these systems, as well as some specific practices, such as those related to seasonal livestock movements (transhumance) that still continue in some parts of Spain (Oteros-Rozas et al. 2014).

The maintenance of *local traditional knowledge and practices* is also perceived to be a benefit. Specific practices that are valued in the montado and dehesa include tree pruning techniques, firewood extraction, canopy clearing to maximize acorn production from holm oaks and cork extraction, cattle ranching, goat, sheep and swine herding (Garrido et al. 2017). Cork debarking and tree pruning are still done in traditional ways by skilled local people with high levels of expertise. Social recognition of traditional knowledge and practices makes an essential contribution to rural development as this knowledge and these practices can help to support or create local food brands, generate tourism income and be transmitted to future generations.

In general, practices based on local knowledge are recognized to represent a profound form of ecological knowledge, about natural resources and possible responses to disturbance, which may prove helpful in identifying adaptive strategies to cope with global environmental changes (Oteros-Rozas et al. 2013). Local expertise is also highly valuable for a co-construction of knowledge that incorporates local and scientific knowledge to find appropriate management interventions (Guimarães et al. 2018).

Studies in both Spain (Garrido et al. 2017) and (Surová et al. 2018), show that those who work or manage the dehesa and montado and contribute to the maintenance of the silvopastoral systems view this as a privilege, which contributes to their feelings of leading a *meaningful life*. In addition, these systems provide people with feelings of tranquility and retreat, which could be classed as *spiritual values*, Land managers of public areas in the dehesa highlighted local festivities and celebrations that take place there as signs of spiritual and religious values (Gaspar et al. 2016).

Alongside identity values, the *aesthetic quality of the landscape* of silvopastoral systems is highly valued by society. The visual aspect owes itself to a combination of particular colours, trees with irregular shapes, planted in diverse densities, grazing animals and undulating terrain that allow different panoramic views across an extensive savanna-like landscape. This diversified landscape with trees is highly favoured (Gómez-Limón and Lucio Fernandez 1999; Surová et al. 2014) and the movement of livestock across it adds a dynamic quality to the landscape that is positively evaluated by different user groups (Surová and Pinto-Correia 2008; Serrano-Montes et al. 2019). In general, the public finds an agricultural landscape with livestock, a component of this system that has a long tradition, more attractive than without.

The aesthetic quality of the landscape of is closely related to the areas' *recreation and tourism* potential (Table 6.2). The silvopastoral systems are appreciated as providing excellent conditions for hunting activities, eco-tourism, agro-tourism, hiking, bird watching, or just making contact with nature (Almeida et al. 2016). Most of these activities have been increasing in recent years or are already established as economic activities on private properties, especially in the Spanish dehesa (Moreno and Pulido 2009). However, there is no reliable or comprehensive data about how much revenue these activities generate.

6.3.2 Economic benefits

Our literature review also showed that the material products of these landscapes are also important to society. Silvopastoral systems are recognized not only for the *material products* from trees, i.e. cork, acorns, firewood and charcoal, but also for numerous products from the understorey, i.e. livestock feed, Iberian pork, lamb, quality beef of endogenous breeds, cheese and other milk products, sheeps' wool, cereals, game, honey and its by-products and mushrooms (Oteros-Rozas et al. 2014; Gaspar et al. 2016; Surová et al. 2018).

Society acknowledges and values the *diversity of goods* that originate from these silvopastoral systems (Jørgensen and Quelch 2014). A diversity of products leads to a diversity of farm income and thus contributes to farms' stability (García de Jalón et al. 2018). Furthermore, most of the products mentioned above have very particular characteristics and that are specific to their niche in the ecosystem. As research shows, this *exclusivity of products* is explicitly valued by people. For instance, the montado has become an essential symbol of the areas where it exists and, even without a unifying brand, diverse economic activities benefit from using this concept in their own promotion. The montado has an appeal and can contribute to tourism attraction, as well as giving value to non-agricultural products that would not exist without this "montado" symbolism (Surová et al. 2018). Gaspar et al. (2016) found that Spanish consumers highly value local products originating from the dehesa, considering them to be quality products.

6.3.3 Environmental benefits

From the environmental point of view, the silvopastoral systems are overall very positively evaluated. Due to their *adaptation to local edapho-climatic conditions, high biodiversity, low chemical treatments and low-energy inputs,* they are considered as healthy environments that produce quality food and secure biodiversity (Pinto-Correia and Carvalho Ribeiro 2012).These systems are also seen as more resistant to wild forest fires than landscapes with other tree species and contribute towards the conservation of soil and natural resources (Surová et al. 2018). Garrido et al. (2017) found that the stakeholders who they interviewed saw trees as playing an essential role in preventing soil erosion and generating a local micro-climate that supports the existence of specific plant communities (this soil quality benefit is supported by research). It is a tangible output even at the farm level (e.g. improved productivity and reduced soil management costs) as well as the more comprehensive landscape-scale (e.g. less flooding and water purification costs). It can be a useful indicator of the long-term productivity of farms (García de Jalón et al. 2018). When livestock density is kept low, this has positive impact on trees' regeneration (Garrido et al. 2017).

6.4 Perceived challenges facing the silvopastoral systems

Studies addressing perceived challenges that the silvopastoral systems face nowadays are less common than those looking at the perceived benefits, yet the research suggests that society recognizes that these highly preferred landscapes are facing numerous challenges. Most of the perceived issues relate to economic inefficiencies and the environmental vulnerability of the systems (Table 6.3), although societal challenges are also recognized, mainly related to societal priorities and policy incentives.

6.4.1 Perceived socio-cultural challenges

As Table 6.3 shows, there is awareness about the continuous disappearance of local traditional knowledge and skills (Surová et al. 2018). The personal interactions of today's society with silvopastoral systems is becoming increasingly occasional and causes apparent *detachment* between landowners and the land, as well as between local communities and the land (Surová et al. 2018). However, the social-ecological nature of the silvopastoral systems requires the continued maintenance of their human component. If these landscapes are to continue providing territorial identity, the human dimension is crucial, mainly through the building of community linkages. In this way the diverse and meaningful links between modern societies and silvopastoral systems should be encouraged.

A significant change has also occurred in terms of public access to formerly publicly open silvopastoral areas (Carolino et al. 2011). Nowadays, many pasture fields are fenced, many gates are closed, old dirt roads and paths are blocked and, in many places, public access is restricted. According to surveys,

Table 6.3 The perceived challenges faced by silvopastoral systems

Socio-cultural challenges	Literature references
Disappearance of local traditional knowledge and practices	Garrido et al. 2017, Surová et al. 2018; Pinto-Correia et al. 2019
Detachment between landowners and the land, but also between local communities and the land	Surová et al. 2018
Restricted public access	Surová and Pinto-Correia 2009, Carolino et al. 2011; Fagerholm et al. 2016, Torralba et al. 2017
Unclear common goods rights	Surová and Pinto-Correia 2009
Tourism development	Gaspar et al. 2016, Surová et al. 2018
Heterogeneous societal expectations	Gómez-Limón and de Lucío Fernández 1999; Surová and Pinto-Correia 2008; Pinto-Correia et al. 2011; Carolino et al. 2011; Barroso et al. 2012; Almeida et al. 2014; Surová and Pinto-Correia 2016; Garrido et al. 2017; Surová et al. 2018; Pinto-Correia et al. 2019
Economic challenges	
Low employment	Surová et al. 2018
Required high efficiency in management due to the extensiveness of the system	Surová et al. 2018
Complexity of the field work management due to the multi-functionality	García de Jalón et al. 2018
Complicated administration	García de Jalón et al. 2018
Poor market competitiveness	Surová et al. 2018
Ill-fitting policy incentives	Gaspar et al. 2016, Surová et al. 2018, Pinto-Correia et al. 2019
Environmental and regulating challenges	
The systems' vulnerability - tree regeneration, intensive pasture effects, long-term thinking needed	Garrido et al. 2017, García de Jalón et al. 2018, Surová et al. 2018, Guimarães et al. 2018, Pinto-Correia et al. 2019

visitors would appreciate signposting of walking paths and the provision of tourist and educational information including information about the presence of wild cattle, hunting zones or fire risks. These improvements would allow occasional visitors a higher degree of independence and make them less reliant on being part of large guided groups which could increase the quality of their experiences (Surová and Pinto-Correia 2009). Moreover, accessibility impacts upon people's perceptions of the provision of benefits and influences peoples' ability to explore and experience the landscape (Torralba et al. 2017), as well as their willingness to pay for the benefits, as would be required if new market products are to be created.

Additionally, there are practices which were formerly considered as *common goods rights* with social and traditional meanings, specifically collecting mushrooms, asparagus and aromatic plants, and beekeeping (Surová and

Pinto-Correia 2009). These activities integrate productive and cultural aspects of people's perceptions and experiences of silvopastoral systems, but require formal or informal agreements between users and land owners. This relates to property and use rights and is addressed in other chapters of this book.

Tourism in the silvopastoral systems is as yet an underexploited opportunity. Agritourism, active tourism (e.g. hiking, horseback routes, touristic cycling) and hunting tourism could be feasible ways to develop the regions where these systems are located. Researchers argue that these activities have the potential to create new employment opportunities and diversify and increase farm income (see, for example, Gaspar et al. 2016; Surová et al. 2018).

Numerous studies reveal *heterogeneous societal expectations* concerning the silvopastoral systems (Surová and Pinto-Correia 2016; Garrido et al. 2017; Surová et al. 2018). Figure 6.2 illustrates the main focus of different societal

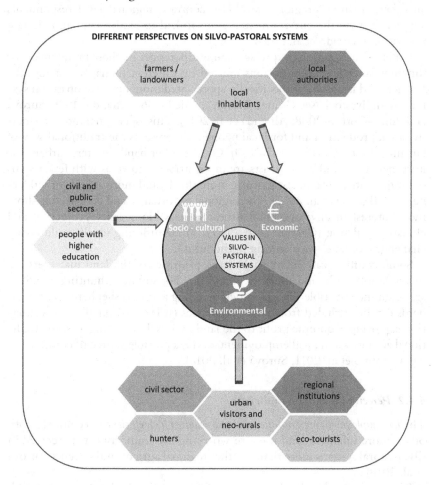

Figure 6.2 Different perspectives on the silvopastoral systems identified based on the review of the up-to-date literature.

groups according to literature review described above about the heterogeneous societal expectations.

Some differences have been observed not only between different user groups, but also between groups of people with different socio-demographic background as well as between different governance levels and sectors. This heterogeneity, combined with the different management models currently applied to these systems, represents a clear challenge in relation to public policies or market mechanisms to support their multiple values.

The socio-cultural dimension of the silvopastoral systems is increasingly recognized by people in both the private and public sectors who work at the local governance level (Garrido et al. 2017) and by people with higher education levels (Surová and Pinto-Correia 2016). The economic aspects connected to providing services are more recognized at the local governance level than the regional level. By contrast, regional level respondents appreciate more the environmental gains linked to biodiversity and climate regulation (Garrido et al. 2017).

Different user groups expressed some differences in their preferences for different landscape compositions created by different management models. Farmers and landowners specifically appreciated more open silvopastoral systems with livestock (see Gómez-Limón and de Lucío Fernández 1999; Surová and Pinto-Correia 2008; Almeida et al. 2014). This aspect guarantees farmers livestock production and for local people, it represents the traditional way of farming (Pinto-Correia et al. 2019). On the other hand, hunters, urban visitors, neo-rural and eco-tourists are more attracted to areas with fewer signs of human intervention, less pasture, more shrubs and more aspects of wilderness (see Barroso et al. 2012). The regional governance and civil sectors show more interest in environmental issues, such as biodiversity protection and climate regulation (Garrido et al. 2017). These differing ideals could create challenges for land use planning in the future.

Some conflicts can emerge when the priorities of the land managers are incompatible with land users' expectations. For instance, hunting activities are often incompatible with livestock breeding and thus shepherds with their herds can be excluded from some properties (at least seasonally). This situation can bring economic profit to the landowners, but endangers some traditional activities and local employment, and is a possible source of conflict (see e.g. Carolino et al. 2011; Surová et al. 2018).

6.4.2 Perceived economic challenges

The *low employment opportunities* and the *required high efficiency* to compete with other, more intensive land uses are worrying economic aspects perceived in silvopastoral systems more than in other regional agricultural systems (Surová et al. 2018).

The *complexity of fieldwork management* and *administration* in comparison with conventional agriculture, due to the combination of the forest and pasture components, was also mentioned by some stakeholders. Some land managers in Spain claimed difficulties in getting permission for pruning, access for transhumance and getting financial and administrative support for the multi-functional system (García de Jalón et al. 2018).

In addition, the products and services of the silvopastoral systems still have, most frequently, relatively *weak market competitiveness* and there is a need for them to be better promoted and marketed. The economic potential of such products as quality labels seems to have potential.

While tourism in the Spanish dehesa seems to provide modest economic benefits, especially hunting, the Portuguese montado still has a long way to go in this direction. This could be an area that could benefit from attention from academics who could help identify opportunities for landowners deriving more economic benefits from tourism in ways that are in harmony with system's sustainability and enhance the areas' reputations (Surová et al. 2018).

Furthermore, existing agricultural *policy incentives* do not seem to be well-tailored to extensive silvopastoral systems (Surová et al. 2018). Although there have been small changes introduced, the implementation of the Common Agricultural Policy in Spain and Portugal still encourages farmers to invest in intensification of cattle production, with costs for the long-term resilience of the whole system (Pinto-Correia and Azeda 2017). An opinion prevails that owners should receive *compensation* through subsidies or tax reductions for the ecosystem services they supply to society as a whole (Gaspar et al. 2016). However, some stakeholders think that a *dependence on subsidies* is a negative factor for the preservation of these systems (Gaspar et al. 2016; Surová et al. 2018).

6.4.3 Perceived environmental challenges

From the environmental point of view, the *vulnerability* of the silvopastoral systems is also known (Table 6.3). The struggle to regenerate tree cover is an issue that is widely discussed (García de Jalón et al. 2018; Surová et al. 2018) and the consequences of pasture intensification depleting the soil is also recognized (Garrido et al. 2017). The system needs continuous human intervention through adequate land stewardship to secure its sustainability and future continuity (Pinto-Correia et al. 2019).

6.5 Lessons learned

Based on the empirical evidence, gathered through a variety of methods, we conclude that silvopastoral landscapes in south-western Iberia are recognized as essential in terms of the social-cultural, economic and environmental dimensions of local communities' well-being. Our literature review shows that members the contemporary society living close to these systems perceive that they receive multiple benefits that are far more complex than those resulting from their management for production purposes. Montado and dehesa are perceived being of outstanding importance for cultural heritage and ecological balance in the regions in which they exist. They are not valued by all stakeholders in the same way, but they do satisfy multiple societal expectations and are generally highly valued. The disappearance or degradation of these systems would mean a loss of unique benefits, many of which would disappear if the land-use system were to significantly change.

In general, the evidence from these studies reveals that society expects the persistence of the silvopastoral systems in a way that balances their socio-cultural, economic and environmental benefits. The society also recognizes that these landscapes are facing several challenges, on multiple fronts, if they are to continue rendering the same benefits to society.

Yet, it should also be highlighted, that citizens may not be aware of all the benefits they receive from these systems and the challenges that the systems are facing. The public does seem to be, in general, aware of the socio-cultural benefits of the silvopastoral system, so focusing on these identity values could play an important role in supporting a bottom-up approach to protecting these systems. These values could also be enhanced by increasing social awareness about the cultural and environmental importance of these systems. Additionally, local action groups could spread and communicate the ethnological heritage associated with montado and dehesa.

As a whole, society does not seem familiar with the complete commercial possibilities and difficulties of these land-use systems. Moreover, the seriousness of specific environmental threats, such as diseases, and the potential role of these systems in mitigating the consequences of climate change, is only partially recognized. In this regard, public awareness could be improved to support the maintenance of these systems and the benefits that they are perceived to bring.

Undoubtedly, help is needed from specific public policies that relate to these silvopastoral systems in order to maintain and improve them and the benefits they bring to society. At the public policy level, the support given to silvopastoral systems should recognize their particularities. The highly valued multi-functionality of the silvopastoral systems suggests that policy incentives intended to promote individual functions are insufficient or even counter-productive. The historical evidence indicates that silvopastoral systems were most often managed for multiple products at the same time (see Box 6.1). Specific supportive schemes could be developed to support management that simultaneously addresses the productive, ecological and social-cultural dimensions of silvopastoral systems. Results-based payments, a new tool included in the menu of the future Common Agricultural Policy, maybe a place to start, at least in an experimental phase. Moreover it is clear that strategies that exclude human use are not suitable to safeguard the full range of the benefits of these silvopastoral systems. Chapter 14 focuses specifically on this issue.

TEXT BOX 6.1

The remarkable plasticity and ability of silvopastoral systems to provide multiple values

Mario Torralba, University of Kassel, Germany

Key points – Due to their capacity to provide multiple values, silvopastoral systems have survived pressures for land-use change by adapting their role to the changing societal needs.

Silvopastoral systems are still very important across large parts of Europe. Developed by almost every European society at some point in history, silvopastoral landscapes have proved themselves to be resilient towards social-ecological changes. Due to their plasticity and ability to provide multiple benefits, silvopastoral models have been progressively adapted and moulded to changing societal needs. Over time, that has resulted in a great heterogeneity of silvopastoral systems within Europe. They all share similar structural characteristics and might look relatively alike. However, they are all very different in relation to their main uses, and especially in relation to their societal role.

Take the examples of Swedish pedunculated oak wood pastures and Spanish holm oak dehesas (Figure 6.3). In the past, in both cases, these systems were

Figure 6.3 Swedish pedunculate oak wood-pasture (left) and Spanish holm oak wood-pastures (right).

central elements within a multifunctional agro-silvopastoral landscape fulfilling multiple roles. In Sweden, pedunculated oaks were a key resource providing animal fodder and refuge in the snowy winters, while in Spain holm oaks did so in the warm and dry summer.

However, societal needs evolved differently in Spain and Sweden and the role of silvopastoral systems diverged in each country. In both cases, silvopastoral systems remained valuable, but for very different reasons. In Sweden, agricultural systems were heavily intensified and food production was de-localized in the market economy, the productive role of silvopastoral systems was almost lost. They have remained in patches as a relict system from the former agro-silvopastoral landscape. Currently, they are highly appreciated for their cultural and identity value and managed primarily for recreational purposes. In Spain, the added value that holm oak plays in the production of high-quality meat products still makes these dehesas profitable systems. Its products rank among the most emblematic in Spanish gastronomy, while holm oak is one of its icons. As such, dehesas are also appreciated, but primarily managed for productive purposes.

We would see similar differences if we looked at silvopastoral systems in other regions, which only highlights the remarkable plasticity of silvopastoral systems and their ability to provide multiple values to society.

To support new tools that can face the challenge of supporting these multi-functional systems, research is needed on policy construction and on land owners willingness to change their business model according to policy interventions.

From another angle, innovative market mechanisms should be created, which would compensate land owners for providing some of the amenities that society values. Research on these mechanisms, and the scale at which they need to be implemented and the values that can legitimately be assigned market value (and those that cannot) is highly needed.

Considering the expectations of multiple actors at different levels of governance and simultaneously reinforcing the ecological, socio-cultural and economic dimensions of the dehesa and montado will be challenging, but is really needed. Failing to do would show an incomplete misunderstanding of what people care about, and undermine the future management and sustainability of these areas. At the farm management level, knowledge about societal expectations makes it possible for decision-makers to assess what impacts the present and planned management will have on users' satisfaction with the resulting landscape. New business models need to make use of this knowledge, with simple indicators and thresholds. Innovative and integrative research is highly needed for this development. This knowledge is likely to be highly relevant at the single farm or local landscape level, especially in regard to hunting and other recreational activities, Moreover, it also opens up the possibility of informing users about where, and to what extent, a given landscape may meet their expectations.

Acknowledgements

The first author, Diana Surová, would like to thank funds from the International mobility grant CZ.02.2.69/0.0/0.0/16_027/0008366, provided by the European Union and Ministry of Education in the Czech Republic.

References

Almeida, M., Loupa-Ramos, I., Menezes, H., Carvalho-Ribeiro, S., Guiomar, N. and Pinto-Correia T., 2016. "Urban population looking for rural landscapes: Different appreciation patterns identified in Southern Europe". *Land Use Policy* 53: 44–55,

Barroso, F. L., Pinto-Correia, T., Ramos, I. L., Surová, D. and Menezes, H., 2012. "Dealing with landscape fuzziness in user preference studies: Photo-based questionnaires in the Mediterranean context". *Landscape and Urban Planning* 104(3-4): 329–342.

Carolino, J., Primdahl, J., Pinto-Correia, T. and Bojesen, M., 2011. "Hunting and the right to landscape. Comparing the Portuguese and Danish traditions and current challenges". In: Egoz S., Makhzoumi J. And Pungetti G. (Eds.), *The Right to Landscape: Contesting landscape and human rights*, Ashgate, London, pp. 99–112

Carvalho-Ribeiro, S., Incerti, G., Miggliozi, A. and Pinto-Correia, T., 2013. "Placing land cover pattern preferences on the map: bridging methodological approaches of landscape preference surveys and spatial pattern analysis". *Landscape and Urban Planning* 114: 53–68

Fagerholm, N., Oteros-Rozas, E., Raymond, C. M., Torralba, M., Moreno, G. and Plieninger, T., 2016. "Assessing linkages between ecosystem services, land-use and wellbeing in an agroforestry landscape using public participation GIS". *Applied Geography* 74: 30–46

García de Jalón, S., Burgess, P. J., Graves, A., Moreno, G., McAdam, J., Pottier, E., Novak, S., Bondesan, V., Mosquera-Losada, M., Crous-Duran, J., Palma, J. H. N., Paulo, J. A., Oliveira, T. S., Cirou, E., Hannachi, Y., Pantera, A., Wartelle, R., Kay, S., Malignier, N., Van Lerberghe, P., Tsonkova, P., Mirck, J., Rois, M., Kongsted, A. G., Thenail, C., Luske, B., Berg, S., Gosme, M. and Vityi, A., 2018. "How is agroforestry perceived in Europe? An assessment of positive and negative aspects by stakeholders". *Agroforestry Systems* 92(4): 829–848.

García-Llorente, M., Martín-López, B., Iniesta-Arandia, I., Lóopez-Santiago, C. A.,Aguilera, P. A. and Montes, C., 2012. "The role of multi-functionality in social preferences toward semi-arid rural landscapes: An ecosystem service approach". *Environmental Science & Policy* 19-20: 136–146.

Garrido, P., Elbadkidze, M., Angelstam, P., Plieninger, T., Pulido, F. and Moreno, G., 2017. "Stakeholder perspectives of wood-pasture ecosystem services: a case study from Iberian Dehesas". *Land Use Policy* 60: 324–333,

Gaspar, P., Escribano, M. and Mesias, F. J., 2016. "A qualitative approach to study social perceptions and public policies in Dehesa agroforestry systems". *Land Use Policy* 58: 427–436.

Gómez-Limón, J. and de Lucío Fernández, J. V., 1999. "Changes in use and landscape preferences on the agricultural-livestock landscapes of the central Iberian Peninsula (Madrid, Spain)". *Landscape and Urban Planning* 44: 165–175.

Guimarães, M. H., Guiomar, N., Surová, D., Godinho, S., Pinto Correia, T., Sandberg, A., Ravera, F. and Varanda, M., 2018. "Structuring wicked problems in transdisciplinary research using the Social-Ecological systems framework: An application to the Montado system, Alentejo, Portugal". *Journal of Cleaner Production* 191: 417–428.

Hartel, T. and Plieninger, T. (Eds)., 2014. *European Wood-pastures in Transition. A social-ecological approach.* Routledge, London.

Jørgensen, D., Quelch, P., Hartel, Tibor and Plieninger, T., 2014. "The origins and history of medieval wood pastures" in Hartel, T. and Plieninger, T. (Eds) *European wood-pastures in transition: a social-ecological approach.* Routledge, New York, NY, pp. 55–69.

Martín-López, B., Iniesta-Arandia, I., Garcia-Llorente,M., Palomo, I., Casado-Arzuaga, I., Del Amo, D.G., Gomez-Baggethun, E., Oteros-Rozas, E., Palacios-Agundez, I., Willaarts, B., Gonzales, J.A., Santos-Martin, F., Onaindia, M., Lopez-Santiago, C. and Montes, C., 2012. "Uncovering eco-system service bundles through social preferences". *PLoS ONE* 7 (6): e38970.

Moreno, G. and Pulido, F., 2009. "The functioning, management, and persistence of dehesas". In: Rigueiro, A., McAdam, J. and Mosquera, R. (Eds.), *Agroforestry in Europe*, Springer, Amsterdam, pp. 127–160.

Nieto-Romero, M., Oteros-Rozas, E., González, J. A. and Martín-López, B., 2014. "Exploring the knowledge landscape of ecosystem services assessments in Mediterranean agroecosystems: Insights for future research". *Environmental Science & Policy* 37: 121–133.

Oppermann, R., Beaufoy, G. and Jones, G. (Eds.), 2012. *High Nature Value Farmland in Europe*, Verlag Regionalkultur, Heidelberg-Basel.

Oteros-Rozas, E., Martín-López, B., González, J. A., Plieninger, T., López, C. A. and Montes, C., 2014. "Socio-cultural valuation of ecosystem services in a transhumance social-ecological network". *Regional Environmental Change* 14(4): 1269–1289.

Oteros-Rozas, E., Martín-López, B., López, C. A., Palomo, I. and González, J. A., 2013. "Envisioning the future of transhumant pastoralism through participatory scenario planning: A case study in Spain". *The Rangeland Journal* 35: 251–72.

Pinto-Correia, T. and Carvalho Ribeiro, S. M., 2012. "High nature value farming in Portugal". In: Oppermann, R., Beaufoy, G. and Jones, G. (Eds.), *High Nature Value Farmland in Europe*, Verlag Regionalkultur, Heidelberg-Basel, pp. 336–345.

Pinto-Correia, T. and Azeda, C., 2017. "Public policies creating tensions in Montado management models: Insights from farmers' representations". *Land Use Policy* 64: 76–82.

Pinto-Correia, T., Barroso, F., Surová, D. and Menezes, H., 2011. "The fuzziness of Montado landscapes: Progress in assessing user preferences through photo-based surveys". *Agroforestry Systems* 82: 209–224.

Pinto-Correia, T., Guiomar, N., Ferraz-de-Oliveira, M.I., Sales-Baptista, E., Rabaça, J., Godinho, C., Ribeiro, N., Sá Sousa, P., Santos, P., Santos-Silva, C., Simões, M.P., Belo, A., Catarino, L., Costa, P., Fonseca, E., Godinho, S., Azeda, C., Almeida, M., Gomes, L., Lopes de Castro, J., Louro, R., Silvestre, M. and Vaz, M., 2018. "Progress in identifying High Nature Value Montados: relating biodiversity to grazing and stock management". *Rangeland Ecology and Management* 71: 612–625.

Pinto-Correia, T., Guiomar, N., Guerra, C. and Carvalho-Ribeiro, S., 2015. "Assessing the ability of rural areas to fulfill multiple societal demands". *Land Use Policy* 53: 86–96.

Pinto-Correia, T., Muñoz-Rojas, J., Thorsøe, M. H. and Noe, E. B., 2019. "Governance discourses reflecting tensions in a multifunctional land use system in decay; tradition versus modernity in the Portuguese montado". *Sustainability* 11(12): 3363.

Plieninger, T., Hartel, T., Martín-López, B., Beaufoy, G., Bergmeier, E., Kirby, K., Jesús Montero, M., Moreno, G., Oteros-Rozas, E. and Van Uytvanck, J., 2015. "Wood-pastures of Europe: Geographic coverage, social-ecological values, conservation management, and policy implications". *Biological Conservation* 190: 70–79.

Serrano-Montes, J.; Martínez-Ibarra, E. and Arias-García, J., 2019. "How does the presence of livestock influence landscape preferences? An image-based approach". *Landscape Online* 71: 1–18.

Silva Pérez, R. and Fernández Salinas, V., 2015. "Claves para el reconocimiento de la dehesa como 'paisaje cultural' de UNESCO". *Anales De Geografía De La Universidad Complutense* 35(2): 121–142.

Surová, D. and Pinto-Correia, T., 2016 "A landscape menu to please them all: Relating users' preferences to land cover classes in the Mediterranean region of Alentejo, southern Portugal". *Land Use Policy* 54: 355–365.

Surová, D., Surovy, P., Ribeiro, N. and Pinto-Correia, T., 2011. "Integration of landscape preferences to support the multifunctional management of the montado system". *Agroforestry Systems* 82(2): 225–237.

Surová, D. and Pinto-Correia, T., 2008. "Landscape preferences in the cork oak montado region of Alentejo, Southern Portugal: Searching for valuable landscape characteristics for different user groups". *Landscape Research* 33(3): 311–330.

Surová, D. and Pinto-Correia, T., 2009. "Use and assessment of the 'new' rural functions by land users and landowners of the Montado in southern Portugal". *Outlook on AGRICULTURE*, 38(2): 189–194.

Surová, D., Pinto-Correia, T. and Marušák, R., 2014. "Visual complexity and the montado do matter: Landscape pattern preferences of user groups in Alentejo, Portugal". *Annals of Forest Science* 71: 15–24.10.1007/s13595-013-0330-8

Surová, D., Ravera F., Guiomar, N., Martínez Sastre, R. and Pinto-Correia, T., 2018. "Contributions of Iberian silvopastoral landscapes to the well-being of contemporary society". *Rangeland Ecology & Management* 71: 560–570.

Torralba, M., Oteros-Rozas E., Moreno, G. and Plieninger, T., 2017. "Exploring the role of management in the coproduction of ecosystem services from Spanish wooded rangelands". *Rangeland Ecology & Management* 71(5): 549–559.

Section B

Institutions and the institutional framework

7 The spatial, temporal and social construction of the concept of silvopastoral systems

The case of montado and dehesa

José Ramón Guzmán-Álvarez and
Maria Helena Guimarães

7.1 Introduction

Silvopastoral systems are the sum of their component parts, which can exist in different combinations. They are multifunctional. Each land manager can follow a model that puts more emphasis on one or more of its components or products. This is the case in the montado and dehesa. Such diversity not only emerges from different business models but also when looking at how different scientific disciplines and regulatory instruments define them.

The multiple difficulties in reaching a common definition have to do with one of the determinant traits of dehesas and montado: their characteristics which mix and link forestry and agriculture (mainly livestock) within a single spatial and management unit. This way, what can be seen as an ecological and even socio-economic strength, has become, paradoxically, its main weakness because of the absence of policy coherence and coordination, which is largely due to the fragmentation and compartmentalization of the regulatory framework for rural land-use.

Any attempt at the definition of the dehesa and montado implies the establishment of boundaries: this is the first source of divergence in the different montados and dehesas we find in our vocabulary. And these boundaries, as we will point out later in this chapter, need to be established across two quite different dimensions: one related to intangible aspects (e.g. property and use rights, the visual attributes, etc.), the other with tangible items (e.g. the physical elements as tree species, tree cover, etc.).

The final chosen definition will never be neutral. It will classify some territories as dehesa and montado land and exclude other territories. Equally important: it will determine our personal and social expectations of what should or could be dehesa or montado.

This chapter deals with montado and dehesa as concepts that evolve along with changes in the social context. In this chapter, we present an overview of this diversity and discuss the implications of this rich picture in terms of governance. This chapter starts by providing some examples of definitions

DOI: 10.4324/9781003028437-8

used in the scientific arena and, then, at the regulatory and policy levels. Later we discuss the spatial and temporal dimension of the concepts, the social construction of the dehesa and montado and draw some conclusions regarding how such diversity affects the governance of these systems.

7.2 The multiple definitions of montado and dehesa in the scientific arena

Dehesa and montado are polysemic terms, having different meanings when they are distinctively considered as an agroforestry system, a farming system, a juridical or socio-political subject, an enterprise or a historical entity with deep roots in the past (see for instance, Joffre et al., 1988; Plieninger, 2007; Guzmán-Álvarez, 2016). The literature (Blondel and Aronson, 1999; Joffre et al., 1999; Castro, 2009; Huntsinger et al., 2013) emphasizes that, in both montado and dehesas, the territory is used simultaneously for multiple purposes, and involves a complex entwinement between quasi-natural ecosystems and extensive agricultural use.

According to a recent definition (Moreno and Cáceres, 2015, p. 3), dehesa and montado are silvopastoral systems, specifically "wood pastures where trees, native grasses, crops, and livestock interact positively under specific management practices (...) result from a simplification, in structure and species richness, of Mediterranean forests and shrublands, and are attained by clearing of evergreen woodlands, reducing tree density, eliminating shrub cover, and favouring the grass layer by means of grazing and crop culture."

It's also illuminating to introduce the definitions derived from two documents elaborated in Portugal and in Spain that collected discussions and reflections on the nature of the montado and dehesa. The Green Paper on Montado (Livro Verde do Montado) describes the montado as a heterogeneous production system of non-woody forestry, based on the exploitation of *Quercus* species, usually cork and holm oaks. Such forestry component is combined with a non-intensive use of the understory for farming, livestock or hunting (Pinto-Correia et al., 2013). The definition of the dehesa in the Green Paper of Dehesa (Libro Verde de las Dehesa) states that they are a functional system of stockbreeding and/or hunting exploitation in which at least 50% of the surface area is occupied by pastureland with adult scattered trees which produce acorns and with a proportion of the canopy cover of from 5% to 60% (Pulido and Picardo, 2010, English translation in Sánchez Martín et al., 2019).

To this, we need to add that for the more extended vision the tree cover is specific: holm oaks (*Quercus rotundifolia or Quercus ilex*), cork oaks (*Quercus suber*), and, occasionally, mountain oaks (*Quercus pyrenaica*), although in some definitions, as in the Andalusian Law of Dehesa (ley 7/2011), the range can include other species as wild olive trees, and occasionally other types of trees. While the dominant species in montados is the cork oak – *Quercus suber* (Godinho et al., 2016), in the case of dehesas it is the holm oak – *Quercus Ilex* (Elena-Roselló et al., 2013). Reality, however, is much closer to a regional and local mosaic with many nuances across both Spain and Portugal, in which the productive system has been historically more oriented towards grazing where holm oaks prevail, and to cork-production in cork oak-dominated systems.

According to Moreno and Cáceres (2015), the dehesa occupies 2.3 million ha in Spain and the montado 0.7 million ha in Portugal. As we will later detail, estimates of the size and distribution of these areas can vary depending on the definition used. Over time, there have been different estimates of the total area occupied by these two silvopastoral systems across Spain and Portugal. Costa (2009) suggests that Portugal has 800,000 ha of montado, mainly in the region of Alentejo (91%). Different figures for the Spanish dehesas have been proposed, ranging from 2,000,000 ha to 3,000,000 ha. Olea and San Miguel (2006) argue that the joint size of dehesas and montados is today around 3,500,000 to 4,000,000 ha, although some authors give a higher figure of up to 6,000,000 ha (Gaspar et al., 2007). This lack of consensus, even in terms of basic surface data, is directly linked to the lack of agreement on the definition of the system itself and is illustrative of the lack of a single land-use category for dehesa or montado in official statistical databases relating to Spanish or Portuguese agriculture data. This is a key challenge that public policy institutions in both countries have so far been unable to resolve.

Both systems are characterized by the rearing of traditional livestock breeds (cattle, sheep, pigs and goats) at low stocking densities. As a result of the management practices, montado and dehesa are usually structured in two plant layers: the herbaceous (mostly natural pasture, but sometimes crops and improved or sown pastures) and low to medium density and dispersed tree component which has shelter and forage functions and conservation values. Shrub understory is also common, although is usually heavily managed to enhance pastures, although it has to be pointed out the nutritional value of this layer and its role in trees' natural regeneration when shrub encroachment occurs.

The definition of dehesa from the Spanish Society for the Study of Pastures states it has a "surface with a more or less sparse tree canopy and a sound developed grassland layer from which scrubs have been largely removed; encompassing agriculture and livestock production. Its main production is extensive or semi-extensive livestock, feeding in grassland and in branches and fruits as well", (Ferrer et al., 2001). This has points in common with the definition proposed by Moreno and Cáceres (2015), although subtle but noticeable differences can be detected. It pays specific attention to the origin (agricultural and stockbreeding) and purpose (extensive or semi-extensive grazing, using grasses, browsing pastures and fruits of trees) of the system. This pasture approach is quite widespread in the scientific literature, stressing the savannah-like appearance of the whole system where the Mediterranean evergreen oaks (mainly *Quercus ilex* and *Q. suber*) are the dominant tree cover (i. e. San Miguel, 1994; Joffre et al., 1999).

In the case of montado, we can exemplify the diversity in the concept by referring to some of the definitions used in different fields of study. Godinho et al. (2016) estimate 1 million ha of montado in the Alentejo region in 2006, by defining a threshold canopy coverage of 10%. Studies focusing on the economic viability of the montado refer to the existence of about 2 million ha, of which 1 million ha is forest of *Quercus[ilex] rotundifolia* and *Quercus suber* (e.g. Fragoso et al., 2011). In such economic analyses, the agroforestry production systems include other land uses such as olive groves, vineyards and cereal systems (Campos et al., 2009). Pereira et al. (2015), when focusing on the relationship between birds and the montado, define the system by three main components: tree composition, tree density and production system. After revising what they also described as a diversity of definitions, proposed by other authors, they advance their work using a definition that describes the montado as system dominated by *Quercus* species with a tree cover equal to or above 10% (≥ 20 trees/ha) where human intervention is occasional or frequent, with a multifunctional character and where one or more forestry, pastoral and hunting/tourism activities take place. They found a higher consensus regarding the tree composition and production systems components of this definition than in regard to the minimum and maximum density of trees. A higher agreement was found on the minimal percentage of tree cover of around 10% and a higher discrepancy when trying to define the maximum percentage of tree cover (e.g. Pulido and Picardo 2010 (60%); Pinto-Correia and Almeida 2013 (100% in a dense montado and 20% in a sparse montado)).

These are some of the possible definitions of montado and dehesa. As it has been pointed out they are concepts with multiple meanings, somehow differing in subtle details, but others being apparently sharply different (Silva Pérez, 2010; Rodríguez Estévez et al., 2012; Pereira et al., 2015; Silva Pérez and Fernández Salinas, 2015; Guzmán-Álvarez, 2016; Guimarães et al., 2018). This is partly a consequence of the historic evolution of the concepts and parts of the different approaches that can be taken when dealing with this type of productive farm-state, ecological system or landscape (Silva Pérez, 2010; Guzmán-Álvarez, 2016). However, the lack of consensus in the definitions leads to difficulties in quantifying how extensive these systems are or in applying specific public politics.

7.3 The temporal evolution of the concepts

Looking at the intangible perspective of dehesa and montado concepts, history shows that both were born in medieval times as terms conceived for legal purposes related with the appropriation of the right to exploit some product in a limited territory. The word dehesa came from the concept 'to defend or to protect', in the sense of removing land from common use reserving it for some specific social actors, such as private owners, municipalities or ecclesial properties (Cabo, 1998; Martín Vicente and Fernández Alés, 2006). This led to the creation of fenced areas of different categories in each village called *dehesa de caballerías* (for war horses), *dehesa boyal* (for draught oxen), *dehesa yegual* (for mares and their young) or *dehesa carniceras* (for butchers' herds). The word *montado* refers to the tax (in Spain this tax was called *montazgo*) or the payment made by the livestock breeder to the landowner – often the municipality (Pinto-Correia and Fonseca 2009).

When property rights stabilized and communal lands lost importance, these concepts evolved to name an unit of management (a particular piece of land that maintained the ancient denomination of 'dehesa' or 'montado' and was managed as a unit) and lastly, in some areas and circumstances, to name an estate or unit of property (reducing the significance of the concept dehesa or montado to something like a label, without linkages to the ancient definition based on the kind of rights to use the territory). In summary, throughout the Middle Ages the concept evolved into having different meanings.

From XVIII century onwards, new semantic speciation started to rise in the dehesa area, first hesitantly and then becoming more widespread: dehesa was identified as a *Quercus* tree-layer open – pastureland. Possibly this emerging particularization was supported by the fact that the area where the dehesa concept was most strongly rooted coincided with that of the prevalence of Mediterranean oak-forest, that when cleared, gave rise to a valued grazing land with the complement of acorns.

The result is that nowadays dehesa and montado are usually identified as being characterized by a tree layer of *Quercus* species and an understorey of grass. However, as we have previously exposed talking about the non-tangible components, maps of the Iberian Peninsula show that there are many place names of montados and dehesa in areas without *Quercus* species, fossil records of territories where oaks have disappeared or where they were never present.

From medieval times until today the cutting of *Quercus* in dehesa and montado has been regulated. However, despite the protection of the tree component, several historical milestones have contributed to its decline in covered area and tree density. The regime designated by the Cereal Law of Portugal that started in 1889, the existence of a *latifundia* property structure, the wheat campaign (1929–1949), the Civil Spanish War (1936–1939) and the post-war two decades, the Agrarian Reform in Portugal (1975–1979), the integration of Portugal and Spain into the European Union (EU) and the 1992 revision of the CAP are all historic milestones in the evolution of the montado and dehesa. Factors such as the relative lack of fertility of the soils and the *latifundia* farm structure, which is associated with a web of actors and influencing lobbyists, led to the continuation of traditional use of the montado and dehesa, whereas others shifted towards cereal production, were converted into irrigated areas or saw an intensification of livestock production, all of which diminished the area of montado and dehesa (Campos, 1983; Guimarães et al., 2018).

A further complication when we try to elaborate the concepts of dehesa and montado arises when seeking to link the intangible and the tangible components. At this point, we observe a historical semantic divide. The first emerges when the concepts dehesa or montado matured as a synonym of management or property unit in the *Quercus* distribution area. As a result, dehesas and montados become estates or management units characterized mostly by the presence of *Quercus* sp. open forest pastureland, in which other productive land uses, such as olive groves, vineyards and cereal systems might have been included.

The other side of the semantic divide emerges when the property or management perspective is not taken into account. From this perspective, the characteristics of the dehesa and montado are considered solely in terms of their physical components or visual attributes. This result in parcels of land sharing the tangible attributes, that fulfill the definition (i.e. number of trees per hectare, tree cover, etc.), independently of intangible issues as property or, even, management.

7.4 Dehesa and montado as explicit geographical concepts

Whichever definition we use, we can conclude that dehesa and montado are geographic entities: areas with physical and functional characteristics (tangible or intangible) that differentiate them from other areas. Any definition should distinguish the distinct assemblages of features and phenomena that allow us to assign the label of dehesa or montado to a territory. These features can be a limited or an expansive set of characteristics or attributes that need, at least to some extent, to be fulfilled (Joffre et al., 1988; Carruthers, 1993; San Miguel, 1994; Ferrer et al., 2001; Pulido and Picardo, 2010).

One essential feature of geographical entities is that it should be possible to map them, drawing boundaries that separate one entity from another. In this sense, the list of characteristics required to draw these boundaries can represent an important source of divergence about where to draw the separation line (see, for instance, Carruthers, 1993). These features can take a qualitative style, as in the European Union's definition of Habitat 6310 Dehesa with evergreen *Quercus* spp. (European Commission, 2007): "a characteristic landscape of the Iberian peninsula in which crops, pastureland or Meso Mediterranean arborescent matorral, in juxtaposition or rotation, are shaded by a fairly closed to a very open canopy of native evergreen oaks (*Quercus suber, Q. ilex, Q. rotundifolia, Q. coccifera*). It is an important habitat of raptors, including the threatened Iberian endemic eagle (*Aquila adalberti*), of the crane (*Grus grus*), of the endangered felid (*Lynx pardinus*)."

The definitions of montado and dehesa tend to be quite precise in quantifying at least some of its attributes, such as the density of trees or the canopy cover. Nevertheless, some of these attributes are clearer and easier to observe than others when trying to define the boundaries of dehesa/montado areas. Some can be described in binomial terms i.e. values of 0/1 (the boundaries of a property, for instance), and others with discrete expressions (tree species that may be considered as components of the tree-layer, see Vieira Natividade, 1950; Montoya, 1989; Ferrer et al., 2001). Quantitative traits (either continuous, such as tree cover, or discrete, such as tree density), are usually a matter of controversy because these limits or boundaries must be fixed.

Legal definitions of dehesa and montado have to be geographically explicit. If we go back beyond some of the more recent ones, we find two distinctive laws in Spain, established in two of the regions with a large surface of dehesa. Both share a farming perspective that identifies dehesa as a type of estate, but the characteristic features in each of them are quite different. The one from the Extremadura Law of Dehesa of 1986 focuses on the size and the productive function ("it is an estate with an area of more than 100 ha where the most suitable agrarian use would be livestock extensive grazing", in Ley 1/1986, de 2 de mayo, sobre la Dehesa en Extremadura); in this definition, the spatial representation must be obtained case by case. Alternatively, the Andalusian law, approved in 2010, is much more detailed and includes a two-step definition (Ley 7/2010, de 14 de julio, para la Dehesa). The first level is the concept of vegetation structure using the term *formación adehesada* ('open-woodland dehesa area'), a specific type of open woodland forested land characterized by a tree stratum with a soil surface area covered by the projection of the tree crowns in between 5% and 75%, mainly comprising of holm oaks (*Quercus ilex*), cork oaks (*Quercus suber*), Portuguese oaks (*Quercus faginea*) or wild olive trees and occasionally other types of trees, which allow the development of a herbaceous layer. The second level is the management dimension through the introduction of the concept *dehesa* itself, defined as a spatial unit of property constituted for the most part by *formación adehesada*, subject to a system of land use and management based mainly on extensive livestock that uses grass, fruits and browsers, as well as obtaining another forestry, hunting or agricultural products. The concept *formación adehesada* is based on tangible components and can be directly mapped, even, from a theoretical point of view, directly from the appropriate source of information, although there would possibly be misclassifications due mainly to the quality of the data. The second concept, dehesa, includes the intangible attribute of property, the limits of which can be drawn on a map with precision.

At the national level in Portugal, the policy instruments that explicitly refer to the montado are the Basic Law for Forestry Policy, the 29/26 law and the regulatory instruments designated Decreto-lei (DL)169/2001 which were replaced by the DL 155/2004. The definitions used in these regulatory instruments diverge to a certain extent from the definitions found in the academic literature. At the start of the DL 169/2001 montado is defined in plural (i.e. montados) as systems used for agri-silvopatoral activities. In a later section of the document the definition used refers only to the tree component and describes the montados as stands of cork or holm oak trees or a mixture of both in a density equal or higher than 10%. Analyzing such definitions Pinto Correia et al. (2013) concluded that an important area of montado is excluded from such regulatory instruments. The montado area is split between different policy instruments, some are within the forestry sector (stands of more than 45 trees/ha) while another part is considered to be in the agricultural sector, as pastures. Such a classification collides with the agro-forestry nature of the montado, which can contain different land use classes within the same farm unit.

The agroecosystem-based definitions discussed previously in this chapter are examples of scientific approaches to the dehesa and montado which differ from the legal perspective. One of the main differences is that the former approach considers dehesa and montado as systems (Gastó Cordech et al., 2009). This approach has prevailed in recent decades and views dehesa and montado as paradigmatic examples of agroforestry systems in Europe (Moreno and Pulido, 2009; Pinto-Correia et al., 2011). Yet these ecological definitions of dehesa and montado must also be mappable as discrete units with boundaries that separate them from distinct neighbouring areas or systems, which raises the challenge of translating the attributes of the definition (frequently vaguely expressed) into figures and spatial boundaries. When issues such as the holistic functioning of the system, the existence of emergent properties or the integration of the components are included in the ecological definition of dehesa and montado, the challenge is even greater.

It is usually more difficult to define and geographically represent dehesa and montado in the scientific arena than it is when using regulatory or policy definitions. The definitions derived from ecological, landscape, forest or cultural perspectives imply conceiving the existence of entities that are distinguished one from another in their structural components and process. Once a definition is accepted, the challenge of delineating the spatial distribution of the traits then emerges, which is necessary to establish the boundaries and unambiguously map them. This implies the individual who draws the boundaries making decisions, some of which are bound to be subjective as it is not possible to arrive at a universal consensus regarding the criteria for boundaries. Farmers and politicians have criticized the ecological or cultural definitions as being overly biological and physical, considering that they are is more visual and phenomenological than based on processes. They argue that following such definitions can lead us to forget the fundamental role of human beings and livestock management in the functioning of dehesa and montado.

Silvopastoral systems such as dehesa and montado can be found worldwide as depicted in the following text box (Box 7.1).

TEXT BOX 7.1

Silvopastoral systems in the world

Gerardo Moreno, Forestry School, INDEHESA - Institute for silvopastoralism Research University of Extremadura

Key points: Silvopastoral systems can be found all over the world and play a valuable role.

Silvopastures are frequently diffuse transitions from agricultural to forest lands, which makes it difficult to produce precise estimates of their size or to map their locations and boundaries. Dehesa and montado, defined as grazed lands with scattered oak trees in the southwest of the Iberian Peninsula are estimated to occupy around 3.5 million ha (Moreno and Rolo, 2019). Den Herder et al. (2017) estimate that there are up to 4.3 million ha of grazed wood pastures on the Iberian Peninsula, increasing up to 6.6 million ha if other grazed lands (grasslands and scrublands) with sparse tree cover are included. While dehesa and montado form the main core of silvopastoralism system in Europe, silvopastoral systems are also common in other Mediterranean and Eastern European countries. Den Herder et al. (2017) estimated a total of 15.1 million ha in the European Union, which corresponds to about 3.5% of the territorial area, and up to 35% of the grazed lands in these 27 countries. Most European wood pastures are based on scattered oak and resemble, in some way, the Iberian dehesa and montado, but other systems are also common (e.g. reindeer pastoral woodland with birch and pine in subarctic Europe).

Grazed oak woodland that resembles the Iberian dehesa and montado can also be found in northern Iran, with > 5 million ha (Valipour et al., 2014), and California, with 1.9 million ha owned by ranchers out of 3.4 million ha of total oak woodland area (Huntsinger et al., 2013). Porqueddu et al. (2016) also reported the importance of silvopastoral systems in other Mediterranean regions, such as the 'WANA' region (Western Asia and Northern Africa), although no figures about the totals were reported. In South America, the Espinales, with about 2 million ha are the most widespread agroforestry system of the Mediterranean climate region of central Chile (Ovalle et al., 1990), while native forests under silvopastoral use cover ~6,8 million ha in Argentina (Peri et al., 2016). Many other examples document the importance of silvopastoralism across the world. Nair (2012) gives a rough estimation of 450 million ha of silvopasture in the world, mostly semi-arid and sub-humid lands in Africa, India and the Americas (for more details see Moreno and Rolo, 2012).

7.5 The dehesa and montado as social representations of changing relationships with nature and the need for consensus

People involved with the montado and dehesa will have quite different perceptions of it depending on their activity or discipline: for some, they may be just commonly used words, fixed in the dictionaries, for others an administrative or legal term, a geographical entity, a source of livelihood or a set of environmental attributes assigned by the natural sciences. But whatever focus we begin with, it is important to adjust our own lens to account for other people's perceptions, thereby expanding our own.

Several studies of the montado report the absence of a common mental model that is used to define it (Guimarães et al., 2018). Although the montado is considered to be central to regional development in the Alentejo Region, there is a lack of consensus over which activities could enhance its future sustainability (Guimarães et al., 2018, 2019). Guimarães et al. (2019) describe a desire to combine development strategies that might be difficult to reconcile with each other as they include maintenance of the montado, intensification of farming production and the expansion of the area of irrigated crops. In a study conducted by Surová and Pinto-Correia (2008), different preferences for the montado landscapes were identified depending on who is using the system and for what purposes. For instance, hunters prefer an open montado (i.e. low tree cover) whereas beekeepers favour dense montado (i.e. high tree cover) with scrub. Guimarães et al. (2018) show the diversity of the words that users of the montado link to the term. The words were categorized into the following groups: resources (i.e. provisioning and supply of regulating services), sensations and feeling (i.e. stimulus derived), heritage/belonging (i.e. attributed cultural values), appreciation (i.e. quality recognition), disfavour (i.e. negative features).

Although it has not been studied in such detail, we consider that the case of the dehesa would be quite similar: for instance, whereas most common definitions adopted by environmental NGOs tend to highlight the natural components, farmers' reports tend to focus their attention on production functions and services. From a stakeholder's perspective, the definition adopted by FEDEHESA (Federación Española de la Dehesa, http://fedehesa. org/concepto-de-dehesa/), linked to the one of the Spanish Grassland Study Society (Ferrer et al., 2001) is closer to a livestock-centred point of view. A World Wildlife Fund (WWF) report on the future of dehesa (Hernández, 2014) adopts the basic components of the FEDEHESA definition, adding a complementary one focussing on its agro-silvopastoral dimension and its sound ecological and economical results under appropriate management. WWF identifies dehesas as a High Natural Value System where "extensive livestock raising plays a leading role in the creation, use and maintenance of grazing lands".

Lay knowledge may bring other representations. In the case of the dehesa, Acosta (Acosta, 2008) has pointed out that the local population cognitively views individual estates as the central unit, irrespective of the plant cover. While local people distinguished differences in vegetation cover between different patches of territory, these differences were not as relevant as the unit of land ownership.

The temptation is to consider dehesa and montado as fuzzy concepts whose boundaries of application can vary according to the context or conditions, but this is perhaps more useful in Alice´s Wonderland than in our real world. While concepts of the dehesa and montado should be as clear and unambiguous as possible, they will inevitably be fuzzy at the edges to some extent.

Fortunately, not everything is fuzzy. All the definitions of dehesa and montado share some attributes in common. One, as we have shown, is that they are spatial entities, though the particular limits of these entities vary according to each definition.

We could represent the dehesa and montado as the whole space that fulfils the requirements of all their possible definitions. This space could be divided into subareas through the use of Venn diagrams corresponding to each specific definition. Theoretically, it would be possible to identify a global intersection zone, corresponding to some kind of the median of the social representations of montado and dehesa at any given time. In our sheet, outlying areas might be residues from the past (fossilized concepts in the process of vanishing, as the first definition in the Diccionario "Tesoro de la lengua castellana o española" of Sebastián de Covarrubias (1611), being dehesa just an herb field where livestock grazes) or emerging new patterns, which may be subtle nuances, or be the result of a disruption or a new vision (as the concept of dehesa as a type of agroecosystem, in Acosta (2014, p. 305)). These outlying areas might be either shrinking and expanding over time in accordance with semantic evolution.

Each of us will may align ourself with one, or maybe several, subareas in the diagram that matches our specific conception of montado or dehesa, which we would take as our reference point. Usually, we would share the same conception of dehesa and montado with our peers or colleagues. For example, we would expect that urban inhabitants would have a different representation of dehesa and montado than the rural population living in, and from, the dehesa and montado area. Finally, we could obtain a typical representation of the image of dehesa and montado for different social groups (farmers, birdwatchers, foresters, cultural heritage experts and so on).

It is possible that, these different spatial representations of dehesa and montado may coexist without disturbance. A birdwatcher´s perception, for instance, may not be affected by a property registrar's conception. But things are not really like that. The spatial semantic representation of dehesa and montado are an arena of social interactions. Farmers, or foresters, or ecologists, or whoever, assume that the social archetype is theirs, sowing the seeds for possible conflicts (see Chapter 11). Such conflicts are difficult to resolve because they are the result of the mix of distinct semantic dimensions: the different groups may not reach an accord because they do not share the same concepts, the baseline on which begin discussions. Such differences strengthen the argument for having a specific definition, as the alternative is confusion.

The lack of a proper definition of dehesa and montado has contributed to difficulties in the conservation of these systems. A recent declaration by dehesa representatives emphasizes that dehesa (and, by extension, the montado) have been mistreated and unprotected by public policies in the EU because their characteristics are not properly recognized (Sánchez Fernández et al., 2018). According to this claim, the lack of existence of an administrative definition of dehesa and montado has helped to prevent proper treatment by CAP subsidies and regulations. As an example, Sánchez Fernandez et al. (2018) point out that dehesa and montado are regarded as an aggregation of components (croplands and permanent pastures), and when applying the CAP rules to calculate the surface declared eligible for payment rights, the surface covered by the *Quercus* tree canopy and the shrubs were ignored.

To obtain a shared definition for a particular aim, we should look together at the core intersection area in our theoretical Venn diagram. As a social concept, we will need debate and discussions to achieve a consensus about this common area of representation of dehesa and montado for major social interactions. This may involve each of the groups to modify parts of their own definitions and reshape their semantic map. Only after this effort, we will obtain a social concept of dehesa and montado and that is needed for specific purposes, especially in the administrative arena.

References

Acosta, R. 2008. *Dehesas de la sobremodernidad. La cadencia y el vértigo.* Diputación de Badajoz.

Acosta, R. 2014. Dehesa de Tentudia. In: Pardo de Santayana, M., Morales, R., Aceituno, L. & Molina, M. (eds.) *Inventario Español de los Conocimientos Tradicionales Relativos a la Biodiversidad.* Ministerio de Agricultura, Alimentación y Medio Ambiente, Madrid: 305.

Blondel, J. & Aronson, J. 1999. *Biology and Wildlife of the Mediterranean Region.* Oxford University Press, Oxford.

Cabo, A. 1998. Formación histórica de las dehesas. In: Hernández, C. (coord.) *La dehesa. Aprovechamiento sostenible de los recursos naturales.* Editorial Agrícola, Madrid: 15–42.

Campos, P. 1983. La degradación de los recursos naturales de la dehesa. Análisis de un modelo de dehesa tradicional. *Agricultura y Sociedad,* 26: 289–381.

Campos, P., Daly-Hassen, H., Ovando, P., Chebil, A. & Oviedo, J. L. 2009. Economics of multiple use cork oak woodlands: two case studies of agroforestry systems. In: Rigueiro-Rodriguez, A., McAdam, J., & Mosquera-Losada, M. R. (eds.) *Agroforestry in Europe: Current Status and Future Perspectives.* Springer, San Diego, USA: 269–294.

Carruthers, S. P. 1993. The dehesas of Spain – exemplar or anachronisms? *Agroforestry Forum,* 4: 43–52.

Castro, M. 2009. Silvopastoral systems in Portugal: Current status and future perspective. In: Rigueiro-Rodriguez, A., McAdam J. & Mosquera-Losada, M.R. (ed.) *Agroforestry in Europe: Current Status and Future Perspectives.* Springer, San Diego, USA: 111–126.

Covarrubias, S. 1611. *Tesoro de la lengua castellana o española.* Ed. Luis Sánchez, Madrid.

Den Herder, M., Moreno, G., Mosquera-Losada, R., Palma, J. H. N., Sidiropoulou, A., Santiago Freijanes, J. J., Crous-Duran, J., Paulo, J. A., Tomé, M., Pantera, A., Papanastasis, V. P., Mantzanas, K., Pachana, P., Papadopoulos, A., Pllieninger, T. & Burgess, P. J. 2017. Current extent and stratification of agroforestry in the European Union. *Agriculture, Ecosystems & Environment,* 241: 121–132.

Elena Roselló, R., Kelly, M., Martin, A., González-Ávila, S., Sánchez de Ron, D. & García del Barrio, J. M. 2013. Recent Oak Woodland Dynamics: A Comparative Ecological Study at the Landscape Scale. In: Campos Palacín, P., Huntsinger, L., Oviedo, J. L., Starrs, P. F., Díaz, M., Standford, R. B. & Montero, G. (eds.) *Mediterranean Oak Woodland Working Landscapes. Dehesas of Spain and Ranchlands of California*, Springer, Dordrecht, Germany: pp. 427–458.

European Commission. 2007. *Interpretation Manual of European Union Habitats*. DG Environment, Bruxelles.

Ferrer, C., San Miguel, A. & Olea, L. 2001. Nomenclátor básico de pastos en España. *Pastos*, 31: 7–44.

Fragoso, R., Marques, E., Luca, M. R., Martins, M. B. & Jorge, R. 2011. The economic effects of common agricultural policy on Mediterranean montado/dehesa ecosystem. *Journal of Policy Modellling*, 33: 311–327.

Gaspar, P., Mesías, F. J., Escribano, M., Rodríguez de Ledesma, A. & Pulido, F. 2007. Economic and management characterization of dehesa farms: implications for their sustainability. *Agroforestry System*, 71: 151–162.

Gastó Cordech, J., Calzado Martínez, C., Carbonero Muñoz, M. D., de Pedro Sanz, E., Fernández Rebollo, P., Garrido Varo, A., Gómez Cabrera, A., Guerrero Ginel, J. E., Guzmán-Álvarez, J. R., Lara Vélez, P. & Ortiz Medina, L. 2009. *Sostenibilidad de las dehesas. Documento de reflexión*. Grupo de Desarrollo Rural Los Pedroches, Córdoba.

Godinho, S., Guiomar, N., Machado, R., Santos, P., Sa-Sousa, P., Fernandes, J. P., Neves, N. & Pinto-Correia, T. 2016. Assessment of environment, land management, and spatial variables on recent changes in montado land cover in southern Portugal. *Agroforestry Systems*, 90: 177–192.

Guimarães, M. H., Guiomar, N., Surova, D., Godinho, S., Pinto Correia, T., Sandberg, A., Ravera, F. & Varanda, M. 2018. Structuring wicked problems in transdisciplinary research using the Social Ecological systems framework: An application to the montado system, Alentejo, Portugal. *Journal of Cleaner Production*, 191: 417–428.

Guimarães, M. H., Esgalhado, C., Ferraz de Oliveira, I. & Pinto-Correia, T. 2019. When does innovation become custom? A case study of the montado, southern *Portugal. Open Agriculture*, 4: 144–158.

Guzmán-Álvarez, J. R. 2016. The image of a tamed landscape: Dehesa through history in Spain. *Culture and History*, 5. doi: http://dx.doi.org/10.3989/chdj.2016.003

Hernández, L. 2014. *Dehesas para el futuro. Recomendaciones de WWF para una gestión integral*. WWF/ADENA, Madrid.

Huntsinger, L., Campos, P., Starrs, P. F., Oviedo, J. L., Díaz, M., Standford, R. B. & Montero, G. 2013. Working landscapes of the Spanish Dehesa and the California Oak Woodlands: an introduction. In: Campos P., Huntsinger, L., Oviedo, J. L., Starrs, P. F., Diaz, M., Standiford, R. B. & Montero, G. (eds.) *Mediterranean Oak Woodland Working Landscapes: Dehesas of Spain and Ranchlands of California*. New York, Springer: 3–23.

Joffre, R., Vacher, J., De Los Llanos, C. & Long, G. 1988. The dehesa: An agrosilvopastoral system of the Mediterranean region with special reference to the Sierra Morena area of Spain. *Agroforestry System*, 6: 71–96.

Joffre, R., Rambal, S. & Ratte, J. P. 1999. The dehesa system of southern Spain and Portugal as a natural ecosystem mimic. *Agroforestry Systems*, 45: 57–79.

Martín Vicente, A. & Fernández Alés, R. 2006. Long term persistence of dehesas. Evidence from history. *Agroforestry Systems*, 67: 19–28.

Montoya, J. M. 1989. *Encinas y encinares*. Mundi Prensa, Madrid, Spain.

Moreno G & Rolo V. 2019. Agroforestry practices: Silvopastoralism. In: Mosquera-Losada, M. R. & Prabhu, R. (eds.) *Agroforestry for Sustainable Agriculture.* Burleigh Dodds Science Publishing, Cambridge, UK: 1–46.

Moreno, G. & Pulido, F. J. 2009. The functioning, management and persistence of dehesas. In: A. Rigueiro-Rodriguez, J. McAdam & M.R. Mosquera-Losada (eds.) *Agroforestry in Europe: Current status and future perspectives.* Springer, San Diego, USA: 127–160.

Moreno, G. & Cáceres, Y. 2015. *System report: Iberian Dehesas, Spain.* AGFORWARD. Agroforestry for Europe. https://www.agforward.eu/index.php/en/dehesa-farms-in-spain.html?file=files/agforward/documents/WP2_ES_Dehesa.pdf

Nair, P. K. R. 2012. Climate change mitigation: A low-hanging fruit of agroforestry. In: Nair, P. K. R. and Garrity, D. (eds) *Agroforestry – The Future of Global Land Use. Advances in agroforestry.* Springer, Dordrecht, The Netherlands: 31–67.

Olea, L. & San Miguel-Ayanz, A. 2006. The Spanish dehesa. A traditional Mediterranean silvopastoral system linking production and nature conservation. In: Lloveras, J., González Rodríguez, A., Vázquez Yáñez, O., Piñeiro, J., Santamaría, O., Olea, L. & Poblaciones, M. J. (eds.) *Sustainable Grassland Productivity. Proceedings of the 21st General Meeting of the European Grassland Federation.* Badajoz, Spain: 3–13.

Ovalle, C., Aronson, J., Del Pozo, A. & Avendano, J. 1990. The espinal: Agroforestry systems of the Mediterranean-type climate region of Chile. *Agroforestry Systems,* 10: 213–239. doi:10.1007/BF00122913.

Pereira, P., Godinho, C., Roque, I. & Rabaça, J.E. 2015. *O Montado e as aves e boas praticas para uma gestão sustentável.* LabOr, ICAAM, Universidade de Évora, Câmara Municipal de Coruche, Coruche.

Peri, P. L., Dube, F. & Varella, A. C. 2016. Silvopastoral systems in the subtropical and temperate zones of South America: An overview. In: Peri, P. L., Dube, F & Varella, A. (eds). *Silvopastoral Systems in Southern South America. Advances in Agroforestry, Book-Series 11.* Springer International Publishing, Switzerland: 1–8.

Pinto-Correia, T. & Fonseca, A. 2009. Historical perspective of montado: The example of Évora. In: Aronson, J., Santos Pereira, J., Pausas, J. G. (eds). *Cork Oak Woodlands on the Edge: Ecology, Adaptative Management, and Restoration.* Island Press, Washington: 49–55.

Pinto-Correia, T. & Almeida, M. 2013. Tentative identification procedure for HNV montados In: Pinto-Correia, T., Ribeiro, N., & Ferraz Oliveira, I. (eds.) *ICAAM International Conference, The MONTADO/DEHESA as High Nature Value Farming Systems: implications for classification and for policy support.* http://hdl.handle.net/10174/10353

Pinto-Correia, T., Ribeiro N. & Sá-Sousa, P. 2011. Introducing the montado, the cork and holm oak agroforestry system of southern Portugal. *Agroforestry Systems,* 82: 99–104.

Pinto-Correia, T., Ribeiro, N. & Potes, J. 2013. *Livro verde dos montados.* ICAAM Instituto de Ciências Agrárias e Ambientais Mediterrânica. Universidade de Evora, Evora.

Plieninger, T. 2007. Compatiblity of livestock grazing with stand regeneration in Mediterranean holm oak parklands. *Journal for Nature Conservation,* 15: 1–9.

Porqueddu, C., Ates, S., Louhaichi, M., Kyriazopoulos, A. P., Moreno, G., Del Pozo, A., Ovalle, C., Ewing, M. A. & Nichols, P. G. H. 2016. Grasslands in 'Old World' and 'New World' Mediterranean-climate zones: Past trends, current status and future research priorities. *Grass and Forage Science,* 71: 1–35. doi:10.1111/gfs.12212.

Pulido, F.J. & Picardo, A. (coord.) 2010. *Libro Verde de la Dehesa. Documento para el debate hacia una Estrategia Ibérica de gestión.* Junta de Castilla y León, Sociedad Española de

Ciencias Forestales, Sociedad Española para el Estudio de los Pastos, Asociación Española de Ecología Terrestre y Sociedad Española de Ornitología.

Rodríguez Estévez, V., Sánchez Rodriguez, M., Arce, C., García, A. R., Perea, J. M. & Gómez Castro, G. 2012. Consumption of acrons by finishing Iberian pig and their function in the conservation of the dehesa agroecosystem. In: Leckson Kaonga, M. (ed) *Agroforestry for Biodiversity and Ecosystem Services. Science and Practice.* Intech, Rijeka: 1–21.

Rolo, V. & Moreno, G. 2012. Interspecific competition induces asymmetrical rooting profile adjustments in shrub-encroached open oak woodlands. *Trees-Structure and Function,* 26: 997–1006. doi:10.1007/s00468-012-0677-8.

San Miguel, J. 1994. *La dehesa española.* Escuela Técnica Superior de Ingenieros de Montes, Madrid.

Sánchez Fernández, M. P., Castaño Nieto, J. & Cabello Bravo, R. 2018. *La Dehesa ante la reforma del Reglamento Ómnibus de la PAC.* 2018/03/19, http://observatoriodehesamontado.juntaex.es/index.php?modulo=noticias&pagina=view.php&id=1253

Sánchez Martín, J. M., Blas-Morato, R. & Rengifo-Caballero, J. I. 2019. The dehesas of Extremadura, Spain: A potential for socio-economic development based on agritourism activities. *Forests,* 10(8): 620. https://doi.org/10.3390/f10080620

Silva Pérez, R. & Fernández Salinas, V. 2015. Claves para el reconocimiento de la dehesa como paisaje cultural de UNESCO. *Anales de Geografía de la Universidad Complutense,* 35: 121–142.

Silva Pérez, R. 2010. La dehesa vista como paisaje cultural: fisonomías, funcionalidades y dinámicas históricas. *Ería. Revista cuatrimestral de Geografía,* 82: 143–157.

Surová, D. & Pinto-Correia, T. 2008. Landscape preferences in the cork oak Montado region of Alentejo, southern Portugal: searching for valuable landscape characteristics for different user groups. *Landscape Research,* 33: 311–330.

Vieira Natividade, J. 1950. *Subericultura. Ministério da Economia. Direçao Geral dos Serviços Florestais e Aquícolas.* Lisboa, Portugal.

Valipour, A., Plieninger, T., Shakeri, Z., Ghazanfari, H., Namiranian, M & Lexerd, M. J: 2014. Traditional silvopastoral management and its effects on forest stand structure in northern Zagros, Iran. *Forest Ecology and Management,* 327, 221–230.

8 Property rights and rights of use

Rufino Acosta-Naranjo, Teresa Pinto-Correia and Laura Amores-Lemus

8.1 Introduction

For the institutions that influence the use and management of land, property rights are central. Property rights are directly linked to rights of use, which broadly define who can benefit from and make changes to the resources available on the land. The benefit obtained from the use is thus crucially linked with the concept of property rights. As stated by Ian Hodge (2016), a property right corresponds to "a relationship between people: a right and a reciprocal duty on others not to interfere with that right." Rights over land and resources define how they may be used and how the benefits and costs of resources are to be allocated; these rights also define how others are excluded from benefiting from this land and resources. Land is a material asset and the owner has the right to use this land. There are others, who do not have this right but might derive benefits from the rights of use. The State and formal institutions sanction and protect both property rights and the rights of use. This role is seen as crucial for the normal functioning and development of economic activity (Hodge, 2016).

The acknowledgement of rights of use may also involve informal institutions, formed and maintained in the realm of civil society. The way formal and informal institutions operate and are understood is linked to different understandings of the concept of land (Olwig, 2011). Firstly, land can be seen as a spatial property with related rights of ownership and economic value, or *property rights*, which are clearly defined by legal institutions, as described in the previous paragraph. Secondly, there is land or even at a higher scale, maybe a landscape, as a place that provided different benefits for multiple individuals and communities, related to *rights of use*, which often are customary rights and only secured by informal institutions. Custom is rooted in unwritten practice, informal agreements and not necessarily anchored in legal rights. These rights of use are therefore often more complex to manage and may also be under-valorized when confronted with legal rights, especially in case of conflicts and claims to different incompatible uses.

In order to understand customary rights, we often need to have a look at the past, since many of these rights are embedded in local cultures and

DOI: 10.4324/9781003028437-9

have their roots in past practices, power relations and land use systems. In the light of the increasing societal demand for multi-purpose land use, this understanding is also crucial for the future. It may be the basis for successful and contextually-adapted land governance models for the future, which may involve multiple land managers over the same patch of land (see also Chapter 11) (Jones, 2011).

Both formal and customary rights define rights of use and may exclude other uses. There may be conflicts, particularly when one of the users changes and/or seeks a change in the distribution and balance of rights. Nevertheless, the two types of rights are most often not in conflict with each other. There may be land properties or parcels of these properties where only one use is possible (e.g. a cultivated field), both under the legal frame and under customary law – where only the owner can exercise rights and any other user has to be excluded. There are other types of land where both property rights and multiple rights of use co-exist, with a well-established interplay of both legal regulations and customs (Olwig, 2011).

This is often the case in silvopastoral systems of Iberia. There may be one owner who has property rights over the land, takes care of the trees, pays for its maintenance (pruning) and benefits from the forest products (cork, wood and charcoal). Usually, with some exceptions (public paths, streams etc.), the owner has the full power to decide who can use what. Other individuals can be allowed (for free, or paying for the exclusive use) to use the land to acquire certain goods and/or services (for grazing pastures, for hunting, for collecting wild plants or mushrooms or for bee-keeping, or hikers).

In silvopastoral systems, especially in rural Mediterranean areas, this hybridity is complemented since different actors make use and manage different components of the system (e.g. trees, pastures, shrubs). The complex multi-management structure, where different users use different parts of the montado or dehesa is only possible, without conflict, due to unique governance of the rights of use, informally but consistently linked to property rights.

In this chapter, we explore the complex articulation between property rights and rights of use, in the silvopastoral systems of Spain and Portugal. We review the main rights of use secured by customary law and informal agreements and provide examples.

8.2 Historical background

A brief historical inquest on the rights at the dehesa/montado may be traced back to the times of the Visigoths. There we can track the etymology of dehesa (*defensus*) as a reserve for pasture as opposed to other agrarian uses (Linares, 2012; Guzmán, 2016). The local community had rights of use for the pasture. The same happened in the montado (Pinto-Correia and Fonseca, 2009). That is to say, there was a communalist property concept. There was a collective subject that had property rights and rights of use the land. This was based on a Germanic law, which had continuity (or at least some links) through a communalist tradition that is still alive in certain dehesa areas. It is also reclaimed by some environmentalist approaches, both in academia as well as social movements. In this vein, there has been a rebirth of studies on commons, now revisited from the point of view of ecological economics or anthropology. At the same time, and taking advantage of these studies, social and environmental movements are laying claim for common use and public access to ecosystems to take equal priority to private ownership of the land (Ostrom, 1990; Centemeri, 2018)

As seen in Chapter 1, the Christian conquest of the southern territories of Iberia during the Middle Age and the State financial crisis at the Modern Age, with the consequent sale of large estates to private owners, led to *latifundism* in the dehesa/montado areas, although some rights of peasants, with or without land, to the use of certain resources (mushrooms, wild plants, grazing, dead wood) survived for centuries (Cuéllar Escobar, 1997; Guzmán and Navarro, 2008). The next step towards privatization occurred with the disentitlement at 19th century.

In recent history, the access of the local community to the resources, in the form of wood, pastures, fruits or land for cultivation, on communal lands or on privately owned estates was constrained through various means, including sometimes repression. Governmental policies, administrative structures and legal regulations sought to guarantee land use activities that met the needs of an expanding market for agricultural products, although some research argues that a dynamic land market actually existed before the disentitlement. During the 19th century, liberal governments guaranteed property rights, but since the end of that century, and during the first half of the 20th, a wave of agrarian reforms swept across Europe, which defied the liberal approach towards defending unlimited property rights at all costs, even with repression (Congost and García Orallo, 2018). Despite this, the final result was the consolidation of the owner's full property rights in almost all the places.

This dynamic of the implantation of liberal ideology, economics and legal structures took place in all parts of the Iberian Peninsula. However, the scale of privatization and its consequences varied, depending on whether or not there was, locally, a critical mass of small peasants and their capacity for resistance, both political and economic. Although not widespread, in some places, local communities bought disentitled lands, but this involved a significant number of individuals being able to organize themselves (and raise the funds) in order to bid in the auctions for the large lots. These were subsequently divided into multiple small properties (Acosta-Naranjo, 2002).

Historiography shows the importance of peasant struggles in defence of the common lands, namely out the Guadalquivir Valley in Spain, where the commons had disappeared and the only option for the peasants, reduced to day workers, was to struggle for the fragmentation of the *latifundio* (Acosta Ramírez, Cruz Artacho and González de Molina, 2009).

Political and ideological contestation have been constant features in the "non-integrated agrarian societies" (González de Molina, 2014) of the Iberian peninsula, where there was a strong polarization between classes, between the rich and the poor. This was reinforced by the absence of an agrarian middle class, which could have served as the agglutinant of local society and culture. Conflict, either latent or patent, active or passive, has shaped the *latifundist* dehesa/montado territories, both in the *yuntero* (cultivators without land) movement of the Spanish Second Republic and the workers' movement in the 1974 Revolution in Portugal (Picão, 1983). In Alentejo, this culminated in the Agrarian Reform which finally failed shortly after. In Spain, there were also failed small-scale attempts in Extremadura and Andalusia to change the land situation through regional laws (Naredo and González de Molina, 2004).

The process of the total privatization of the dehesa/montado has continued to this day, and currently takes the form of continuing ruptures of the relationship between rural communities with their territory, which occur for various reasons.

Historically, two logics have coexisted in south-western Iberia regarding the right to territory, meaning two forms of territoriality and two types of legality: the legal regulations on the one hand; and on the other, the social uses and the ideological conformations of the rural communities on the territory – as shown in Box 8.1. Even when private ownership of the land was clearly established, that exclusive right was nuanced by the territoriality of the local communities. As an entity, these communities delimited their space and maintained a material, economic but also a moral relationship with what they considered 'their territory'. It was usual that villagers (but not outsiders) could take their livestock and take advantage of the remains of the harvest and the pastures on private land (Carolino et al. 2011). The same can be said of hunting and gathering wild plants and mushrooms, or scrub and certain forest products with no critical importance to the owners (Acosta-Naranjo, 2002; Carolino et al. 2011). Indeed, these activities arguably brought some benefits for the owners, since grazing field crops after harvesting fertilize the fields with animal manure and leave the field cleaner for the next sowing. Similarly, collecting scrub and forest products keep the forest more open and accessible. There were many win–win arrangements behind those uses.

The process of privatizing the dehesa and montado has been influenced by growing privatization of economic resources on a global scale. In the dehesa and montado, this process, which has separated rural communities from their territory and resources, has been strongly influenced by changes in the agrarian sector itself, especially in the *latifundist* domain. As described in Chapter 3, manpower has been drastically reduced on the farms, so that very few people now have a continuous relationship with the territory through their work, and are thus missing the main channel of connection with the land and its resources. This reinforces the privatization of all dimensions of the silvopastoral land.

Another case of public access to estates is transit on private farms, for example for recreational, aesthetic or ritual uses (see Box 8.1). On most estates, the passage is not legally prohibited, but the enclosure of plots (fences) for livestock management has made access difficult. Except for public ways, many owners

TEXT BOX 8.1

La Matilla, a 680 Ha holm oak Dehesa estate. Puebla del Maestre (Extremadura, Spain)

Rufino Acosta-Naranjo (University of Seville), based on an interview with Sebastián Acosta Naranjo, president of the Pallares' Hunters Asociation April 2020.

Key points: This textbox highlights the various rights within a single dehesa farm and the relationship between the landowner and the community.

This estate is unique in the region, as it is the only privately owned one where the owners do not have the right to cultivate. The Earl of the village, La Puebla del Maestre, had jurisdictional lordship during the Modern Age. With disentitlement he became the private owner of the farm but the residents kept the exclusive right to cultivate, which takes place every eight years. Today the residents have no economic interest in the crops but some of them continue cultivating with the sole objective of the village maintaining this right. This is a collective symbolic reference, a way of claiming the territory and an emblematic element of local identity (Acosta-Naranjo, 1992).

Like all Dehesa, the farm is fenced off with wires for livestock management. It is a private hunting reserve, although the hunting uses have been controversial. While the farm is located in the municipality of Puebla del Maestre, it is much closer to another village, Pallares, which also considers the estate as part of its territory. Years ago, hunting rights were transferred to the local hunting society of Pallares which created unrest in Puebla del Maestre.

This farm, whose farmhouse is a 16th-century Renaissance palace, has had symbolic importance for the two villages, who both consider it part of their history and identity. Until the 1970s they jointly held the *romería* (feast in the countryside) of San Isidro on the property, with permission from the owners. Since the 1980s, the village of Puebla del Maestre celebrates alone elsewhere. Pallares continued to celebrate it on the farm, although this has caused conflicts with the new owners, upset by the celebrations, especially when they started to last several days. Finally, Pallares decided to move its celebration to another place closer to the village. *Calderetas* (collective meals of groups of men) were also held there until the 1980s, next to a spring that has local historical and cultural significance.

One of the local reference points for the people of Pallares is the Cerro Castillo, the second-highest hill of its surroundings where there are remains of an old Arab fortress, and from which there are excellent views of the village and its surroundings. It is common for people to go up to Cerro Castillo, at least once in their lives, which in the past was regularly used by children as a play area. The people of Pallares do an annual hike, that sometimes reaches the top of Cerro Castillo, crossing part of the estate (Figure 8.1). There is a disused drove road, part of which people can still use as a public right of way, although part of it has been fenced off and appropriated by the owners. Although the farm is fenced off, it is still common for the villagers of Pallares to go there to gather asparagus as well as some other wild plants such as *Crataegus monogyna* or *Nasturtium officinale*.

Figure 8.1 The announcement of the organized hiking tour, with the view of Pallares from Cerro Castillo.

forbid access to their farms. This is more and more frequent as the farms are now emptier of people. In this way, both physically and symbolically, the process of deprivation of the ecological and moral relationship with the territory is being completed, thereby also deteriorating local knowledge about the environment (Acosta-Naranjo, 2008; Carolino and Pinto-Correia, 2011).

In this context, hunting and gathering, especially of asparagus and mushrooms, becomes a pretext, a kind of moral safe-conduct for accessing private properties, using the pretext of ancestral practices (Carolino et al. 2011; Acosta-Naranjo, Guzmán-Troncoso and Gómez-Melara, 2020), while at the same time not harming the interests of the owners. Nevertheless, damage to fences does occur and is used as an argument by the owners to restrict access to their land. Hunting also represents a symbolic and cognitive appropriation by the local community of the privately owned territory.

8.3 Today: the most common rights of use

We begin with **hunting,** in which rights over game resources are not fully regulated. There is some ambiguity about rights to "wild" animals (or birds). Some understand that hunting is *res nullius* (Cuéllar Montes, 2015), as the property owners are not the owners of wild animals, which move through different spaces (Macaulay, Starrs and Carranza, 2013). The landowners, by contrast, consider they own everything on their farm and can prevent others from having access to game whilst it is on their property. The problem takes on another dimension when the owners of farms and hunting preserves produce the game, breed it, buy-in stock to improve the genetic pool and/or to be hunted, or feed them during the hungry seasons. This they argue is closer to a husbandry activity. A paradoxical situation occurs when, for example, a deer or a wild boar breaks into a road and causes an accident that damages a vehicle. There have been some cases in which the civil responsible has been the owner of a hunting estate from which the animal left but in other cases, this was not the case. There are many controversies over hunting rights. In some cases, these have become so charged that hunters have caused forest fires in protest of their deprivation of hunting rights.

In Portugal, there are four modalities of managing hunting rights:

1   Tourist reserves – Owners of large properties are allowed to close their property for hunting, and have the sole right to use the game in their land – and also the responsibility. They can sell the hunting rights in the form they prefer, on a daily, annual or other basis. They often introduce game to secure abundance, and land may be managed to favour game species. A few montado- properties, in the most marginal areas, have hunting as their main income.

2   Associative reserves –An association makes agreements with several landowners of contiguous properties, who transfer all hunting rights to the association. Only the association's hunters have the right to hunt in these areas. The landowners may or may not be members of these associations.

3   Non-hunting areas – If there is livestock production that is not compatible with hunting activities, the landowner may require his/her property to be free from hunting; in such properties, no one can hunt, and this means the right to hunt is non-existent. These first three modalities require a license by the State for a fixed period of time. They are the most common in montado.

4   Municipal hunting areas – In all other areas, normally small-scale farms; the landowners do not have the right to the game nor can he/she decide on the distribution of these rights. Most of these municipal areas are managed by local hunting associations. In many situations, hunting rights are separated from land ownership, and conflicts may emerge, mainly in municipal areas. A similar situation can be found in Spain, which also has private, social and state hunting areas.

Anglers' access to water bodies is also controversial. In Spain, every watercourse that crosses more than one property can be used by anybody. Owners cannot forbid access, nor are they obliged to facilitate it; people are permitted to walk along the line of the watercourse.

The case of **gathering** is different. The general improvement of living standards and changes in food habits have meant that interest in wild plants has evolved from being an essential service to a cultural service, especially recreational (Reyes-García et al. 2015; Acosta-Naranjo, Guzmán Troncoso and Gómez-Melara, 2020; Chapter 5). The social conflicts which existed in the past over access to resources by the most disadvantaged people in local communities have vanished. In the past, it was common, for example, for dayworkers to steal acorns from large farms. This was a crime and the rural police arrested the perpetrators, and beatings were frequent (Alagona et al. 2013). Despite this, there was a moral understanding on the part of the local community, that if acorn collecting took place on *latifundios* where acorns were abundant, and the workers who did this illegal collecting were frequently much in need it was acceptable (Acosta-Naranjo, 2002).

The gathering of mushrooms, asparagus, or even acorns in Portugal, is increasingly popular as a leisure activity (see Chapter 3). In Portugal, and in certain regions in Spain, (e.g. Castilla y León) a license fora "reserved mushroom collection area" can be obtained by the owners, and the concerned plots can be closed from free mushroom picking. Otherwise, access has been traditionally free, for mushrooms. When access to areas where mushrooms are known to grow is unregulated, owners often complain about damage to fences or disturbances to livestock. There are also local/outside tensions. Traditionally, the collectors have been the people of the village or surrounding area, but increasingly people are coming from outside the local area. This is true both for asparagus (Bermúdez, 2018), but also mushrooms, as there has been a profusion of mycological societies in cities. New communication technologies and geo-location devices enable the circulation of information about the territory and its resources, without the need for pickers to have local knowledge or to be linked with the territory. In certain places, this is leading to tensions between local people and outsiders (Bermúdez, 2018).

Access to blossom by **beekeepers** is regulated by informal agreements with the landowners, as described in Chapter 3. This is a customary right that is largely maintained: beekeepers are allowed free access to their hives as needed. They usually make a gift of honey to the landowner. Other rights and uses have disappeared, such as gleaning (allowing livestock to browse cultivated fields after the harvest) (Acosta-Naranjo, 2002).

As shown in this section, there are traditional rights of use, some of which are still maintained today, which do not coincide with ownership rights. This is particularly Mediterranean and brings in particular dimensions of co-involvement of different people, in the everyday management of the silvopastoral systems.

8.4 New uses and controversial rights – societal demand for public goods

Currently, a multitude of other actors, concerned with the **cultural services** of agroecosystems and the struggle for rights of access to the montado and dehesa are becoming more prominent. Besides gathering vegetal resources and bee-keeping, there are now multiple other uses that can be framed as countryside consumption (Pinto-Correia et al. 2016). This includes hikers, rural tourists, naturalists, birdwatchers or nature photographers (as referred to in Chapter 3 and in Text Box 8.1 in this chapter). The hikers, or the companies that offer them services, claim the right of way through the dehesa/montado. As it is increasingly difficult to question access to private estates, one growing subject of conflict is the right of passage on public roads and the designation of which are and are not public roads (and therefore can or cannot be closed by landowners). Public paths and drover routes are the subjects of dispute between different agents: owners, users and administrations. Drover roads became meaningless with motorization and the end of transhumance and *trasterminancia* (short-range transhumance, between nearby towns). There is increasingly an appropriation of these trails by the owners, for different reasons: the trails are not in use anymore and the owners incorporate them as part of production land; or there is growing pressure on the use of the trails by multiple users, especially in dehesa/montado that is close to urban areas, and the landowners react by limiting access. The long-term inaction of the State reinforces this problem. However, there are few cases in which there is a claim and fight for the recovery of user rights, especially in places related to rituals, festivals and ceremonies. In Extremadura and Andalucía, tours that use (disputed) public roads (sometimes abandoned for years) to places of pilgrimage have proliferated, in a symbolic way of communities reattaching themselves to the local territory and making a claim of the right to use it.

In other cases, tourists, cyclists or hikers, their associations and environmental groups are claiming the recovery of livestock routes. In other (fewer) cases, there are groups such as landless goatherds who carry out acts of protest against the appropriation of roads and paths.

8.5 The role of the States

The State is a leading actor in the struggle for rights. In fact, the State regulates – or is expected to regulate – the relations between the actors that define the protection of the common good and the common interest of society by limiting and protecting the rights of different actors. But, in addition to acting as a referee, the State is also the owner of rights. Although now reduced in size, Spain still has public dehesa properties, usually owned by municipalities (Nevado Peña and Rodríguez Cancho, 1994). In Portugal, this aspect is less relevant – there are very few public properties, and they are managed in a comparable way to private properties.

Fundamentally, the State has the capacity to regulate and limit private property rights. Both the Spanish and the Portuguese Constitutions explicitly state that private property also has a social purpose. This has to do with the possibility of expropriating land, as occurred in the case of the Agrarian Reform in *latifundio* areas in Spain and also in Portugal (both failed, as said earlier). The same is true of certain works or activities of social interest, which allow the expropriation of land, although, in practice, this right is not exercised, except for major public works, such as roads, railways, dams, etc.

At the same time, rivers, their channels, adjacent areas as well as the sub-soil resources, including water, are public domains, which are owned by the State, which grants use rights to private owners. This is very relevant today in situations of water scarcity and the increase in livestock stocking and climatic irregularities.

But today, perhaps the most relevant issue is the capacity for surveillance, control and sanction that the State has when it comes to controlling activities on farm estates and other farm units. Although this has always been the case, today it acquires a new dimension and becomes more pronounced with emerging environmental issues. The regulation of the activities that private owners carry out on farms is a priority issue. The property right of the farms refers to discrete units of territory, while the ecosystems are open: many elements of them cannot be confined to spatial limits, they are systems of dynamic and open relationships. The farms are not merely private spaces even if they have an owner; since their use affects everyone even if they are not owners. The new conception of the commons is based on the collective dimension and consideration of existing and future generations, since these activities affect the air, water, soil and biodiversity, or what nowadays is called nature. Citizens are increasingly claiming a right to a healthy environment and its sustainability against the absolute right of the owners (Scheidel et al. 2018).

To address such environmental problems, the State must act and set limits on the freedom of action of the owners of dehesa/montado. Many of the frictions between owners and the State today have to do with limitations on the management of soil, trees or livestock, especially in Protected Areas (Coca and Quintero Morón, 2018). These might include land-use changes and plowing permanent pastures; and limitations on fencing that block the entry of wild fauna into grazing areas (despite the sanitary problems that this may pose for livestock). Sometimes, these conflicts are not directly about the sovereignty of the owners over their land, but about other assets or sources of income, created by the State, such as subsidies, which today are the main regulatory instrument of agriculture, as discussed in Chapter 10. Public goods that are paid for and associated with the provision of ecosystem services should also be included in this mix in the future. This is an import issue, as a significant part of the market price of a dehesa/montado farm is related to the associated right that the farm has, to claim public subsidies for its production. Despite the long-standing tradition of several differentiated users of resources in silvopastoral land which belongs to one single owner, the new demands of society create new tensions and new challenges, which are still partially resolved. More and more explicit, societal expectations can be expected to emerge in the future. In addition to the traditional social function of the land, its environmental function is being added as a central function in the ecological transition scenario. The agrarian question has been transmuted into a socio-ecological question.

8.6 Common lands – a vanishing collective governance model

The communal lands represent a particular case of rights over the dehesas and are particularly significant especially in the parts of Spain bordering Portugal, in La Raya/A Raia. In Portugal common lands in the montado existed in some limited areas, but were abolished decades ago.

As previously mentioned, communal land ownership does not fit well with the founding principle of liberalism, as the owners are not individuals, institutions or companies, but neighbours, the community, those born and as yet unborn in a territory. Local people use various formulas, using the gaps or empty spaces left by the new legality, to maintain their forms of collective ownership and management.

Beyond legal issues, the collective ownership and management of the dehesa have had significance in local politics, economy and society, not only as a source of resources for locals but also as a collective memory, a mentality and a praxis that facilitates governance. Public involvement in ecosystem management is a practice that is embedded in the local tradition of communal dehesa territories. Moreover, it's also part of the identity of communities and the basis for forms of collective social action and social economy (Rangel Preciado and Parejo Moruno, 2017; Beltrán Tapia, 2018). In many of these places, there is still a distinction between rights to the *suelo* (ground) and the *vuelo* (the right to pick the fruits and trees), which for example allows those who have the right to acorns to continue to pick them by hand and sell them, something which is unusual in other part of the dehesa.

8.7 Conclusions: what can we learn for the future governance of silvopastoral systems

The review of the way property rights and rights of use have been evolving in the montado and dehesa shows us, on one hand, that there are multiple uses which can be made of the resources on the same land; and on the other, that the right to these multiple uses can be appropriated by different users, and that landowners, throughout history as well as today, combine the use that they make of the land with allowing (or tolerating) it's use by others. This complexity makes sense in the Iberian context. Iberian people understand it and, in most cases, still feel confident with the resulting arrangements. It is unique and may be hard for an external observer to understand. In recent years, societal demand for different uses of these systems has changed (and will continue to change), with increasing interest for rights related to leisure, quality of life and well-being (Surová and Pinto-Correia, 2016). This increases the importance of finding new governance models that maintain this hybridity. Due to the historical roots of the montado and dehesa, they are part of the Iberian culture which makes them even more precious.

Smart solutions for new governance models for the rights of use are required, that will support the reconnection of people with their territory and their landscape. It is important to recognize the various groups possessing or claiming use-right of the dehesa and montado as stakeholders in governance processes, alongside private or communal owners. Today's culture and connections are increasingly global and virtual but, paradoxically, well-being is increasingly linked to close associations and bindings at the local level. The montados and dehesas are landscapes with immense importance for recreation and connections to nature and culture, and these types of rights of use should be afforded the uppermost priority.

References

Acosta-Naranjo, R. 1992. "La siembra de La Matilla. Un derecho histórico de Puebla del Maestre". Saber Popular 6: 9–25.

Acosta-Naranjo, R. 2002. *Los entramados de la diversidad. Antropología social de la dehesa.* Badajoz: Diputación de Badajoz.

Acosta-Naranjo, R. 2008. *Dehesas de la sobremodernidad. La cadencia y el vértigo.* Badajoz: Diputación de Badajoz.

Acosta-Naranjo, R., A. J. Guzmán-Troncoso and J. Gómez-Melara. 2020. "The persistence of wild edible plants in agroforestry systems: the case of wild asparagus in southern Extremadura (Spain)". Agroforestry Sytems, 94: 2391–2400 doi.org/10.1007/s10457-020-00560-z

Acosta Ramírez, F., S. Cruz Artacho and M. González de Molina. 2009. Socialismo y democracia en el campo (1880-1930). Los orígenes de la FNTT. Madrid: Ministerio de Medio Ambiente.

Alagona, P.S., A. Linares, P. Campos and L. Huntsinger. 2013. "History and Recent Trends", in: Campos, P. Huntsinger, L. Oviedo J. L., Starrs, P. F., Díaz M., Standiford R. B. and Montero, G. (eds.): *Mediterranean Oak Woodland Working Landscapes. Dehesas of Spain and Ranchlands of California.* Dordrecht: Springer, 25–58.

Beltrán Tapia, F. J. 2018. *En torno al comunal en España: una agenda de investigación llena de retos y promesas.* Sociedad de Estudios de Historia Agraria, Documentos de Trabajo.

Bermúdez, V. 2018. *La Recolección de Espárragos ante la Despoblación en Valverde de Burguillos (Badajoz).* Master's Thesis. Universidad de Sevilla.

Carolino, J. and T. Pinto-Correia. 2011. "Material landscape, symbolic landscape, and identity in the municipality of Castelo de Vide". *Analise Social,* 46(198): 89–113.

Carolino, J., J. Primdahl, T. Pinto-Correia, and M. Bojesen. 2011. "Hunting and the right to landscape: Comparing the Portuguese and Danish traditions and current challenges", in: Egoz, S., Makhzoumi, J. and Pungetti, G. (eds.): *The Right to Landscape. Contesting Landscape and Human Rights.* Farnham: Ashgate. 99–112.

Centemeri, L. 2018. "Commons and the new environmentalism of everyday life. Alternative value practices and multispecies commoning in the permaculture movement". *Rassegna Italiana di Sociologia,* 64 (2): 289–313.

Coca, A. and V. Quintero Morón. 2018. "Otro mundo es posible, o el movimiento (ambiental) de los corcheros y arrieros en Andalucía", in: Cortés Vázquez and Beltran (coords.): *Repensar La conservación. Naturaleza, mercado y sociedad civil,* Barcelona: Universitat de Barcelona Edicions, 179–196.

Congost, R. and R. García Orallo. 2018. "¿Qué liberaron las medidas liberales? La circulación de la tierra en la España del siglo XIX". *Historia Agraria*, 74: 67–102. DOI 10.26882/histagrar.074e03c

Cuéllar Escobar, S. 1997. "Los baldíos de Alburquerque". *Revista de Estudios Extremeños*, 53(1): 157–176.

Cuéllar Montes, T. 2015. *La Naturaleza del Derecho de Ocupación y su Repercusión en la Propiedad de las Piezas de Caza*. Ph. D. Thesis (University of Extremadura).

González de Molina, M. 2014. "La tierra y la cuestión agraria entre 1812 y 1931: latifundismo versus campesinización", in: González de Molina (coord.): *La Cuestión Agraria en la Historia de Andalucía. Nuevas perspectivas*, Sevilla: Fundación Pública Andaluza Centro de Estudios Andaluces, Consejería de la Presidencia, Junta de Andalucía, 21–60.

Guzmán, J. R. 2016. "The image of a tamed landscape: Dehesa through history in Spain". *Culture & History Digital Journal*, 5(1).

Guzmán, J. R. and R. M. Navarro. 2008. "De las dehesas del pasado a las dehesas del futuro: reflexiones sobre la evolución de un concepto pastoral", *Actas de las Jornadas sobre Dehesas y Mundo Rural en Andalucía*. Sevilla: Consejería de Medio Ambiente. Junta de Andalucía.

Hodge, J. 2016. *The Governance of the Countryside. Property, planning and policy*. 1st ed. Cambridge, UK: Cambridge University Press.

Jones, M. 2011. "Contested Rights, Contested Histories: Landscape and Legal Rights in Orkney and Shetland", in: Egoz, S., Makhzoumi, J. and Pungetti, G. (eds.): *The Right to Landscape. Contesting Landscape and Human Rights*. Farnham, Surrey: Ashgate, 71–83.

Linares, A. 2012. "La evolución histórica de la dehesa: entre la persistencia y el cambio", in: Linares, L. and L. Pedraja (eds.): *Santiago Zapata Blanco: Economía e Historia Económica*, Cáceres: Fundación Caja de Extremadura, 11–36.

Macaulay, L. T., P. F. Starrs and J. Carranza. 2013. "Hunting in Managed Oak Woodlands: Contrast Among Similarities", in: P. Campos, L. Huntsinger, J. L. Oviedo, P. F. Starrs, M. Díaz, R. B. Standiford and G. Montero (eds.): *Mediterranean Oak Woodland Working Landscapes. Dehesas of Spain and Ranchlands of California*. Dordrecht: Springer, 311–350.

Naredo, J. M. and M. González de Molina. 2004. "Reforma Agraria y desarrollo económico en la Andalucía del siglo XX", in: de Molina González and J. Parejo Barranco (eds.): *La Historia de Andalucía a Debate. Vol. II: El campo Andaluz*. Barcelona: Anthropos, 88–16.

Nevado Peña, A. and M. Rodríguez Cancho. 1994. "Las dehesas boyales cacereñas". *Actas del VII Coloquio de Geografía Rural*, 96–102.

Olwig, K. 2011. "The right rights to the right landscape?", in: Egoz, S., Makhzoumi, J. and Pungetti, G. (eds.): *The Right to Landscape. Contesting landscape and human rights*. Farnham, Surrey: Ashgate, 39–50.

Ostrom, E. 1990. Governing the Commons: *The Evolution of Institutions for Collective Action*. Cambridge: Cambridge University Press.

Picão, J. S. 1983. *Através Dos Campos. Usos e Costumes Agrícolo-Alentejanos*. 1st ed. Lisbon: Dom Quixote.

Pinto-Correia, T. and A. Fonseca. 2009. "Historical Perspective of Montados: The example of Évora", in: Aronson, J., Santos Pereira, J. and Pausas, J. (eds.): *Cork Oak Woodlands on the Edge: Ecology, Adaptative Management, and Restoration*. Washington, DC: Island Press, 49–54.

Pinto-Correia, T., N. Guiomar, C. A. Guerra, and S. Carvalho-Ribeiro. 2016. "Assessing the ability of rural areas to fulfil multiple societal demands". *Land Use Policy* , 53:86–96.

Rangel Preciado, J. F. and F. M. Parejo Moruno. 2017. "El origen y desarrollo del nego-cio corchero extremeño. La contribución de la Comarca Sierra Suroeste", in: Segovia Sopo, R. (coord.): *Arqueología e Historia en Jerez de los Caballeros y su entorno. I Jornadas de Historia en Jerez de los Caballeros*, Xerez Equitum y Diputación de Badajoz, 405–425.

Reyes-García, V., G. Menendez-Baceta, L. Aceituno-Mata, R. Acosta-Naranjo, L. Calvet-Mir, P. Domínguez, T. Garnatje, E. Gómez-Baggethun, M. Molina-Bustamante, M. Molina, R. Rodríguez-Franco, G. Serrasolses, J. Vallès and M. Pardo-de-Santayana. 2015. "From famine foods to delicatessen: Interpreting trends in the use of wild edible plants through cultural ecosystem services". *Ecological Economics*, 120: 303–311.

Scheidel, A., L. Temper, F. Demaria and J. Martínez-Alier. 2018. "Ecological distribu-tion conflicts as forces for sustainability: An overview and conceptual framework". *Sustainability Science*, 13: 585–598.

Surová, D. and T. Pinto-Correia. 2016. "A landscape menu to please them all: Relating users' preferences to land cover classes in the Mediterranean region of Alentejo, south-ern Portugal". *Land Use Policy*, 54: 355–365.

9 Regulatory and institutional frameworks of silvopastoral systems in Tunisia

Mariem Khalfaoui, Ali Chebil, Hamed Daly-Hassen and Aymen Frija

9.1 Introduction

Tunisia has a total forest area of 1.15 million ha in 2015 which represents 8% of the total surface of the whole country (ONAGRI 2019). About 0.78 million ha (60% of the forest area) in Tunisia are open for grazing in silvopastoral systems that combine forest trees, forage and other cash crops, in addition to animal production on the same unit of land. Tunisian silvopastoral systems derive their importance from their natural, economic and social value. From a natural perspective, they constitute a platform for biodiversity preservation and for the conservation of water and soil against erosion and desertification, especially given that 94% of the silvopastoral systems' surface is located upstream of mountainous areas in northern side of the country (ONAGRI 2019). They also provide a multitude of goods and services to the local and national population. It is estimated that the total economic value of the forest sector contributed 0.3% to the gross domestic product (GDP) in 2012 (Croitoru and Daly-Hassen 2015). The social contributions of silvopastoral systems are also highly important. It has been estimated that around 750,000 people (around 8% of the country's population) live within Tunisia's silvopastoral systems (FAO and DGF 2012). This importance is not restricted to Tunisia, but also exists in several other Maghrebian countries as shown in Text Box 9.1

Despite their importance, silvopastoral systems in Tunisia are subject to several pressures and facing important challenges. The combination of the effects of climate change and human pressure coming from overgrazing and the excessive extraction of silvopastoral produce by the local population is creating an alarming situation that requires serious intervention at different technical, institutional and policy levels (Ben Mansoura et al. 2001; Campos et al. 2007; Tardieu et al. 2018).

The legal framework that regulates silvopastoral systems is the Forest Code (law) and its subsidiary texts. It covers three main areas: (1) application of the forest regime (including the organization of forest users, ownership of forest products and a regime for temporary occupations and forest concessions) and silvopastoral management, (2) hunting management and the conservation of game, and (3) the protection of nature, wild flora and fauna.

DOI: 10.4324/9781003028437-10

TEXT BOX 9.1

Socio-economic contributions of the Agdal institution, a traditional agro-silvopastoral communal governance system of the Maghreb

Pablo Dominguez, Laboratoire de Géographie de l'Environnement (GEODE), UMR-5602 CNRS Université Toulouse 2, France.

Key points: Agro-silvopastoral commons are a positive management regime in social terms and in terms of sustainable environmental governance, but such systems are undergoing swift degradation.

This makes it important to draw public attention to the need to support such systems.

Local livelihoods in the mountain territory of Yagur (situated between 2,000 and 3,600 masl in the High Atlas of Marrakech, Morocco) are highly dependent on the role of the local silvopastoral system which had been maintained for thousands of years by natural resource governance system of the *agdal* (Figure 9.1). The Yagur is managed by this institution of the *agdal*, a system for communal governance of natural resources of the Amazigh people (a Berber ethnic community). Key points of this system include a ban on herding during three months, mainly in spring, to allow vegetation to regenerate, as well as relatively equal access to it for all households. At the same time, the integration of the traditional *agdal* within local religious practices and their associated ethics entails a local conservationist and egalitarian set of human values. Specifically, the economic pastoral contribution of the Yagur to the population of the Ait Ikiss belongs to the Mesioua

Figure 9.1 View of the lake Yagur at 2,465 masl.

Photo credit: Pablo Dominguez.

tribe. One year of fieldwork and cohabitation with the Ait Ikiss, which covered the entire cycle of agro-pastoral seasons, allowed the author to calculate that the Yagur provides up to 18% of its users' annual agro-pastoral gross monetary income through its fodder contribution. This is probably an underestimate since much of the community's livelihoods are non-monetized and cannot be readily or rigorously economically accounted for. Complementary qualitative data give a better understanding of all the agro-economic and eco-anthropological dimensions of the *agdal*. These findings lead us to conclude that, even if highly underestimated by the national administration, the agro-silvopastoral *agdals* of Morocco and more broadly the Maghreb are a key practice for the continuity of agro-silvopastoral systems and their economic productivity, as well as being a very important tool for maintaining local communities' social and spiritual well-being.

The Forest Code provides a clear definition of the types, properties and management of the pastoral and forest domain. About 95% of the silvopastoral systems area is under public (state) ownership (World Bank et al. 2016). The state is engaged in the conservation and management operations of these lands and gets some commercial benefits (through public tenders), mainly from cork, wood, mushrooms, myrtle and game. The local population also has specific rights of access and use. The Forest Code recognizes the fulfilment of the basic needs of the local population and its definition of open access rights allows for livestock grazing in specific areas, the collection of deadwood and brushwood and the non-commercial use of other forest products. Other Non-Wood Forest Products (NWFP) such as aromatic shrubs are harvested by local communities through specific authorizations provided by the National Forestry Department against established tariffs.

As mentioned above, silvopastoral systems in Tunisia are currently subject to high anthropogenic pressures and are overexploited by local communities. This phenomenon worsened after the Revolution in 2011 with weaker enforcement of laws and sanctions. The consequences of overexploitation are soil degradation, reduction of forest density and the lack of forest regeneration.

The objective of this chapter is to provide an analysis of the regulatory and institutional frameworks of silvopastoral systems in Tunisia. In the following sections, we will focus on the description of the main silvopastoral systems, then a description of the main institutional and management aspects, finalizing the challenges and perspectives for better regulation of these resources.

9.2 Tunisian silvopastoral systems

The silvopastoral systems of Tunisia are mainly located in the northern and central-western regions of Tunisia (Figure 9.2). These regions are bordered by the Mediterranean Sea on the north and east, Algeria on the west, and by

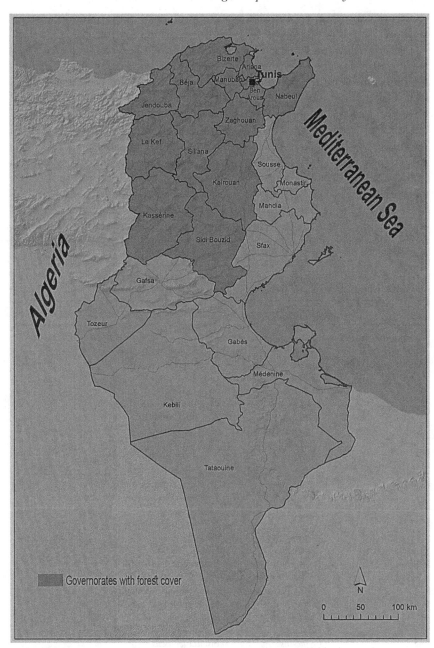

Figure 9.2 Location of major silvopastoral systems in Tunisia.

Source: Edited by authors based on DGF (2010).

the steppes of central Tunisia in the south. About 94% of the total silvopastoral areas are located in 12 (out of 24) governorates (DGF, 2010).

The dominant trees and livestock species of the different silvopastoral systems in Tunisia vary according to the bioclimatic conditions. Aleppo pine is the most dominant species, occupying around 35% of this total silvopastoral area. Other important forest species include cork oak (*Quercus Suber*), wild olive (*olea europea*), *maritime pine*, oak zeen (*Quercus canariensis*), and *Pinus nigra*. Table 9.1 provides details of the main understory species in the silvopastoral systems in the north and central-west regions of the country.

Livestock production is an important part of the silvopastoral systems in Tunisia. Silvopastoral systems have always been directly related to forestry, targeting the production of woody material (trees) and fodder, as well as livestock farming. The Forest Code recognizes livestock grazing among the list of usage rights that can be freely practiced by local communities. Livestock represents the primary source of revenue for many of these marginalized

Table 9.1 Main species in silvopastoral systems of Tunisia

Climate	Main tree species	Main understory	Main livestock

Humid-sub humid :
Annual rainfall
  between 500 and
  1500 mm
	Qurcus Suber
	Maritime pine
	Quecus Suber
	Quercus Canariensis
	Olea Europea
	Eucalyptus
	Acacia Mimosa
		Erica arborea
		Pistacia lentiscus
		Myrtus communis
		Quercus coccifera
		Cistus salvifolius
		Calicotome villosa
		Genista tricuspidata
		Phillyrea media L.
		Phillyrea angustofilia
		Cistus monspeliensis
		Lavandula stoechas L.
		Trifolium Subterraneum
		Trifoluimrepens
		Ceratonia siliqua
		Hydesarum coronarium
			Goats, cattle

Arid-semi arid :
Annual rainfall
  between 250 and
  550 mm
	Aleppo pine
	Pinus nigra
	Pinus pinea
	Tetraclinus articulata
	Tetraclinis occidentalis
	Acacia
	Eucalyptus
	Casuarina
	Cypressus
	Erable de Montpelier
		Rosmanirus officinalis
		Quercus coccifera
		Rosmarinus tournefortii
		Phillyrea angustifolia
		Ampelodesma mauritanica
		Cistus villosus
		Lonicera inplexa
		Pistacia terebinthus
		Pistacia lentiscus
		Coronilla valentina
		Hedysarum
		Spinosissimum
		Atriplex subsp
			Cattle, sheep

Source: Own elaboration.

communities. It has been estimated that silvopastoral systems in Tunisia contain respectively about 80%, 70% and 50% of total cattle, sheep and goat flocks in the country. Rangelands in the central and southern parts of the country are more devoted to goats and sheep and are suffering continuous degradation, mainly due to overgrazing.

A good example of silvopastoral systems in Tunisia is the cork oak-based system which dominates most of the northwestern landscapes. This system plays a key role for the local population and also for the State. The main commercial benefits from the cork oak woodland area derive from multiple final products, including cork, mushrooms, berries, essential oils, licences for hunting wild boar and firewood collected for domestic use. Local communities mainly benefit from traditional livestock rearing, firewood from cork oaks (only dead wood lying on the forest floor is allowed to be collected and used) and shrub biomass, which is used for fodder, firewood, charcoal and shelter, as well as seasonal forestry jobs. Local families are also allowed to cultivate subsistence crops grown in small treeless plots within or bordering the cork oak woodlands (Ben Mansoura et al. 2001). Grazing and crops (forages and residues) grown in these agroforestry areas are one of the main sources of livestock feed for many of the local communities living in the border of forests. However, the current institutional arrangement seems to be failing to ensure the sustainability of forests where high levels of degradation are observed mainly due to overgrazing, clearing, and ploughing. In the 1950s, cork oak forests occupied more than 100,000 ha in Tunisia. Recently, the forest inventory reported a total area of 70,000 ha (ONAGRI 2019).

Unrestricted grazing, increasing demographic pressure, and lack of proper surveillance have led to the degradation and destruction of natural resources throughout the cork oak woodlands of the country (Ben Mansoura et al. 2001; Daly-Hassen et al. 2009). The objective of the forest administration is to generate woodland capital income for the government and, simultaneously, to develop local householders' subsistence economy, while respecting the Tunisian Forest Code regarding forest resource usage rights and restrictions. The regulations for silvopastoral systems are considered weak given the specificity and complexity of those systems that involve different types of embedded natural resources and related property and usufruct rights.

9.3 The institutional aspects and management of silvopastoral systems

Many national and regional institutions are involved in the management and exploitation of Tunisia's silvopastoral systems. At the national level, the Ministry of Agriculture (MA), through its various departments, is the principal manager. However, other ministries, such as the Ministry of Environment and Sustainable Development, Ministry of State Property, etc. are also involved in some land tenure and management aspects. The main central departments of the MA that are directly involved in silvopastoral systems management are:

- The General Directorate of Forestry (DGF), which is responsible for the application and implementation of the Forest Code and for managing, protecting, and developing public forests and other woodlands and rangelands. This administration is specifically in charge of (1) forest and rangeland protection, conservation and surveillance, (2) the definition of the forestry and pasture strategy, and (3) the preparation and implementation of forestry and rangeland projects.
- The Directorate of Forest Exploitation is responsible for the exploitation of forest products, namely cork, logging and timber sales. It is the only entity within the Forestry Administration that is authorized to market forest products.
- The General Directorate of Soil and Water Conservation, which is responsible for soil erosion control and water catchment protection.
- The Livestock and Rangelands Office, which is a public agency working on the promotion and development of the livestock sector and related feed and pasture resources.
- The North-West Sylvopastoral Development Office, which works on the promotion of agro-sylvopastoral development in the mountainous and forest areas of the five governorates of the northwest of the country (Béja, Bizerte, Jendouba, Le Kefand Siliana). The mission of this public agency is to implement integrated rural development projects within its areas of intervention.
- At the regional level, the DGF's activities are coordinated by provincial forestry services which are hosted at the regional agricultural administrations.
- Local communities. These may or may not be organized in community-based organizations called "agricultural development groups."

Upon independence in 1956, the application of forest code was strictly oriented towards the protection and preservation of these ecosystems, restricting the access and use of local communities. However, the strict application of restrictions, the high poverty rate and lack of employment among local communities led to conflicts between the conservation of these silvopastoral systems and local householders' subsistence.

Institutional regulatory changes have been undertaken in 2012 to deal with the high pressure and continuous degradation of silvopastoral systems in Tunisia. This was based on the clear conviction that parts of the current Forest Code of Tunisia was failing to deliver effective protection of the resources on the ground. The institutional changes cover four major themes (Daly-Hassen 2012):

1   The consolidation of usage rights of local communities with the objective of better involving them in setting and deciding about forest management

plans. This happened through providing the legal framework for the organization of these communities into groups with the aim of facilitating their participation and integration in the territorial management with the aim of improving their socio-economic conditions.

2 Additional incentives for the encouragement of private initiatives and investments. Examples are the replacement of the standard administrative authorizations with simple specifications for the purpose of harvesting private forest lands.

3 Ensuring the compliance of Tunisian forest laws with international conventions signed and ratified by Tunisia on all issues related to environmental conservation and sustainable development.

4 The decentralization and devolution of part of the administrative activities and the transfer of some forest areas to be managed by farmers' organizations.

Several similar forest strategies and programs were implemented within this new national regulatory framework relying on higher community involvement oriented to sustainable management, taking into consideration the existence of the local population, naming::

1 National Programme for Forest Protection against Forest Fires (1992)[1]
2 National Master Plan for Silvopastoral Systems Development (1997)[2]
3 National Strategy for the Development of the Forest Sector (2001–2011)[3]
4 National Strategy for Forest Conservation (2001)[4]
5 National Forest Programme (NFP) (2007)[5]
6 The Five Year Development Plan (2010–2014)[6].

According to State of the World's Forests 2016 (SOFO), early forest strategies in Tunisia have helped in increasing the extent of forest cover from 2.2% in 1956 to 5.9% in 1990, representing an average annual increase of 8,000 ha or 0.09% p.a. over a 34- year period. Similarly, the implementation of the revised policies since 1992 has allowed the forest cover to attain a level of 8.0% in 2010, representing an average annual increase of 17 000 ha or +0.10% annually. However, this increase of the forest areas hasn't resulted in positive impacts on the livelihood of the neighbouring communities. Contrary, the number of illegal activities (cleaning, logging, ploughing, etc.) increased fourfold between 1992 and 2014 (FAO 2016).

The most recent strategy for the development and sustainable management of forests and rangelands, 2015–2024 (MA 2014), acknowledges the previously stated limits. Its approach follows the Mediterranean Forests Strategies, oriented towards the sustainable management and the reinforcement of forests' role in rural development (DGF 2015). The strategy is based on four objectives: (1) further adaptations of the institutional and legal framework and capacity building, (2) optimization of the contribution of the forestry sector to the socio-economic development of local communities and at the national level, (3) maintaining and improving forest ecosystem services and, (4) the consolidation, restoration and improvement of the forests' natural capital. Despite its relevance and wide thematic coverage, the application of this strategy was expected to be challenging since the development of appropriate regulations for silvopastoral systems in Tunisia needs to be aligned to the specificity of these systems that involve different types of natural resources, rights of use and tenure systems. The current regulation is not explicit about the role of specific types of resources in relation to each other (the example of rangelands and forest which are two different types of resources, involving different types of land tenure, but are considered in the same Forest Code). The control of the resources and law enforcement through appropriate executive agencies and organs is another major problem related to State intervention. Furthermore, most of the institutional instruments designed to manage natural resources are based on regulations with low usage of other (economic) instruments such as incentives and taxes. These failures in the regulatory framework are leading to wide governance failures in Tunisia's silvopastoral systems.

9.4 The challenges facing the regulations in silvopastoral systems and ways forward

The major challenge facing silvopastoral systems in Tunisia is related to the identification of appropriate ways to reconcile the production of environmental services on one hand, and the socio-economic development of local communities on the other hand. Besides all the limits and regulations stated in the Forest Code, the presence of local communities with access rights to livestock grazing, collection of deadwood and brushwood and non-commercial use of other forest products without explicit property rights has been always conflictual, taking into account the fact that the high poverty rate among this population, 46% in 2010 (FAO and DGF 2012) compared to 25% average poverty rate in the country (INS[7]).

Considering the fragility and vulnerability to climate change, the conflictual situation with the local population and the deficiency of certain aspects of the legal framework, a positive change of the legal framework for better governance of the silvopastoral systems is much needed.

The first step towards such institutional change is to determine the original reasons behind the legal framework's failure. Three main factors can be identified in this regard: (1) unclear rights of use, (2) the exclusion of the local communities from decision making for forest management and (3) the socio-economic context, specifically the economic vulnerability of the local population.

Regarding the local population, the DGF has always sought short term and temporary solutions, such as the increase of occasional employment as the destructive behaviour (clearing, logging, grazing in protected areas, ploughing, etc.) of the local population is related to their need for additional income, while overgrazing is to livestock production being the only profitable activity that can be carried out in the forest regions. To avoid such behaviour, their involvement in decision procedures regarding the forests and the provision of income resources would probably be a great help.

Drastic changes should be integrated into the institutional framework including new governance models based on participatory and integrated approaches. Such approaches will favour more involvement of the local population in the management of silvopastoral resources. The adoption of new co-management tools, combined with a reinforcement of the existing legal framework may increase the population's access to forest and pastoral products in a more efficient way. Supporting capacity-building for administration and local organizations will be also required to ensure effective co-management. The reform of regulations to facilitate access to forest and pastoral resources and services for local communities will indirectly increase their income and decrease their destructive behaviour, especially if such access is well structured and controlled by collective community-based organizations.

A more optimistic idea can be based on the integration of new projects in forest regions: Projects such as eco-tourism and establishing certified labels for locally produced NWFP. Such actions will encourage the local population to adopt more conservation oriented behaviour with regards to the forest and increase their involvement in the preservation of the forests.

9.5 Conclusion

This chapter has reported on the different regulation instruments used for State intervention in silvopastoral systems in Tunisia and suggests guidelines and recommendations for enhancing the current regulatory framework for these systems.

The complexity of the silvopastoral system situation is mainly due to their vulnerably to climate change, human pressure and inadequacy of the legal framework. The lack of compliance of the Forest Code by the local population and their economic dependence on forest products for their subsistence is leading to destructive behaviour and constitutes a handicap to efficient management of these silvopastoral systems. For better governance, Tunisian decision-makers need to move towards a co-management approach and the promotion of public-private partnerships. To do so, there is an urgent to modernize the Forest Code, to integrate the local population into decision making and to increase the flexibility of the laws with regards to access to forests. The Forest Model described in the Text Box 9.2 might be a good alternative to the structures that currently exist in Tunisia.

Another possible alternative is the valorization and exploitation of Non-Wood Forest Products, especially the most abundant species, by local population organized in associations. Such actions may lead to increases in

---

TEXT BOX 9.2

Model Forest: a multifaceted governance approach for Mediterranean silvopastoral systems

Gerardo Moreno, Forestry School, INDEHESA – Institute for Silvopastoralism Research University of Extremadura.

Key points: Model Forests can be an important approach to dealing with governance issues of

Mediterranean silvopastoral systems. Current applications show promising outcomes.

A Model Forest is a working landscape of forests, farms and stakeholders, who define what, sustainability means in their own context, identify a common vision and set of goals, devise a governance structure and strategic plan. They then work together to achieve the goals set out in that plan. Currently, the International Forest Model Network (https://imfn.net/) includes 60 territories in 37 countries managing > 73 million ha all around the world, with nine cases in seven countries of the Mediterranean Basin.

A good example of silvopastoral territory is the Moroccan Ifrane Model Forest, where livestock grazing in rangelands (within and outside forests) plays a considerable socioeconomic role in the region. It is the main source of income for more than 8,245 households that practice transient grazing over 364,358 ha, with people and herds moving seasonally between the plateaus and the higher mountains to exploit well-defined pastoral areas. The forest regulations set out the tribes and tribal fractions who may enjoy these rights, which are not transferable. However, changes in the practice of transhumance and the use of established paths had led people to bypass traditional pastoral agreements and to neglect the laws governing the use of common land, threatening the local forest of *Cedrus atlantica* (Navarro-Cerrillo et al. 2013). To halt the degradation, the Ifrane Model Forest structure was created in 2009 to define an Action Plan that includes forestry and pasture management activities, valorization of products, such as medicinal plants and wool, and awareness-raising actions on the value of the environment. The Silvopastoral Management Association and the National Association of Sheep and Goat Producers participate in the governance structure of the Ifrane Model Forest. Since its creation, this is contributing to improving the living conditions of the inhabitants of the area (Qarro et al. 2014).

The Bucak Model Forest in Turkey, where multiple non-wood forest products provide many jobs for local people, have demonstrated that the forums have a greater likelihood of succeeding if they involve the informed participation of all stakeholders and that it is important to consider the capacity of local organizations, not as an alternative but as an important complement to both central and local bureaucratic institutions (Tolunay et al. 2014).

---

household income from forests without the destruction of silvopastoral systems, the improvement of the social and economic situation of the local population. These however require a new legal framework.

Acknowledgements

This work was partly supported, through the staff time of the co-authors, by the CGIAR Research Programme (CRP) on Institutions, and Markets (PIM) led by the International Food Policy Research Institute (IFPRI), and by the CRP Livestock (led by ILRI).

Notes

1. Ministère de l'Agriculture, Direction Générale des Forêts (1999) Plan National de la Protection des Forêts contre les Incendies. Ministère de l'Agriculture, des Ressources Hydrauliques et de la Pêche Maritime, Tunis.
2. Ministère de l'Agriculture, Direction Générale des Forêts (1999) Stratégie de développement des ressources forestières et pastorales. Ministère de l'Agriculture, des Ressources Hydrauliques et de la Pêche Maritime, Tunis.
3. Ministère de l'Agriculture, Direction Générale des Forêts (2002) Stratégie Nationale pourla consérvation des forêts. Ministère de l'Agriculture, des Ressources Hydrauliques et dela Pêche Maritime, Tunis.
4. Ministry of Agriculture (MA) (2014) National Strategy for the Development and Sustainable Management of Forests and Rangelands. 2015 –2024. Tunis.
5. National Strategy for the Development of the Forest Sector (2001–2011). Ministère de l'Agriculture, des Ressources Hydrauliques et de la Pêche Maritime, Tunis.
6. Ministère du développement économique (2010) Plan de Développement Economique et Social 2010-1014. Ministère de l'Agriculture, des Ressources Hydrauliques et de la Pêche Maritime, Tunis.
7. INS Institut National de Statistique. http://www.ins.nat.tn/. http://www.ins.nat.tn/

References

Ben Mansoura A, Garchi S & Daly-Hassen H (2001) "Analyzing forest users' destructive behavior in Northern Tunisia". *Land Use Policy* 18:153–163. https://doi.org/10.1016/S0264-8377(01)00004-7

Campos P, Daly-Hassen H & Ovando P (2007) "Cork oak forest management in Spain and Tunisia: two case studies of conflicts between sustainability and private income". *International Forestry Review* 9:610–626.

Croitoru L & Daly-Hassen H (2015) *Analyse des Bénéfices et des Coûts de la Dégradation des Forêts et Parcours*, Banque Mondiale, Washington DC.

Daly-Hassen H, Campos P & Ovando P (2009) "Economic analysis of cork oak woodland natural regene- ration in the region of Ain Snoussi, Tunisia". In: *Cork Oak Woodlands and Cork Industry: Present, past and future* (Zapata B., ed). Girona, Museu del Suro de Palafrugell.

Daly-Hassen H (2012) *Financing for Sustainable Forest Management in Tunisia*. Helsinki, Finland.

Direction Générale des Forêts (DGF) (2015) *Stratégie de Développement Durable des Forêts et des Parcours en Tunisie*. 2015–2024 Ministère de l'Agriculture, des Ressources Hydrauliques et de la Pêche Maritime. Tunis, Tunisia.

Direction Générale des Forêts (DGF) (2010). *Résultats du Deuxième Inventaire Forestier et Pastoral National*. Tunisie. 180.

FAO & DGF (2012) *Evaluation Economique des Biens et Services des Forêts Tunisiennes*. Edition: Direction Générale des Forêts, Tunisie.

FAO (2016) *State of the World's Forests: Tunisia Case Study by Hamed Daly-Hassen In "The State of the World's Forests (SOFO), Forests and agriculture: land-use challenges and opportunities"*. 25, Rome.

Navarro-Cerrillo R M, Manzanedo R D, Bohorque J, et al. (2013) "Structure and spatio-temporal dynamics of cedar forests along a management gradient in the Middle Atlas, Morocco". *Forest Ecology and Management*, 289, 341–353.

ONAGRI (2019) *Indicateurs Clés sur la Forêt, les Produits et Services Forestiers en Tunisie*. Observatoire National de l'Agriculture Tunis.

Qarro M, Valbuena P & Segur M (2014) "Managing cedar forests in Morocco's Middle Atlas Mountains". *Unasylva* (English ed.), 65(242), 40–44.

Tardieu L, Muys B, Tuffery L, et al (2018). "Human needs and ecosystem services". In: *State of the Mediterranean*. FAO and UNEP/Plan Bleu, 331.

Tolunay A, Türkoglu T, Elbakidze M & Angelstam P (2014) "Determination of the support level of local organizations in a model forest initiative: Do local stakeholders have willingness to be involved in the model forest development?" *Sustainability* 6(10):7181–7196.

10 The complexity of public policies in Iberian montados and dehesas

José Muñoz-Rojas, José Ramón Guzmán-Álvarez and Isabel Loupa-Ramos

10.1 Introduction

The overarching public policy framework for European agriculture, forestry and sustainable rural development is set by the common agricultural policy (CAP) and by complex panoply of other policies at EU level (Jordan and Lenschow, 2010). This common framework is downscaled across the Member States via a wide variety of policies and instruments. Lastly, common principles, including subsidiarity, rule of law and sustainability, are set via the EU's and global (e.g. FAO's) long-term agendas, which national and regional policy visions and plans need (and are encouraged) to be in line with. This all results in a complex picture that is the subject of lively societal and scientific debates.

This complexity in policy settings strongly influences the governance of silvopastoral systems, along with the private actors and levels examined in Chapter 11 of this book. Policy frameworks and regimes generally encompass a diversity of instruments that differ widely in their purposes, enforcement mechanisms, timescales, agency and policy objectives. When examined as a whole, policy frameworks and regimes can be defined by differences in their degrees of internal coordination, coherence and integration, both horizontally (across policy areas) and vertically (across institutional levels from the local to the international).

By *policy coordination* we refer to the joint consideration, under a unified policy agenda, of the effects of policy instruments that may be either mutually conflicting or synergistic. *Policy coherence* is understood as the compatibility between individual policy purposes and objectives. Lastly, *policy integration* is the process of implementing joint strategic and administrative decisions aimed at tackling complex challenges that cannot be effectively tackled addressed through individual policies (Cejudo & Michel, 2017). According to these same authors, poor levels of policy coordination, coherence and integration fragment policy frameworks and regimes, which is a major barrier towards achieving the increased levels of sustainability and resilience sought by current international rural policy objectives (Boulanger & Messerlin, 2010).

DOI: 10.4324/9781003028437-11

Silvopastoral systems are particularly vulnerable to the perils of policy fragmentation because of their complex and multi-functional nature (Campos et al, 2013). The synergistic properties of these systems require integrated management (Campos et al, 2013), and yet this requirement does not fit well with the sector-oriented policy framework that still dominates EU policies (Faludi, 2010), in which agriculture, forestry, rural development, biodiversity and landscape are too frequently considered as separate and independent policy areas (see Figure 10.1; see also Text Box 10.1 later in this chapter).

In the case of Iberian silvopastoral systems, improvements are clearly required in the coordination, coherence and integration of public policy institutions and instruments. To tackle such needs, actions should be undertaken that embrace multiple spatial-temporal scales and institutional levels, types of actors, policy areas and decisions.

In this chapter, we describe, characterize and compare the current policy frameworks affecting silvopastoral systems in Alentejo and Andalucía. The chapter begins by introducing the diverse types of public policies (Section 10.2). The legal definitions of montado and dehesa are then presented (Section 10.3), followed by a description of the specific frameworks in

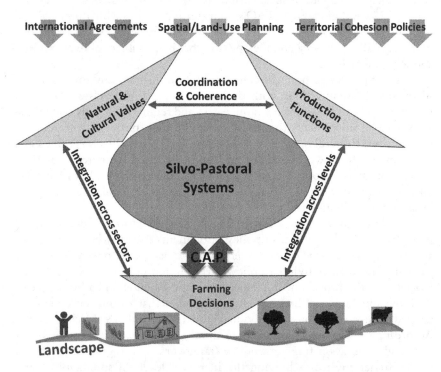

Figure 10.1 Illustration of the role and place of policy integration, coherence and coordination in the governance and planning of silvopastoral systems across levels of governance and subsidiarity of the EU.

Andalucía and Alentejo, and how these relate with the common overarching international policy framework, with a focus on the EU level (Section 10.4). This is intended to shed light on how different national and regional public administrations can act differently even when dealing with similar systems and challenges, and sharing a common underpinning policy framework (in this case, EU and other international policy).

We close the chapter by discussing the relevance of our findings to facilitate progress towards improved levels of policy coordination, coherence and integration for Mediterranean silvopastoral systems (Section 10.5). Finally, provide some retrospective reflections of the key lessons learned from this analysis and the future options available (Section 10.6).

10.2 Key aspects of rural public policies for the governance of silvopastoral systems

In this section, we propose criteria for identifying key aspects of rural public policies. While these criteria are overarching, they are of specific relevance for silvopastoral systems.

10.2.1 Purpose of the policy: strategic or action-oriented

Firstly, the distinction arises between strategic and action-oriented public policy instruments. Whilst the former are aimed at setting the basic principles representing political and social goals, the latter's main intent is facilitating the implementation of such principles at the appropriate institutional levels which, in compliance with the subsidiarity principle, should as close to the citizens as possible.

An example of direct relevance for silvopastoral systems is provided by the 1992 UN's Convention of Biological Diversity, hereinafter referred to as CBD[1], which contains the main strategic principles and targets for biodiversity conservation set worldwide, and that has been then transposed into the EU's Biodiversity Strategy[2], and also into a specific Portuguese Biodiversity Strategy and Action Plan (in the form of a National-level law – Rcm 55/2018, Presidência do Conselho de Ministros Governo da República de Portugal 2018). The principles in these strategies are expected to be enacted through specific actions for valuable habitats, to be identified and monitored at the local level. In the specific case of the Portuguese policy framework, the strategic plans and policies driving land-use changes become legally binding when integrated into municipal master plans (MMP) and can help shape the distribution of CAP funding. The same logic and procedures also apply in Andalucía, where the CBD is effectively downscaled via the Natural Heritage and Biodiversity National Strategic Plan[3] and the Andalusian Biodiversity Integrated Management Strategy[4] to then become effective at local scales through a combination of land-use planning and schemes to support rural development.

10.2.2 Enforcement mechanisms: incentives and regulations

A further distinction needs to be made between voluntary incentives and compliance that is enforced through regulations. Incentives are delivered through monetary payments or tax breaks and aim to compensate farmers for any financial loss (e.g. caused by setting aside land for conservation), or to encourage them to make certain decisions. In contrast, laws and regulations are compulsory and secured through fines and criminal convictions. A combination of these "carrot and stick" approaches is most commonly used in the search for more salient, legitimate and efficient mechanisms to achieve policy goals. In both Portugal and Spain, the silvopastoral systems are generally less affected by enforcement-oriented instruments (e.g. via planning) and more so by voluntary payment schemes (e.g. via CAP).

10.2.3 Timescales: long-term programmes and short-term actions

Generally speaking, higher institutional levels and their strategic instruments are enacted at longer timescales, a key example of this being the six-year programmes under which the CAP is managed. Lower scale, implementation-oriented, policy instruments that set out concrete actions to be implemented by farmers or other actors on the ground tend to be shorter term. This allows them to be tweaked on the basis of how effective and efficient they prove to be in practice. This rule is nonetheless far from universal, for example, a municipal spatial plan can be in place for up to 20 years, whilst some concrete CAP measures can be revoked or updated following a mid-term review of the CAP six-year cycle, as happened in 2003.

Incentives and plans are both generally set at the timescales at which the higher-level strategic policies and programs are defined (e.g. six years in the case of the CAP). This is particularly relevant for silvopastoral systems in which long-term cycles of plantation and renovation dominate and can have around a ten-year cycle. An example of this is the nine-year cycle of cork harvesting, a key (if infrequent) source of income for farmers in Iberian montados and dehesas. This results in a timescale miss-match between agroforestry production cycles, and the national policy (four-year) and CAP budget (six-year) cycles.

10.2.4 Agency: bottom-up initiatives and top-down regulations

In accordance with the principle of subsidiarity that underpins the Europe 2020 strategy[5], policy decisions across all sectors should be made at the closest possible level to citizens and secure the highest possible levels of public participation. This can be considered as a "wicked challenge" (*sensu* Duckett et al, 2016) in the case of Iberian dehesas and montados, where multi-functionality, complexity and uncertainty are all key characterizing features (Pinto-Correia et al, 2011).

Despite the administrative differences encountered between Spain (with administrative regionalization) and Portugal (more centralized), achieving more bottom-up decisions still seems a long way off in both the montado and the dehesa.

10.2.5 Policy object: sectoral or integrative

Lastly, a key problem for silvopastoral systems is related to the prevalence of sectoral policies that do not effectively address complex territorial systems and landscapes as a whole. This negatively impacts the potential for achieving more effective policy coordination, coherence and integration (Cejudo & Michel, 2017), raising multiple challenges related to both the coordination of policy areas and cooperation amongst actors and institutions, both horizontal and vertical, in silvopastoral systems.

10.3 The existing policy framework

10.3.1 The EU and international regulatory approaches to silvopastoral systems (including dehesas and montados)

The CAP is the single largest element of EU expenditure, representing 37.8% of the total EU budget for the period 2014–2020 (down from 66% in the early 1980s) and totalling 408.3 bn Euros. The main bulk of the CAP budget is financed by the European Agricultural Guarantee Fund (EAGF) under Pillar I, which secures that direct payments are channelled through national or regional administrations to farmers.

Alongside the CAP, the Policy for Regional and Territorial Cohesion also operates two main funds: the European Regional Development Fund (ERDF) and the Cohesion Fund (CF) (Faludi, 2010). For 2014–2020 these two funds accounted for €351.8 bn (32.5% of the EU's 2014–2020 budget). Together with the European Social Fund (ESF), the European Agricultural Fund for Rural Development (EAFRD) and the European Maritime and Fisheries Fund (EMFF), they make up the European Structural and Investment (ESI) Fund, which amounts to €641.9 bn total budget, with a contribution from the EU of €460.5 bn and the remainder being covered by national or regional governments. There is partial overlap amongst these funds, notably between EAFRD (Pillar II) and the CAP budget. Such overlap makes it difficult to obtain a clear-cut picture of the overall expenditure.

EU funding requires that national policies are compliant with both CAP regulations and other relevant EU policy instruments and standards. These include the Water Framework Directive (2000/60/EC)[6], the Habitats Directive (92/43/EC)[7], the Nitrates Directive (91/676/EEC)[8] and the Birds Directive (2009/147/EC)[9]. The EU lacks a common forestry policy, although forest lands and products have been incorporated within the Pillar II scheme.

Competencies for spatial planning lie at national, regional or local levels, depending on the administrative structure of each EU Member State. This is also the case for landscape policies, which are not legislated for at the EU level, although some guidance was provided by the European Council, through the European Landscape Convention (EC, 2000)[10]. Despite this, EU policies can strongly influence the possibilities for rural and local landscape changes that are central to the governance of silvopastoral systems.

Along with EU policies, there are other international levels and institutions that are also relevant for policies relating to silvopastoral systems, including those related to the natural and cultural heritage conservation, IUCN and UNESCO respectively, and also those related to food and agriculture (FAO).

There is no unified legal and widely accepted definition of silvopastoral systems at the EU level, leading to difficulties in regulating the Iberian dehesas and montados or in setting common principles and targets.

In such a complex policy arena, dehesa and montado farmers frequently face the challenge of different regulations that apply to different parts of their own farm estates, which might make their decisions mutually incoherent, if not incompatible. For instance, the UE Regulation 2013/1305[11] on support for rural development by the European Agricultural Fund for Rural Development establishes a specific aid to enhance agroforestry systems (Article 23). Agroforestry systems are defined in this regulation (Pillar II of the CAP) as "land use systems in which trees are grown in combination with agriculture on the same land". This definition is further extended in the CAP sub-measure 8.2;

> land-use systems and practices where woody perennials are deliberately integrated with crops and/or animals on the same parcel of land management unit without the intention to establish a remaining forest stand. The trees may be arranged as single stems, in rows or in groups, while grazing may also take place inside parcels (silvoarable agroforestry, silvopastoralism, grazed or intercropped orchards) or on the limits between parcels (hedges, tree lines)
>
> (cited in Mosquera–Losada et al, 2016 & 2017).

This scheme originally aimed to establish new agroforestry systems, so it became difficult to apply it to dehesas and montados (the most widespread agroforestry system in Europe, according to Moreno & Pulido, 2009), which are long-established systems, and where the more acute problems are often related to their ageing. Because of this policy instrument's focus was on planting a specific number of trees per hectare on existing agricultural land, it proved to be hardly applicable to montados and dehesas. A few years later, following national and regional claims, the regulation was modified, adding the possibility of supporting the regeneration or renovation of existing silvopastoral systems (Regulation (EU) 2017/2393)[12].

As another example of the challenges posed by EU policies, CAP EU schemes and rules include the consideration of grazing areas as one of the main rural land-use categories in need of urgent policy reform across Europe. However, as it has been pointed out by representatives of extensive livestock farmers, the CAP has been more tailored towards herbaceous pastures and discriminates against wood pastures or grassland pastureland under tree cover.

Legally defining grasslands is a difficult task, as they may be conceived in terms of land use or land cover. As a recent overview has shown (Velthof et al, 2014), there are several definitions of grasslands that currently co-exist in the grey and policy literature. This hampers the potential for setting policy targets and objectives for dehesas and montados.

A further relevant EU-level policy for these systems comes from the EU Habitat Directive[7], which addresses biodiversity and nature conservation priorities. This Directive designates dehesas and montados with evergreen *Quercus* spp., as having an importance for nature conservation. Habitat type code 6310, states that they are: a characteristic landscape of the Iberian peninsula in which crops, pastureland or Meso-Mediterranean arborescent matorral, in juxtaposition or rotation, are shaded by a fairly closed to very open canopy of native evergreen oaks (*Quercus suber*, *Q. ilex*, *Q. rotundifolia*, *Q. coccifera*). It is an important habitat for raptors, including the threatened Iberian endemic eagle (*Aquila adalberti*), for the crane (*Grus grus*), for large insects and their predators and for the endangered lynx (*Lynx pardinus*).

The agriculture/forest divide that permeates much of EU policies may be appropriate for northern Europe's rural landscapes but in the more complex and hybrid rural landscape mosaics of the Mediterranean, it is much more difficult to discern whether certain patches ought to be considered as forestry or agriculture. This results in management, policy and funding miss-matches and in challenges for policy coordination and integration. Too frequently the legal distinction is based on tree density. At other times, the main point of distinction is made around the concept of naturalness. Though dehesas and montados are cultural systems, they also hold relevant natural values that can result in habitat designation and protection and in related policy and management conflicts. These are alternately considered as cultural and purely natural systems (see Figure 10.1), yet the conservation of the landscapes' character and ecosystem values depends on management practices which have yet a third set of objectives, inextricably linked to human activities, such as agriculture.

In summary, the complexity of Iberian agro-silvopastoral systems makes it rather difficult to accommodate them under a single policy area, with agriculture, forestry and nature conservation policies all playing key (and often contradictory) roles in regulating their management.

10.3.2 Dehesas in Andalucía

10.3.2.1 National level

Over the last two decades in Spain several policy reports and documents have addressed the multiple challenges facing the dehesa, most of which highlight the absence of a unique definition underpinning its legal status and requirements for protection. This has implications for progressing towards more sustainable management and governance models for dehesas. The following documents are especially relevant:

- The Spanish Senate Report on the Dehesa (Study Paper on the Protection of Dehesa ecosystem (Senado, Gobierno del Reino de España, Boletín Oficial de las Cortes Generales, 24/1/2010).
- The Dehesa Green Book (Pulido and Picardo, 2010). A similar document also exists for the Portuguese montado (Pinto-Correia, Ribeiro & Potes, 2013).

The Spanish national forest regulation (Gobierno del Reino de España, 2003, Ley 43/2003, de Montes) considers the dehesa as a particular case of forest land-use characterized by its mixed agro-silvopastoral nature in which livestock production shares a common piece of land with forestry. Being under the umbrella of forest regulation implies that dehesas are subject to the regime of protection and conservation of forest lands. Nevertheless, since no other references to this complex system are made in generic forest law, it is difficult to establish a proper legal and instrumental definition of the system, including the land-cover and land-use types.

A unique exception is found in Spanish forest statistics where the item "dehesa" is acknowledged, whereby it is defined as "a man-made forest system whose main trait is its multiple-use, and which is composed mainly by Holm, Cork and Deciduous Oaks, Wild Olives and Ash tree (and occasionally by other tree species); its vegetation structure facilitates the development of a herbaceous layer (grassland), which is commonly used for livestock production or for hunting activities of animal species that feed on branches and fruits as well" (Ministerio de Agricultura, Pesca y Alimentación, 2017, p. 11)[13].

The Iberian Pork Production Regulation establishes its own definition of dehesas for the purpose of defining quality standards (Ministerio de Agricultura, Alimentación y Medio Ambiente. Gobierno del Reino de España. Real Decreto 4/2014), whereby it defines dehesa as a geographic area with the predominance of an agroforestry system mainly used for extensive livestock production over a continuous grassland cover with a density of Mediterranean tree species (fundamentally *Quercus* species) of, at least, 10 trees/ha, and where human action is required for its conservation and maintenance.

The lack of a common perspective that defines dehesas as complex agroforestry systems is clearly reflected in the social visions held across Spain of this system, which expand well beyond the mere administrative framework. In this sense, most common definitions adopted by environmental NGOs tend to highlight the natural components, whereas reports by farmers or landowners tend to focus on production functions and services. Nevertheless, over the last few years, both perspectives have been coming closer together, possibly as a result of the need to join efforts in order to reach a common diagnosis about the crisis that is impacting the system as a whole.

From a stakeholder's perspective, the definition adopted by Federación Española de la Dehesa[14] (FEDEHESA) puts the focus on the productive point of view, thus assuming the one included in the nomenclator of the Spanish Grassland Study Society (Ferrer, San Miguel & Olea, 2001):

> surface with a more or less sparse tree canopy and a sound developed grassland layer from which scrubs have been largely removed, encompassing agriculture and livestock production. Its main production is extensive or semi-extensive livestock, feeding in grassland and in branches and fruits as well.

A World Wildlife Fund (WWF) report on the future of the dehesas (Hernández, 2014) adopts the basic components of the former definition, adding a complementary aspect focussing on its agro-sylvopastoral dimension and its sound ecological and economical results under appropriate and efficient management options. WWF thus identifies dehesas as a High Natural Value Systems whereby "extensive livestock raising plays a leading role in the creation, use and maintenance of grazing lands".

10.3.2.2 Regional level

Concerns for the conservation of dehesas in Andalusia originally led to a multi sectorial agreement in 2005 called *Pacto Andaluz por la Dehesa* (Andalusian Agreement for the Dehesa, Junta de Andalucía, 2005). Due to the wide degree of representativeness achieved (it was signed by more than 800 individuals and entities), this can be understood as a bottom–up governance initiative that includes a diagnosis and objectives set to achieve the sustainable development of the Andalusian dehesas.

As a direct result of the *Pacto Andaluz por la Dehesa*, in 2010 the Andalusian Regional Law for the Dehesa was approved (Gobierno del Reino de España 2010). Several factors lay behind this, including: the social relevance of the Andalusian dehesas, the large amount of land surface occupied by this agro-ecosystem in the region (more than 1,000,000 ha, roughly covering 12.5% of the whole region), the presence of the UNESCO Biosphere Reserve of Dehesas de Sierra Morena comprising an area of 424,000 ha, and the relevance of dehesas as natural designated areas for nature and landscape conservation in Spanish regulations and the EU Habitats Directive (Council Directive 92/43/EEC)[7].

Dehesas in Andalucía are legally considered as forest land and therefore come within the regulatory framework of Spanish and Andalucía Forest Laws. In this sense, one of the main targets of the Andalusian Dehesa Law is to enhance coordination and coherence among the different agricultural and forestry policies and activities in place, rather than merely establishing new regulations.

For the purpose of achieving its goals, the Andalusian law includes some relatively novel regulatory instruments. At the regional level, it establishes the Master Plan for the Dehesas of Andalusia, which was approved in 2017 (Decreto 172/2017, de Andalucía, 2017), as a strategic planning instrument. At the farm level, it fosters integrated management plans to facilitate more efficient management of the dehesas, focussing on acknowledging and protecting multi-functionality, and on promoting their sustainability. Improvements in research and capacity building are promoted under this Law. From a governance perspective, the Andalusian regulation includes an Administrative Coordination Commission (Comisión Andaluza para la Dehesa), and a Steering Committee for the Master Plan of the Dehesa. This is a step forward towards improving policy coordination and social participation in order to design more integrated public policies and apply them to the Andalusian dehesas.

From a spatial planning perspective, the Andalusian Regional Land-Use Plan (de Andalucía, 2006), based on Law 1/1994, Spatial Planning of the Autonomous Community of Andalusia) explicitly considers the dehesa as one of the main territorial assets in the region. However, this plan does not establish any specific legal indications for its conservation and sustainability. Local planning in Andalucía does not specifically consider the dehesa, since municipal level policy instruments largely have an urban planning orientation (Law 7/2002, Urban Planning of the Autonomous Community of Andalusia, de Andalucía, 2002).

According to the Master Plan for the Dehesa of Andalucía (Document of annexes, p. 135), 319,884 ha of open-woodland dehesas ("formación adehesada") are included in areas designated as Natural Parks across Andalucía This means that roughly 28% of the area covered by dehesas falls under specific nature conservation regulations (based on Natural Resources Management Plans and Master Plans for Use and Management, as defined in Regional Law 2/1989 de Andalucía, 1999).

In some of these Natural Parks dehesa is the main ecosystem and the more common landscape type. With the Natural Parks of Andalucía now designated as Sites of Community Importance (EU Habitat Directive, 1992), Andalusian dehesas have become part of the Natura 2000 European Ecological Network. In order to better integrate and coordinate diverse regulations, Habitat Class nº 6310 in Andalusia has become equivalent to a "formación adehesada" (open-woodland dehesa area).

10.3.3 Montados in alentejo (Portugal)

Portugal equally lacks a unified and operational definition of its montados and there is poor coordination and integration between the range of governance scales and institutional actors that influence the management and protection of this threatened and valuable land-use system.

Unlike in Spain, there is no fully evolved regional level for policy making for land-use (including the montados) in Portugal. Although some regulations addressing specific aspects and activities in the montado such as forestry have been in place for more than 100 years[15], these are of little relevance for the current challenges faced by the montado. Our analysis looks at the current legal instruments that control the governance of these unique silvopastoral systems.

• Law for the Protection of Holm and Cork Oaks - Decreto-lei de 30 de Junho 155/2004 (Presidência do Conselho de Ministros, Governo da República de Portugal, 2004).

This is a legal instrument that is rooted on previous legislation originally dating back more than a century (Decreto de 24 de Dezembro de 1901) and that has now been partially superseded by the more specific Forest Regime and Code, from 2009, which nonetheless, does not uniquely target the montado. A 2004 Law is in place specifically and uniquely defining the protection of the trees – Cork and Holm Oaks. This instrument acts through top-down enforcement, in this case prohibiting the cutting of these trees unless a permit is obtained. While this law does prevent an unrestricted degradation of the montado system, it fails to consider its complexities and related multiple interactions.

Focusing on Alentejo, where the majority of montado is located, the following regulatory policy and planning instruments were identified that, directly or indirectly, target the montado;

• Regional Strategy for Smart Specialization in Alentejo (Comissão de Coordenação Regional do Alentejo-CCDRA, 2014a).

This is a strategic planning instrument approved at the regional level that aims at providing a sound strategic vision for the competitive specialization of the region. The strategy identifies a series of key sectors and activities including, in relation to the montado, the production of cork and livestock, prioritized for public support. This approach is in line with the common perception of the montado as the simple sum of commodity-oriented activities. These same strategic objectives are then translated into potential applications in practice through the Operational Programme for the Alentejo Region 2014–2020 (CCDR-A, 2014b) that again focuses on promoting or protecting single elements of the system.

• Regional Spatial Plan for the Alentejo - Resolução do Conselho de Ministros n.º 53/2010 (- Presidência do Conselho de Ministros, Governo de Portugal, 2010).

This is the main spatial planning instrument in Alentejo and embraces a more spatially explicit approach that considers the montado as one of 8

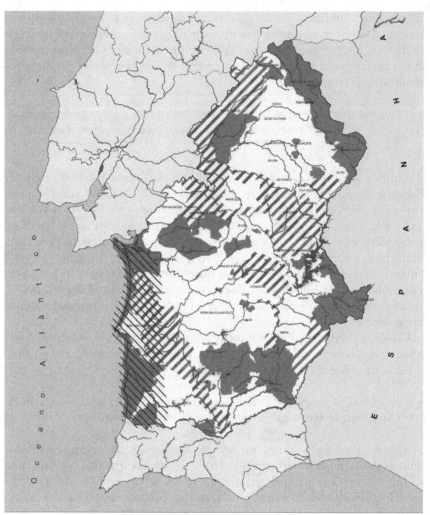

Figure 10.2 Regional spatial structure for the protection and valorization of the environment and coastal green areas of Alentejo, included in the Regional Spatial
Plan (Presidência do Conselho de Ministros. Governo da República de
Portugal, 2010). The areas with diagonal brown stripes mark the zones where
the higher potential for ecological connectivity is to be fostered mainly
through the presence of well-connected corridors and patches of montados.

strategic territorial assets for the sustainable development of the region. The
Regional Spatial Plan defines a corridor of montados that stretches from the
NE to the SW of the region (Figure 10.2) as part of the Regional Green
Infrastructure. Clearly delineated areas are designated as part of these cor-
ridors that should be legally protected and transposed into the Municipal

Zoning Plans (PDM), encouraged to consider how best to protect them through specific regulations and local implementation plans.

- Law for the definition of National Agricultural Reserves-RAN - Decreto-Lei n 199/2015 de 16 de Setembro (Presidência do Conselho de Ministros. Governo da república de Portugal, 2015) and of National Ecological Reserves-REN - Decreto-Lei n 166/2008, de 22 de Agosto (Presidência do Conselho de Ministros. Governo da República de Portugal., 2008).

These two policy instruments only implicitly relate to the montado. In both cases, these are spatially explicit planning policy instruments aimed at protecting biophysical structures and processes by avoiding disturbance or the degradation of their key landscape values. These are instruments that are applied through enactment and enforcement and not via voluntary measures (i.e. incentives).

RAN (National Agricultural Reserves) are designated for the sole purpose of agricultural production by mapping soils with high agricultural potential, whilst REN (National Ecological Reserves) are designated primarily for soil and water conservation purposes. Both RAN and REN result in zoning designations that restrict and permit land-uses that then need to be adapted and enacted by municipalities through the local planning system. As montados are can be found in all types of soils and locations, these regulations play little role in helping their protection.

10.4 Discussion: policy coordination, coherence and integration for dehesas and montados (and beyond)

10.4.1 Policy coordination

The urban-rural policy divide is a key factor explaining the poor levels of policy coordination and coherence. Even though regional and local authorities have the direct competence for planning in their entire territory, the power of agricultural policy, where most funds are allocated, overshadows any other aspects, hampering the multi-functional nature of montados and dehesas. In theory, this could be tackled if the agricultural funds were required to be aligned with spatial planning instruments. A further option could be if local governments had more control over how CAP funds were spent in their own territories. However, in practice this might lead to higher competition for these funds, probably benefitting larger land-owners, as has been the case within some past LEADER initiatives. It remains to be proven whether agroforestry systems might benefit from such arrangements, although the social position of many owners of montados and dehesas seems to suggest that they would benefit.

TEXT BOX 10.1

The EU's novel strategies for biodiversity and a sustainable food system

Jabier Ruiz-Mirazo, European Commission

Key points: A political agreement on the CAP, expected in 2021, and the decisions that Member States make in finalizing their national CAP strategic plans in the following months will become crucial for the future of the EU's food system and the environments on which it impacts. These factors will ascertain whether the ambition to transform EU agriculture, as set out in the Commission's strategies, is achieved and supported by national and regional governments. This could have a large impact on the way farm public subsidies and incentives evolve. Over the coming years CAP, and its implementation at the national and regional level could progressively change the way in which Mediterranean silvopastoral systems operate and are managed, and not necessarily for the better.

The European Green Deal[16] is one of the central political priorities for the European Commission in its 2019-2024 term. With a focus on achieving climate neutrality by 2050, the European Green Deal was presented as a new "green-growth" strategy for the European Union, aiming to address environmental degradation and decouple economic growth from resource use. The European Green Deal was complemented by the proposal of a new European Climate Law and the publication of two flagship strategies: a *Farm to Fork Strategy for a fair, healthy and environmentally-friendly food system*[17], and a new *EU Biodiversity Strategy for 2030*[18].

The scope of these strategies is very broad, stretching well beyond the focus of this book. This text box describes only a few of their main features and how they might change the EU policy framework for Mediterranean silvopastoral systems. Before going into further detail, it is important to clarify the role of such strategies in EU policy-making. Published by the European Commission, and not subject to any co-decision process by the European Parliament or the Council of the EU, such strategies set the political orientation and announce actions to be undertaken by the Commission in the coming years. However, as they are not regulations or directives, they do not have any binding power forcing the Member States to take action.

The EU's Biodiversity Strategy for 2030 announces the intention of expanding protected areas in Europe to cover 30% of land and seas. Additionally, it sets a target to increase the area under *strict protection*, from the current 3% of land (and less than 1% of seas) to 10% (one-third of total designated areas). The European Commission is working to provide guidance on what 'strict protection' means, which could be very relevant for silvopastoral systems if it implies a full exclusion of human activities such as grazing by domestic livestock.

In the EU's Biodiversity Strategy for 2030 section on *Bringing nature back to agricultural areas*, it is striking that no mention is made of farming practices that help preserve biodiversity, of high nature value farming, or the loss of certain habitats and species due to the abandonment of farming. Most of the narrative is on reducing the impacts of intensive farming on the environment, with most agricultural targets for 2030 focused on reducing chemical inputs (fertilizers, antibiotics, pesticides), all issues that are not very relevant in extensively managed silvopastoral systems.

The Biodiversity Strategy also includes an ambitious target on organic farming, aiming to reach 25% of agricultural land in Europe by 2030. Converting to organic production is generally easier on (extensive) pasture-based farms than in arable or permanent crops, and so this target may benefit silvopastoral systems. An additional target of *at least 10% of agricultural area under high-diversity landscape features* could also lead to better policy support for woody vegetation on farmland. However, it will be important that this target embraces the productive role that trees and shrubs play in silvopastoral systems, and not just their biodiversity value.

Another flagship target of the Biodiversity Strategy is to plant *at least 3 billion additional trees in the EU by 2030*. As assessed in a recently published policy brief on the topic by conservationists and land-owners[19], there is potential to grow many of those trees on farmland in a way that is ecologically and agronomically appropriate. Thus, this initiative could help boost the presence of woody vegetation on farms, and channel more support for the establishment or regeneration of trees and shrubs in all types of agroforestry, such as wood pastures.

The main novelty and value of the Farm to Fork Strategy resides in its aim to create a single overarching framework for multiple policy areas that relate to the sustainability of our food system. This covers food production, processing, trade marketing, consumption and waste, so it is a highly ambitious strategy, as proven by the announcement for 2023 of a new legislative framework for sustainable food systems, with the aim to promote policy coherence at EU and national level, mainstream sustainability in all food-related policies and strengthen the resilience of food systems.

The Farm to Fork Strategy adds a few more details on the plans of the European Commission to assist farmers in the transition to sustainability. For instance, carbon farming is identified as a key tool to be further developed, as a way to mitigate climate change while creating a new source of income for farmers. This could be an interesting development for silvopastoral systems, due to their high potential to store carbon. There is also a reference to supporting the most sustainable and carbon-efficient methods of livestock production, although this could operate in favour of intensive animal production if the metrics used are only based on emissions of greenhouse gases per kg of product.

The Common Agricultural Policy (CAP) is covered largely in the Farm to Fork Strategy, as a central instrument to provide funds to implement these ambitions in relation to agriculture. The European Commission considers that the proposed regulations for the next CAP can drive forward the Green Deal but are concerned about certain environmental provisions getting diluted in the negotiation process. Notably, they underline the importance of reinforcing eco-schemes (a novel agri-environmental intervention in CAP's First Pillar) to boost sustainable farming practices such as agroforestry and organic farming. However, eco-schemes could also be focused on achieving the desired reductions in the use of agrochemicals, which could lead to intensively managed farms maintaining or increasing the support that they receive from the CAP.

10.4.2 Policy coherence

The concept of green infrastructures (GI), originating in nature conservation and biodiversity policy, has emerged quite recently. Due to their spatial character, GIs are being increasingly considered as spatial planning tools that can be used across multiple scales, ranging from the EU to the local. They also contribute to the NATURA 2000 network. Some montado-based corridors of Alentejo would directly benefit from these planning instruments.

Further opportunities for future improvement in increasing policy coherence could be enacted through landscape improvement programmes. These have recently been approved in Portugal for highly vulnerable landscapes (Decreto-Lei 28-A/2020 - Presidência do Conselho de Ministros. Governo da República de Portugal, 2020a). Landscape Improvement Programs are novel strategic instruments approved at sub-regional scales which only become legally binding when they are integrated into a local development plan. The main novelty is that they include their own dedicated funding schemes. For the time being these programs are largely focused on wildfire risk zoning (Resolução do Conselho de Ministros n. 49/2020- Presidência do Conselho de Ministros. Governo da República de Portugal, 2020b) but their definition does not exclude them from being expanded to other vulnerable areas. Their success has yet to be demonstrated in practice, but they nonetheless show high potential to improve policy integration in agroforestry systems.

10.4.3 Policy integration

The first point in relation to policy integration is that this is largely dependent on how financial resources are distributed. This creates multiple challenges for implementing effective policy-led actions in multi-functional rural land-use systems, where the policy framework is fragmented, with funds largely targeting only specific elements of the overall system (crops, livestock, trees etc.). In this sense, one logical recommendation would be to design a specific financial scheme in support of the overall agroforestry system, which would help overcome the currently fragmented financial paradigm.

Also, it is clear that the policy fragmentation between forestry and agriculture is not beneficial for Iberian montados and dehesas. This is despite the existence of some specific policy instruments targeting the system as a whole, which are either outdated (in the case of Extremadura) or too strategic and requiring operational guiding actions (as in the case of Andalucía). In Portugal, where forests are 98% privately owned, such policy fragmentation becomes equally apparent when looking at the public forestry and agricultural statistical datasets.

10.5 Conclusion

This analysis has led us to question whether spatial planning should really be implemented as a core public policy tool in agricultural land, including in agroforestry systems, and also whether it should be limited to its current function as a zoning-based land-use permission system. The current planning approach has limitations as regulations are first and foremost based on restrictions and punitive actions. Having a more pro-active planning policy framework would facilitate a deepening of the integration of spatial planning instruments and regional development funding, whilst simultaneously bearing a stronger influence on the definition of a novel CAP and other agricultural and forestry policy measures creating funds that are better coordinated, more cohesive and more closely integrated across diverse levels of land management and decision-making.

Notes

1. https://www.cbd.int/doc/legal/cbd-en.pdf,
2. https://ec.europa.eu/environment/nature/biodiversity/strategy/index_en.htm
3. https://www.boe.es/eli/es/rd/2011/09/16/1274/con
4. http://www.cma.junta-andalucia.es/medioambiente/portal_web/web/temas_ambientales/biodiversidad/estrategia_biodiversidad/estrategia_de%20_biodiversidad.pdf
5. https://ec.europa.eu/eu2020/pdf/COMPLET%20EN%20BARROSO%20%20%20007%20-%20Europe%202020%20-%20EN%20version.pdf
6. https://ec.europa.eu/environment/water/water-framework/index_en.html
7. https://ec.europa.eu/environment/nature/legislation/habitatsdirective/index_en.htm
8. https://ec.europa.eu/environment/water/water-nitrates/index_en.html
9. https://ec.europa.eu/environment/nature/legislation/birdsdirective/index_en.htm
10. https://www.coe.int/en/web/landscape
11. https://eur-lex.europa.eu/legal-content/EN/TXT/?uri=celex%3A32013R1305
12. https://eur-lex.europa.eu/legal-content/EN/TXT/?uri=CELEX%3A32017R2393
13. https://www.mapa.gob.es/es/desarrollo-rural/estadisticas/aef2017_estructuraforestal_tcm30-521520.pdf).
14. http://fedehesa.org/concepto-de-dehesa/
15. http://ww2.icnf.pt/portal/florestas/gf/regflo/enqleg
16. https://ec.europa.eu/info/strategy/priorities-2019-2024/european-green-deal_en
17. https://ec.europa.eu/food/farm2fork_en
18. https://ec.europa.eu/environment/nature/biodiversity/strategy/index_en.htm
19. https://www.wwf.eu/?uNewsID=364674

References

Boulanger, P.H. & Messerlin, P.A. 2010. *2020 "European Agriculture: Challenges and Policies"*. 93 pp. German Marshall Fund of the United States & Sciences Po. Paris, France.

Campos, P., Huntsinger, L., Oviedo, J.L., Starrs, P.F., Díaz, M., Standiford, R.B. & Montero, G. (eds.). 2013. *"Mediterranean Oak Woodlands Working Landscapes. Dehesas of Spain and Ranchlands of California"*. 519 pp. Springer, Dordrecht, Berlin.

Cejudo, G.M. & Michel, C.L. 2017. "Addressing fragmented government action: coordination, coherence, and integration". *Policy Sciences* 50, 745–767.

Comissão para a Coordenação e Desenvolvimento Regional do Alentejo (CCDR-A). 2014.a. *"Estratégia Regional De Especialização Inteligente do Alentejo"*. Évora, December 2014 [https://www.ccdr-a.gov.pt/docs/ccdra/alentejo2020/EREI_Alentejo_vf.pdf].

Comissão para a Coordenação e Desenvolvimento Regional do Alentejo (CCDR-A). 2014.b. *"Programa Operacional Regional do Alentejo"*. Évora, December 2014 [http://www.

aproder.pt/admin/upload/ficheiros/ficheirosMultimedia/PORALENTEJO_2020_vf_THII.pdf].

Duckett, D., Feliciano, D., Martin-Ortega, J. & Muñoz-Rojas, J. 2016. "Tackling wicked environmental problems: The discourse and its influence on praxis in Scotland". *Landscape and Urban Planning*, 154, 44–56.

Faludi, A. 2010. *"Cohesion, Coherence, Co-operation: European Spatial Planning Coming of Age?"*. 226 pp. RTPI Library Series. Routledge, Oxford.

Ferrer, C., San Miguel, A. & Olea, L. 2001. "Nomenclátor básico de pastos en España". *Pastos. Revista de la Sociedad Española para el Estudio de los Pastos*, 31(1), 7–44.

Gobierno del Reino de España. 2003. *"Ley 43/2003, de 21 de noviembre, de Montes"*. «BOE» núm. 280, de 22 de noviembre de 2003, Referencia: BOE-A-2003-21339. Madrid [https://www.boe.es/buscar/pdf/2003/BOE-A-2003-21339-consolidado.pdf].

Hernández, L. 2014. *"Dehesas para el futuro. Recomendaciones de WWF para la gestión integral"*. 48 pp. Ministerio de agricultura, alimentación y medio ambiente & WWF. Madrid. [http://awsassets.wwf.es/downloads/dehesas_savn.pdf].

Jordan, A. & Lenschow, A. 2010. "Environmental policy integration: a state-of-the-art review". *Environmental Policy and Governance*, 20(3), 147–158.

Junta de Andalucía. 1999. *"LEY 2/1989, de 18 de julio, por la que se aprueba el Inventario de Espacios Naturales Protegidos de Andalucía, y se establecen medidas adicionales para su protección"*. Boletín número 60 de 27/7/1989. Sevilla. [https://www.juntadeandalucia.es/boja/1989/60/1].

Junta de Andalucía. 2002. *"Ley 7/2002, de 17 de diciembre, de Ordenación Urbanística de Andalucía. Comunidad Autónoma de Andalucía"*. «BOJA» núm. 154, de 31 de diciembre de 2002. «BOE» núm. 12, de 14 de enero de 2003. Referencia: BOE-A-2003-81. Sevilla. [https://www.boe.es/buscar/pdf/2003/BOE-A-2003-811-consolidado.pdf].

Junta de Andalucía. 2005. *"Acuerdo de 18 de octubre de 2005, del Consejo de Gobierno, por el que se promueve el pacto andaluz por la dehesa"*. Boletín Oficial de la Junta de Andalucía, 78/2006, Sevilla   [http://www.juntadeandalucia.es/medioambiente/site/portalweb/menuitem.7e1cf46ddf59bb227a9ebe205510e1ca/?vgnextoid=931c2709733da010Vgn VCM1000000624e50aRCRD&vgnextchannel=6ce08ad1d6391610VgnVCM 2000000624e50aRCRD].

Junta de Andalucía. (Consejería de Fomento, Infraestructuras y Ordenación del Territorio), 2006. *"Decreto 206/2006, de 28 de noviembre, por el que se adapta el Plan de Ordenación del Territorio de Andalucía a las Resoluciones aprobadas por el Parlamento de Andalucía en sesión celebrada los días 25 y 26 de octubre de 2006 y se acuerda su publicación"*. Boletín Oficial de la Junta de Andalucía259/2006. Sevilla [https://juntadeandalucia.es/boja/2006/250/4].

Junta de Andalucía. 2010. *"Ley 7/2010, de 14 de julio, para la Dehesa"*. «BOJA» núm. 144, de 23/07/2010, «BOE» núm. 193, de 10/08/2010. Sevilla. [https://www.boe.es/buscar/pdf/2010/BOE-A-2010-12891-consolidado.pdf].

Junta de Andalucía. 2017. *"Decreto 172/2017, de 24 de octubre, por el que se aprueba el Plan Director de las Dehesas de Andalucía, se crea su Comité de Seguimiento y se modifica el Decreto 57/2011, de 15 de marzo"*. Sevilla. [https://www.juntadeandalucia.es/boja/2017/207/BOJA17-207-00007-18430-01_00123501.pdf].

Ministerio de Agricultura, Alimentación y Medio Ambiente. Gobierno del Reino de España. 2014. *"Real Decreto 4/2014, de 10 de enero, por el que se aprueba la norma de calidad para la carne, el jamón, la paleta y la caña de lomo ibérico"*. «BOE» núm. 10, de 11 de enero de 2014 Referencia: BOE-A-2014-318. Madrid, 16 pp. [https://www.boe.es/buscar/pdf/2014/BOE-A-2014-318-consolidado.pdf].

Moreno, G. & Pulido, F. J. 2009. "The functioning, management and persistence of dehesas". In: A. Rigueiro-Rodriguez, J. McAdam & M.R. Mosquera-Losada (eds.). *Agroforestry in Europe: Current Status and Future Perspectives*, Springer, San Diego: Springer, pp. 127–160.

Mosquera-Losada, M.R., Santiago Freijanes, J.J., Pisanelli, A., Rois, M., Smith, J., den Herder, M., Moreno, G., Malignier, N., Mirazo, J.R., Lamersdorf, N., Ferreiro Domínguez, N., Balaguer, F., Pantera, A., Rigueiro-Rodríguez, A., Gonzalez-Hernández, P., Fernández-Lorenzo, J.L., RomeroFranco, R., Chalmin, A., Garcia de Jalon, S., Garnett, K., Graves, A. & Burgess, P.J. 2016. *"Extent and Success of Current Policy Measures to Promote Agroforestry across Europe"*. Deliverable 8.23 for EU FP7 Research Project: AGFORWARD 613520. (8 December 2016). 95 pp. Santiago de Compostela.

Mosquera-Losada, M.R., Santiago Freijanes, J.J., Pisanelli, A., Rois, M., Smith, J., den Herder, M., Moreno, G., Lamersdorf, N., Ferreiro Domínguez, N., Balaguer, F., Pantera, A., Papanastasis, V., Rigueiro-Rodríguez, A., Aldrey, J.A, Gonzalez-Hernández, P., Fernández-Lorenzo, J.L., RomeroFranco, R. & Burgess, P.J. 2017. *"How Can Policy Support the Uptake of Agroforestry in Europe"*. Deliverable 8.24 for EU FP7 Research Project: AGFORWARD 613520. Santiago de Compostela.

Pinto-Correia, T., Ribeiro, N. & Sá-Sousa, P. 2011. "Introducing the montado, the cork and holm oak agroforestry system of Southern Portugal". *Agoroforestry Systems*, 82, 99–104.

Pinto-Correia, T., Ribeiro, N. & Potes, J. 2013. *"Livro Verde dos Montados"*. 61 pp. ICAAM, Évora, Portugal. [https://dspace.uevora.pt/rdpc/bitstream/10174/10116/1/Livro%20Verde%20dos%20Montados_Versao%20online%20%202013.pdf].

Pulido, F. & Picardo, A., 2010. *"Libro Verde de la Dehesa"*. 23 pp. Consejería de Medio Ambiente, JCyL, Valladolid, Spain. [https://www.pfcyl.es/sites/default/files/biblioteca/documentos/LIBRO_VERDE_DEHESA_version_20_05_2010.pdf].

Presidência do Conselho de Ministros. Governo da República de Portugal. 2004. "Decreto-lei nº 155/2004 de 30 de Junho que Altera o Decreto-Lei n.º 169/2001, de 25 de Maio, que estabelece as medidas de protecção ao sobreiro e à azinheira". Diário da República n.º 152/2004, Série I-A de 2004-06-30. Lisboa [https://dre.pt/pesquisa/-/search/517471/details/maximized].

Presidência do Conselho de Ministros. Governo da República de Portugal. 2008. *"Decreto-Lei n.º 166/2008 de 22 de agosto que aprova o Regime Jurídico da Reserva Ecológica Nacional e revoga o Decreto-Lei n.º 93/90, de 19 de Março"*. Diário da República n.º 162/2008, Série I de 2008-08-22. Lisboa [https://dre.pt/pesquisa/-/search/453518/details/maximized].

Presidência do Conselho de Ministros. Governo da República de Portugal. 2010. *"Resolução do Conselho de Ministros n.º 53/2010. Aprovação do Plano Regional de Ordenamento do Território do Alentejo (PROTA)* Diário da República, 1.ª série — N.º 148 — 2 de Agosto de 2010". Lisboa [https://dre.pt/pesquisa/-/search/333798/details/maximized].

Presidência do Conselho de Ministros. Governo da República de Portugal. 2015. *"Decreto Lei 199/2015, de 16 de Setembro, que procede à primeira alteração ao Decreto-Lei n.º 73/2009, de 31 de março, que aprova o regime jurídico da Reserva Agrícola Nacional.* Diário da República n.º 181/2015, Série I de 2015-09-16". Lisboa [https://dre.pt/home/-/dre/70309902/details/maximized?p_auth=eVIwl6Va].

Presidência do Conselho de Ministros. Governo da República de Portugal. 2018. *"Resolução do Conselho de Ministros n.º 55/2018 que aprova a Estratégia Nacional de Conservação da Natureza e Biodiversidade 2030"*. Diário da República n.º 87/2018, Série I de 2018-05-07. Lisboa [https://dre.pt/home/-/dre/115226936/details/maximized].

Presidência do Conselho de Ministros. Governo da República de Portugal. 2020a. *"Decreto-Lei n.° 28-A/2020, de 26 de junho, o qual Estabelece o Regime Jurídico da Reconversão da Paisagem"*. Diário da República n.° 123/2020, 1° Suplemento, Série I de 2020-06-26. Lisboa [https://dre.pt/home/-/dre/136678483/details/maximized].

Presidência do Conselho de Ministros. Governo da República de Portugal. 2020b. *"Resolução do Conselho de Ministros n.° 49/2020. Cria o Programa de Transformação da Paisagem"*. Diário da República n.° 121/2020, Série I de 2020-06-24. Lisboa [https://dre.pt/web/guest/home/-/dre/136476384/details/maximized].

Gobierno del Reino de España. 2010. *"Informe de la Ponencia de Estudio sobre la Protección del Ecosistema de la Dehesa*, constituida en el seno de la Comisión de Medio Ambiente, Agricultura y Pesca (543/000009)"*. Serie I: BOLETÍN GENERAL 26 de noviembre de 2010 Núm. 553. Madrid [https://www.senado.es/legis9/publicaciones/pdf/senado/bocg/BOCG_D_09_8_29.PDF].

Velthof, G.L., Lesschen, J.P., Schils, R.L.M., Smit, A., Elbersen, B.S., Hazeu, G.W., Mucher, C.A. & Oenema, O. 2014. *"Grassland Areas, Production and Use. Report Lot 2. Methodological studies in the field of agro-environmental Indicators"*. EUROSTAT & ALTERRA-Wageningen-UR. Netherlands. 155 pp. [https://ec.europa.eu/eurostat/documents/2393397/8259002/Grassland_2014_Final+report.pdf/58aca1dd-de6f-4880-a48e-1331cafae297].

Section C

Governance models

11 Conflict and the governance of Iberian silvopastoral systems

Conflict and interaction among different actors and levels of governance today

Pedro M. Herrera and Fernando Pulido

11.1 Introduction

Conflict, defined as the expression of divergent interests between individuals and/or social groups, has a strong influence on demographic, ecology and evolutionary processes (Behrendorff et al, 2017) and can arise in different contexts, geographic scales and decision-making levels (Byrne and Senehi, 2009). Conflict is considered a structuring element among human and animal communities (Manning and Dawkins, 2012), and its role in the governance and behaviour of complex social-ecological systems had received attention from researchers and practitioners from different perspectives, including political ecology and ecological economics (Martínez-Alier, 2002), social metabolism (González de Molina and Toledo, 2014), geography, sociology and other disciplines. On a wider scale, conflict is an essential part of human social dynamics and a natural driver of change (Bruce and Holt, 2011). While society, academics and experts have often focused on violent conflicts (Stohl et al, 2017), many conflicts involve only symbolic violence or no violence at all. Low-violence conflicts can produce a wide range of both positive and negative outcomes for those involved, as well as third parties, which may influence social-ecological systems (Dahrendorf, 1959). If conflicts are not resolved, they can linger, hidden, with their social, economic and environmental impact unaccounted for (Palomo-Campesino et al, 2018). Latent conflicts, when neglected, tend to expand and escalate, often generating new risks and potentially harmful situations for all parties involved. Moreover, risks also increase as most conflicts usually develop on a subjective level, where perceptions become more important, and more influential, than the original grudge (Maser and de Silva, 2019).

Conflicts that arise are often interlinked to other conflicts, so they are difficult to individualize. They can occur at different political levels (local, county, region, country or international), within reach of different institutions and displayed at multiple scales, with different and simultaneous manifestations at different geographical, political and institutional levels (Aas Rustad et al, 2011).

Most conflicts have some economic logic behind them and are deeply influenced by social, political, environmental and other external factors. However,

DOI: 10.4324/9781003028437-12

environmental issues, which involve a wide range of interest groups, can also lead to conflict (Hipel et al, 2015). Environmental conflicts can and usually do involve both people and institutions, including either the implication of states or their failure (Sandole et al, 2009). The latest changes in global politics and environmental conditions (climate change, urbanization, globalization, etc.) have increased interest in these types of conflicts (Nie, 2003).

This chapter focuses on the influence of low violence conflicts on the governance and use of land and natural resources in southern Iberian silvopastoral systems. Their governance relies on a complex network of practices, institutions and management tools that affect access to and the use of land and resources, as detailed in Chapter 8. As such they are breeding grounds for conflicts, which can emerge whenever people, collectives or sectors hold incompatible goals and develop disagreements as a way of enforcing their position. European countries have secure land ownership rights and, as such conflicts around their governance usually rarely result in violence (Palmer et al, 2009) although they can remain active for a long period of time. Conflicts in silvopastoral systems are dynamic and variable, with different causes and patterns of evolution and can result changes in the existing power balance. Here we address the effect of governance-related factors, such as land rights, access or management decision-making over dehesa and montado (and similar) agroecosystems and how they can damage their governance, leading to sub-optimal performance and, possibly, their degradation. On the other hand, silvopastoral conflicts also can be considered as a creative and active element of innovation, to which participatory methodologies and multi-stakeholder approaches can be applied, limiting the risks and delivering positive outcomes.

11.2 Social-ecological conflicts in Mediterranean silvopastoral systems

Silvopastoral systems are complex, multifunctional and involve multiple actors (Castro, 2009), so they host a great variety of interests that might seed conflict (Plieninger and Huntsinger, 2018). Usually, stakeholders deal with those interests with minimal consequences for the system's performance, although clashes of interest can trigger dangerous confrontations. The list of possible conflicts in silvopastoral systems is long, including the use of infrastructures, access to reserved and protected areas or touristic sites, the use of resources (see Chapters 2 and 3), conflicts between uses (hunting, grazing, hiking or birdwatching), regulations (conservation rules, permissions, approval of new projects) or increasing risks (wildlife as disease vectors, risk of wildfires, working dogs and so on). While these conflicts can be complex and multifactorial, we propose a five tier classification, according to the competing interests and social groups involved in silvopastoral systems:

- Conflicts over land: Land is a limited resource and an economic asset that supports human activities, and it is a key issue for strategy and security. The governance of land holds economic and strategic value, political and cultural significance and great impact on identity, generating tensions and conflicts about ownership, control, use and access (Bruce and Holt, 2011). Pastoral lands often hold nested, flexible and/or customary rights (that often overlap with privative ownership), that can often give rise to specific conflicts (Davies et al, 2016)

- Conflicts linked to landscape change: European landscapes are changing profoundly, driven by combinations of political, institutional, cultural, and natural factors (Plieninger et al, 2016). There is a growing trend towards urbanization in northwestern Europe, generating a polarization of land (Primdahl et al, 2013) where also land abandonment is widespread, particularly in Mediterranean Europe (Sluiter and de Jong, 2007). This situation, often linked to agricultural expansion and intensification along with the industrialization and globalization of food flows, is affecting silvopastoral systems, deepening a wide set of ongoing conflicts and generating new ones. This scenario amplifies the tensions between silvopastoral systems and modern agricultural schemes and threatens to eliminate semi-natural landscapes, leading to the occupation of these tracts of lands and traditional tracks and fragmented access to pasturelands.

- Urban-rural conflicts: Trends towards urbanization have become global (Carlucci et al, 2016), with massive population shifts towards cities and a shift in rural lifestyles into urban ones. Urbanization has become a major public policy issue in southern Europe, as it provokes rural depopulation, undermining the viability of rural communities, changes in land use and habitat fragmentation and the growth of infrastructures and accessibility. The urban and rural worlds operate at different speeds, frequencies and magnitudes which generate many conflicts at the interfaces between the two worlds. The progressive dilution of rural values within an increasingly urbanized European society (Augére-Granier, 2016) leads rural people to feel threatened by urban interests and mistreated by policies, and this often manifests as urban-rural environment conflicts (Herrera et al, 2019), that have a clear influence on the governance of silvopastoral systems.

- Conservation and human/wildlife conflicts: The governance of Mediterranean silvopastoral systems is strongly affected by conservation conflicts which often generate mistrust and confrontation among stakeholders, creating controversy and polarisation. These conflicts can develop when two or more parties strongly disagree over conservation measures, which are perceived as damaging to the interests of one or more of them (Redpath et al, 2013). Human activities increasingly come into conflict with conservation objectives, creating conflicts that not only undermine effective conservation but also prevent the realisation of other social, ecological and economic objectives (Woodroffe, 2005). While all conflicts reflect power relations and differences in decision-making, human-wildlife relationships bring out differences in management that are deeply rooted in social and cultural history (Redpath et al, 2013). As such, protected and High Natural Value areas, such as dehesa and montado, are often become portrayed as arenas of conflict between people and wildlife, or economics and sustainability (Cortés-Vázquez et al, 2014).

• Climate-related conflicts: While the relationship between climate change and violent conflict is still under debate, climate variability does seem to generate conflicts around natural resources (Scheffran et al, 2012). Whereas these conflicts are not often explicitly violent, the influence of climate-related conflict on the governance of land and natural resources has been found to be influential in several case studies (Corbera et al, 2019). As climate change is progressively affecting the Mediterranean agroecosystems, including the dehesa and montado, climate-related conflicts related to water and land use could spread and need to be taken into account.

11.3 Addressing conflict in silvopastoral systems in Extremadura (Spain)

In this chapter, we illustrate some of the conflicts around silvopastoral systems by using a specific case study around conflicts and the sustainability of silvopastoral and High Nature Value Farming (HNV) systems in the county of La Vera, Extremadura (Spain)[1]. As in other silvopastoral areas, the mountains of La Vera are currently struggling with the abandonment of traditional activities and pastoralism, leading to a loss of biodiversity and different ecosystem services, increasing risks such as wildfires, and leading to social and economic losses in terms of employment, land management and sustainability.

The results of this project have been revisited using a conflict approach (Kóvacs et al, 2015) to review the projects' outcomes in order to try to understand how conflict affects the governance of the local silvopastoral systems and reveal the existence of several conflicts within the study area. Table 11.1. displays a matrix of potentially conflicting activities and interests collected from interviews with farmers active in HNV farming systems in and around La Vera. These areas were found to required innovation and improved management to maintain the HNV farmland, particularly to overcome simple conflicts (that can arise between any two parties over specific interests) and complex ones which can affect several parties and divergent interests simultaneously. The parties identified in the matrix correspond to some of the main stakeholders involved in the governance of local silvopastoral systems, which more or less coincide with the main actors described in Section A of this book: landowners, farmers (in charge of agricultural activities on the area, whether or not landowners), workers/employees contracted to work on the estates, other users (including tourists, hikers, naturalists, researchers, sportsmen and women, hunters, etc.), private companies and government bodies.

Farmers are central stakeholders in this matrix, as they are the main people responsible for managing the silvopastoral systems on a day-to-day basis. Table 11.2 characterizes the main conflicts and consequences that affect

Table 11.1 Matrix showing potential conflicts between stakeholders in silvopastoral systems in La Vera (Cáceres, Spain). Columns display the parties engaged in specific activities that may lead to conflict, while the rows display the parties suffering the consequences of such actions. As any party could be imposing or suffering from damage on others, the matrix is asymmetrical. Some of these tensions may not evolve into real conflicts while others (emboldened) are already developed and having strong effects on the abandonment of La Vera's silvopastoral systems

Active party/ recipient party	Landowners	Farmers	Workers/ employees	Users	Private companies	Government agencies
Landowners	Conflicts over boundaries Conflicts overuses	Over-exploitation Land quality loss Conflicts over boundaries	Low responsibility for the land, thereby Increasing risks	**Invasion, unauthorized uses, picking, increasing risks (e.g. wildfires…)**	Overexploitation	**Excessive bureaucracy, permissions, land-use planning, mandatory supervision**
Farmers	**Overpricing land, insecure contracts, land access, conflicts over resources, priority given to high-income activities, transport on public roads**	**Incompatible practices (e.g. GMO vs organic), conflicts over land use and the use of scarce resources**	Low availability, low qualification levels, loss of critical skills and local ecological knowledge	Disturbance, stealing products, unleashed dogs, damaged fences and open gates, general damage	Abusive contracts	**Land use regulations, sanitation campaigns, conservation framework, environmental rules overruling The CAP**
Workers/ employees	Low wages and poor working conditions	Low wages and poor working conditions	Precarious work contracts and a lack of organization	Disturbance	**Low wages and poor working conditions**	Inappropriate rules, inspections and fines

(*Continued*)

Table 11.1 (Continued)

Active party/ recipient party	Landowners	Farmers	Workers/ employees	Users	Private companies	Government agencies
Users	Banning access & fencing, conflicts over access and rights of way	Noise, odour, disturbance	Lack of consideration for other people's needs	Incompatible use, growth in user numbers and misunderstanding of rural ways	**Marketing practices that lessen the values associated with origin and quality**	Ruling access and land uses, banning, fines…
Private companies	Short time leases & insecure contracts	Low degree of formality in trade	Labour & market conflicts	Consumer behaviour	Competence	**Lack of transparency and accountability**
Government agencies	**Unauthorized uses**	**Animal health, and human–wildlife conflicts and unauthorized uses**	Labour & market conflicts	**Access and public use conflicts, unauthorized activities and conflicts over development projects**	Corruption	**Competence conflicts, overruling contradictory policies and a lack of coordination**

Table 11.2 Detailed description of sustainability conflicts in La Vera, focusing on the farmers' perspectives and perceptions

Political conflicts

Conflicts with government bodies related to legal and institutional framework, ruling and application of law

Factors generating conflicts with farmers	*Consequences for farmers and silvopastoral systems*
Legal and institutional framework that is ill-suited to silvopastoral systems	Undifferentiated production, low prices, lack of profitability, pressure towards intensification.
Application of controversial or contradictory biodiversity and nature conservation laws (zoning, banning uses, limiting access, etc.)	Poor access to pastures, under grazing, high costs, little recognition, misunderstanding, lack of acknowledgement of ecological benefits of traditional management practices, abandonment
Sanitary regulations that are not suited to silvopastoral dynamics (sanitation campaigns, cross contagion with wildlife, incompatible schedules...)	Disturbance of production, increase of workload, loss of livestock, pressure towards stabling, abandonment
Overwhelming bureaucracy (disrupting livestock mobility, sales, CAP payments, transport ...)	Less efficiency, focus on payments, difficulties for small farmers
Powerlessness against government actions	Frustration, legal expenses, time wasted, lower profitability
Lack of representation, voice and participation of farmers and the decline of local and traditional land-ruling institutions.	Exclusion from development and land planning regimes, imbalanced grazing (under and overgrazing), bad governance decisions, higher risks

Environmental conflicts

Conflicts with authorities and other users related to conservation, use and access to land and natural resources

Factors generating conflicts with farmers	*Consequences for farmers and silvopastoral systems*
Increasing difficulties of access to land and natural resources.	Increasing costs for farmers, lack of access to resources, lack of profitability, loss of land care
Increasing risks, such as wildfires and other risks	Misunderstanding, incoherence between farming and conservation goals, banning of pastoralism
Bad relationships with wildlife	Livestock predated, frightened or damaged, with associated pain, threat and grievance, lack of compensation and public support, pressure towards intensification
Problems with other outdoor users (hikers, hunters, crop farmers, forest managers, mining companies...)	Rejection by other users, banning or the marginalization of farming activities

(Continued)

Table 11.2 (Continued)

Economic conflicts

Conflicts with markets, consumers and value chain actors over costs, benefits and trade issues

Factors generating conflicts with farmers	*Consequences for silvopastoral systems*
Pressure from conventional markets and industry	Lack of profitability and added value, pressure to intensify, lack of generational turnover, abandonment
Lack of adaptation of market tools, low prices, inappropriate supply chains	Lack of differentiation and recognition, low sales, little interest from institutions and consumers.

Social conflicts

Conflicts with other groups of people about the image, support, culture, welfare, discrimination, etc.

Factors generating conflicts with farmers	*Consequences for silvopastoral systems*
Difficult human relationships (isolation, helplessness, lack of access to public services as health, education …)	Pressure to search for more secure livelihoods
Disputes with other stakeholders and government bodies deriving from the loss of infrastructures, collective work, and technical support	Landscapes become less suitable for silvopastoral systems, increasing pressure for intensification, lack of interest
Structural social factors such as renewal, ageing, loneliness, lack of social organization, inequality, masculinization and lack of gender equality	Loss of social support and influence, less power, weakened governance, lack of generational turnover
Different and poorly compatible lifestyles with urban people, neighbours, newcomers, visitors	Poor image, lack of understanding, marginalization, lack of social status, abandonment

farmers, and consequently the sustainability of silvopastoral systems. To compile this list, farmers were considered as aligned with or against other actors with competing interests. There are also other actual and potential conflicts between other stakeholders that may also affect those systems (e.g. governmental, internal or labour conflicts) although they are beyond the scope of this chapter. Finally, it is important to acknowledge that, although some of the conflicts have not actually become manifest in the La Vera area, some stakeholders feel involved in them (e.g. currently, there is no wolf predation in La Vera, but all extensive livestock farmers feel as harmed by the relationship with wolves and relate this conflict with other conflicts with predators and wildlife.

This list of potential conflicts might look excessively high, although many of them can be managed by farmers and other stakeholders as part of their jobs, thereby reducing their impact. Conversely, other situations can rapidly escalate and will need a more ambitious approach to soften their consequences. Market forces, for instance, drive silvopastoral products to compete undifferentiated with industrial production. Biodiversity rules sometimes limit grazing in conservation areas. Transhumant shepherds are forced into a bureaucratic nightmare. An emerging conflict, which is rapidly growing in intensity, is related to some diseases and sanitary campaigns, specially Bovine Tuberculosis (TB) and other diseases hosted by wildlife and transmitted between wild and domestic animals. This generates a complex situation as wildlife is increasingly expanding due to the abandonment of former grazing and cultivated areas, estates are managed to improve hunting and government action and sanitary campaigns are focused only on domestic animals, with a heavy cost for farmers. Consequently, the problem is growing and conflicts have been escalating for years. This is particularly relevant in areas where silvopastoral systems dominate, which are generally the more remote and marginal areas where land abandonment and shrub encroachment are a growing problem. Text Box 11.1. focuses on animal TB and the solutions that are still needed.

Such situations can lead to conflicts increasing, become aggravated or freezing over time and space (Young et al, 2005). Some of them, such as conflicts of coexistence with wildlife, are widespread, while some others, such as conflicts over sanitation campaigns, are very localized. Some conflicts can be internal to silvopastoral systems, while many others link those systems with the wider world, with non-residents, such as hunters or tourists. But, in general such conflicts undermine the governance and performance of active silvopastoral systems (Herrera et al, 2014), reinforcing the negative effects of current trends and leading these systems to abandonment. Even if the impacts are variable and difficult to account for, most stakeholders believe that they are challenging the sustainability of their livelihoods, while society is losing the public goods and services that they provide (Kóvacs et al, 2015). Changing these trends and restoring silvopastoral systems to a position where they can provide such goods, requires defusing and addressing those conflicts

TEXT BOX 11.1

Animal tuberculosis – the other side of the multifunctionality of *Montado*

Sara Santos and Eduardo Ferreira, University of Évora

Key points: The coexistence of high densities of extensive cattle production and wild ungulates prevents the eradication of animal tuberculosis in Portugal and Spain.

Animal tuberculosis (TB) is amongst one of the most widespread zoonotic diseases. It is caused by *Mycobacterium bovis* that belongs to the *Mycobacterium tuberculosis complex* (MTC) and can be transmitted within and between wildlife species and free-ranging cattle, in a multi-host system. It has high economic repercussions and direct implications on animal and public health. Despite considerable efforts to control and eradicate animal TB in Europe, it has not been eliminated from the Iberian Peninsula, posing constraints on cattle production and trade. The existence of wildlife maintenance hosts has been recognized as the main difficulty for its eradication. Recent studies suggest that the ongoing existence of the pathogen and transmission risk is largely driven by interactions among cattle and wildlife. In the Iberian Peninsula, wild ungulates such as wild boar (*Sus scrofa*) and red deer (*Cervus elaphus*) are considered the main reservoirs, but other potential hosts are also probable: the badger (*Meles meles*) or the red fox (*Vulpes Vulpes*). Thus, the pathogen's persistence and spread are dependent on the density of each potential host species and on inter-species contact rate (which is dependent on the ecology of each species).

In Portugal, the counties with the highest disease prevalence in wild ungulates and cattle are in Idanha-a-Nova and Moura-Barrancos, both within a *Montado* landscape, where large game hunting coexists with extensive cattle production. In these big game hunting estates, wild boar and red deer are maintained in high densities. To achieve desirable kill rates, supplementary feeding and baiting are common practices, increasing the use of shared space between domestic and wild animals. At the same time, the grazing systems are based on extensive regimes, in which cattle are frequently maintained all year round on pastures and supplied with supplementary food and water when needed. This makes these areas particularly attractive for wildlife, fostering increased shared space use and leading to more inter-species interactions.

Conflicts between farmers (especially those managing free-ranging cattle and goat herders), other stakeholders and government around TB has also intensified because farmers feel that they are carrying the main consequences of the disease, suffering eradication campaigns that lead to the slaughter of thousands of animals without accurate detection or adequate compensation payments (although meat from the slaughtered animals still reaches the markets undifferentiated) and having their animals' movements restricted. In the meantime, very few is done about wildlife or the hunting sectors, which manage game products and their activities with lower sanitary restrictions. This conflict is forcing farmers to either abandon or industrialize their farms, generating an acute feeling of mistreatment and defenselessness.

while aligning concerned stakeholders to work together and move in the same direction. This agreement-seeking approach is not the only option to deal with these conflicts, some authors even challenge it and propose alternative perspectives to address natural resource management conflicts, for instance enhancing the role of dissensus or re-adjusting power balance (Mooney and Hunt, 2009). One way or another, addressing these conflicts with a conflict resolution approach rather than addressing strictly technical problems has already achieved some success in La Vera's silvopastoral systems, giving farmers' a sense that their voices are being listened to and improving their social and political relationships.

11.3.1 Applying a conflict-solving approach to silvopastoral systems

Conflicts can be dealt with using different approaches. Table 11.3 displays these mechanisms as a continuum that extends from avoidance to coercion with both extremes representing the greatest losses and risks to the disputing parties. Recent research is currently questioning mainstream participation solely based on consensus and 'win-win' scenarios instead of integrating dissensus as a creative tool for integrative participation (Anderson et al, 2016). This approach is considered complementary to the treatment of conflicts based on agreements between stakeholders used in this case studio.

According to our experience, conflict solving in silvopastoral systems entails properly assessing the situation, improving the communication between stakeholders, maintaining positive attitudes, building capacity and trust and improving relationships between parties and dealing with both subjective (identity, perception, etc.) and objective (material interests, monetary outcomes, etc.) issues. These processes demand sound fieldwork and reliable information to understand local dynamics, especially the value systems of stakeholders. Thus, adopting a governance-based adaptive and integrative approach (McDougall and Banjade, 2015) to conflicts can be dealt with in a positive way for stakeholders and silvopastoral systems. This approach involves the use of participatory tools and evidence-based decision-making throughout the process. Solutions should be implemented under the responsibility and with the commitment of stakeholders, who are more likely to implement durable solutions if they have agreed upon them (Hipel et al, 2015).

Table 11.3 Conflict-solving continuum, adapted from Sidaway (2005)

Avoidance Maintenance of the status quo	Conciliation Compromise between conflicted interests	Decision Adjudication determining who is right	Coercion Adjudication determined by use of power
• Inaction • Indecision • Blockage	• Negotiation • Mediation • Facilitation • Arbitration	• Litigation • Binding arbitration	• Coercion

Negotiated conflict addressing techniques often involves third parties, who should be able to provide expertise, knowledge, and instruments to mediate in the conflict and facilitate solutions. Mediation must hold enough legitimacy, justice, accountability and acceptance to be recognized by the disputants.

Text Box 11.2 gives an example of how participatory approaches can be useful in addressing deeply embedded conflicts related to land and natural

---

TEXT BOX 11.2

Grupo Campo Grande, collaboration for coexistence

Pedro M. Herrera, Fundación Entretantos

Key points: Campo Grande Group, a participatory Spanish nation-wide think tank into the conflicts relating to the coexistence of wolves and pastoralism, has reached a set of agreements between all parties, showing how addressing the social perspective of conflicts can be a valuable approach for dealing with complex environmental problems.

Figure 11.1 Presentation of the Declaration of the Grupo Campo Grande in the Royal Botanic Garden of Madrid.

While there are often problems between people and wildlife, the social conflict arising between the groups of people involved in such conflicts can be stronger, more durable and more violent, to the point that it is damaging all interests involved, including livestock, wildlife and ecosystems. Defusing the conflict is key to any viable long-term solution (Herrera et al, 2019). This is especially the case with wolf-livestock coexistence in Spain which has developed into a polarized social conflict, replete with symbolism that now impacts right across rural society. Worried about this situation, the Entretantos Foundation initiated a social mediation initiative in 2014, whose main focus is the Campo Grande Group (CGG), a participatory, nation-wide, think-tank where people from contending sectors seek collaborative solutions (Figure 11.1). Grupo Campo Grande has used a methodological roadmap to deal with polarized conflicts by drawing priorities around social and subjective aspects. Mediation has been a key, establishing the neutral position of the facilitation team who act as both facilitators and mediators. Actions taken by GCG started with in-depth interviews and content analysis of discourses and continued by establishing a permanent ground for dialogue to pursue collective reflection and dialogue in a solution-oriented process. After almost four years, the group reached a set of agreements endorsed by participants, displaying fair, technically viable and socially acceptable measures helping to de-escalate the conflict (Herrera et al, 2019).

resources, even in complex systems such as silvopastoral ones. International institutions such as FAO are promoting this approach to deal with land tenure issues in pastoral lands (Davies et al, 2016) and institutions like the African Union (2010) advocate the direct participation of pastoralists in order to develop a political framework for pastoralism in Africa, specifically assigning them a role in actively managing the most conflict-ridden aspects, such as mobility and cross-border activities.

11.3.2 New governance for smart fire prevention

Wildfires are one of the biggest problems in Mediterranean ecosystems and can have catastrophic consequences on already declining silvopastoral systems and rural communities living in scarcely populated territories. The way that governments and policies are dealing with the problem is becoming problematic and generating management-related conflicts in the most fire-sensitive areas. The tension between land abandonment, fire extinction policies, lack of prevention and the acute consequences of mega-fires is adding to conflict in the governance of those areas, contributing to a progressive confrontation between local people and government agencies. Though these conflicts may arise in many ways and differ according to social-ecological settings, there is an overwhelming tendency in most territories to allocate most funds to fire extinction at the expense of integrated and participatory prevention measures. Innovative and participatory initiatives such as the Mosaico Project[2] and Ramats al Foc[3] are trying to reshape this situation by developing new governance frameworks for areas at high risk of wildfire in the Iberian Peninsula. The aim is to increase commitment and participation among local stakeholders to prevent the occurrence of large fires. These initiatives use participatory land-use planning and other tools to improve governance and create fire-resistant landscapes. Bottom-up land-use planning approaches that integrate forestry, agriculture and pastoralism into a sound landscape framework seeking to optimize the mix of land needed to maximize ecosystem services while reducing fire risk. In order for these initiatives to be up-scaled, new policy guidelines favouring 'profitable fire prevention' are urgently needed that will encourage farmers to maintain a lower fuel load in the forests

11.4 Conclusion

Conflicts are widespread in silvopastoral systems and disturb or undermine their governance, management, performance and their capacity to provide ecosystem services. Conflicts can often go unattended for a long time, only to escalate under certain conditions, potentially causing real damage. A few remarkable initiatives seek to address the latent conflicts as the first step to building capacity, and improving the resilience and long-term performance of silvopastoral systems. Conflict-solving and participatory approaches have proven to be a good alternative to deal with some of the most pressing problems of silvopastoral systems, by focusing on the confrontation and trying to de-escalate the tension as a first step to applying widely agreed measures. These approaches should be complemented with supportive research and innovative and more inclusive methodologies. Good facilitation and mediation are essential for the success of conflict-solving initiatives. It is also necessary to allocate sufficient resources, staff (teams of skilled, multi-disciplinary and well-trained professionals) and capacity of action to ensure their success. Finally, the communication and dissemination of results should valorize the agreements reached and solutions delivered.

Notes

1. This approach draws on results from a H2020 project (HNV-Link) and other research projects undertaken in the area during the last three years. The HNV-Link project aimed to improve learning and innovation in the area to try to reverse this downward spiral of land abandonment and the loss of HNV agricultural systems. See http://hnvlink.entretantos.org/ or www.hnvlink.eu
2. https://www.mosaicoextremadura.es/
3. https://www.ramatsdefoc.org/es

References

Aas Rustad, S. C., Buhaug, H., Falch, Å., & Gates, S. (2011). "All conflict is local: Modeling sub-national variation in civil conflict risk". *Conflict Management and Peace Science*, 28(1), 15–40.

African Union (2010). *Policy Framework for Pastoralism in Africa.* Department of Rural Economy and Agriculture, African Union Commission, Addis Ababa, Ethiopia.

Anderson, M. B., Hall, D. M., McEvoy, J., Gilbertz, S. J., Ward, L., & Rode, A. (2016). "Defending dissensus: participatory governance and the politics of water measurement in Montana's Yellowstone River Basin". *Environmental Politics*, 25(6), 991–1012.

Augére-Granier, M. (2016) Bridging the rural-urban divide. European Parliament Briefings. EPRS | European Parliamentary Research Service.

Behrendorff, L., Belonje, G., & Allen, B. L. (2017). "Intraspecific killing behaviour of canids: how dingoes kill dingoes", *Ethology Ecology & Evolution*, 30(1) 88–98, DOI: 10 .1080/03949370.2017.1316522.

Bruce, J. W., & Holt, S. (2011). *Land and Conflict Prevention. Colchester Initiative on Quiet Diplomacy.* University of Essex, Colchester.

Byrne, S., & Senehi, J. (2009). "Conflict analysis and resolution as a multidiscipline. A work in progress" in Sandole, D., Byrne, S., Sandole-Staroste, I., & Senehi, J. (eds.). *Handbook of Conflict Analysis and Resolution.* Routledge, London.

Carlucci, M. Grigoriadis E., Rontos, K., & Salvati, L. (2016). "Revisiting a hegemonic concept: Long-term 'Mediterranean urbanization' in between city re-polarization and metropolitan decline". *Applied Spatial Analysis*, 10(3), 347–362.

Castro M. (2009) "Silvopastoral systems in Portugal: Current status and future prospects" in Rigueiro-Rodríguez, A. et al. (eds.). *Agroforestry in Europe: Current Status and Future Prospects.* Springer Science+Business Media.

Corbera, E., Roth, D., & Work, C. (2019). "Climate change policies, natural resources and conflict: Implications for development", *Climate Policy*, 19(1), 51–57, DOI: 10.108 0/14693062.2019.1639299.

Cortés-Vázquez, J., Valcuende, J. M., & Alexiades, M. (2014). "Espacios Protegidos en una Europa en crisis: Contexto para una antropología del eco-neoliberalismo" in Prat, En J. (ed.). *Periferias, Fronteras y Diálogos*, Universitat Rovira i Virgili, Tarragona.

Dahrendorf, R. (1959) *Class and Class Conflict in Industrial Society.* Stanford University Press, Stanford.

Davies, J., Herrera, P., Ruiz-Mirazo, J., Mohamed-Katerere, J., Hannam, I., Nuesri, E., & Batello, C. (2016). *Improving Governance of Pastoral Lands.* FAO. Roma.

González de Molina, M., &Toledo, V. M. (2014). *The Social Metabolism. A Socio-Ecological Theory of Historical Change.* Springer International Publishing, Switzerland.

Herrera, P. M., Alonso, N., Sampedro, Y., Majadas, J., Sánchez, J. A., & Casas, V. (2019) "Social mediation initiative on the coexistence between Iberian wolf and extensive livestock farming". *Carnivore Damage Prevention News*, (18 Autumn 2019), 15–23, https://lciepub.nina.no/pdf/637423424258404840_CDPNews18.pdf

Herrera, P. M., Davies, J., & Baena, P. M. (eds.). (2014). *The Governance of Rangelands: Collective action for sustainable pastoralism.* Routledge, New York.

Hipel, K. W., Fang, L., Cullmann, J., & Bristow, M. (eds.). (2015). *Conflict Resolution in Water Resources and Environmental Management.* Springer International Publishing, Switzerland.

Kovács, E., Kelemen, E., Kalóczkai, Á., Margóczi, K., Pataki, G., Gébert, J., & Mihók, B. (2015). "Understanding the links between ecosystem service trade-offs and conflicts in protected areas". *Ecosystem Services*, 12, 117–127.

Manning, A., & Dawkins, M. S. (2012). *An Introduction to Animal Behaviour.* Cambridge University Press. Cambridge.

Martínez-Alier, J. (2002) *The Environmentalism of the Poor. A Study of Ecological Conflicts and Valuation.* Edward Elgar Publishing, Cheltenham, and Northampton, MA.

McDougall, C., & Banjade, M. R. (2015). "Social capital, conflict, and adaptive collaborative governance: exploring the dialectic". *Ecology and Society*, 20(1),44.

Maser, C., & de Silva, L. (2019). *Resolving Environmental Conflicts.* Taylor and Francis, Boca Ratón.

Mooney P. H., & Hunt, S. (2009). "Food security: The elaboration of contested claims to a consensus frame". *Rural Sociology*, 74(4),. 469–497.

Nie, M. (2003). "Drivers of natural resource-based political conflict". *Policy Sciences*, 36,307–341.

Palmer, D., Fricska, S., & Wehrmann, B. (2009). Towards improved land governance. FAO. Food and Agriculture Organization of the United Nations. United Nations Human Settlements Programme.

Palomo-Campesino, S., Ravera, F., González, J. A., & García-Llorente, M. (2018). "Exploring current and future situation of Mediterranean silvopastoral systems: Case study in Southern Spain". *Rangeland Ecology & Management*, 71(5), 578–591.

Plieninger, T., Draux, H., Fagerholma, N., Bieling, C., Burgid, M., Kizose, T., Kuemmerle, T., Primdahl, J., & Verburgg, H. (2016). "The driving forces of landscape change in Europe: A systematic review of the evidence". *Land Use Policy*, 57, 204–214.

Plieninger, T., & Huntsinger, L. (2018). "Complex rangeland systems: Integrated social-ecological approaches to silvopastoralism". *Rangeland Ecology & Management*, 71(5), 519–525.

Primdahl, J., Andersen, E., Swaffield, S., & Kristensen, L. (2013). "Intersecting dynamics of agricultural structural change and urbanisation within European rural landscapes: Change patterns and policy implications". *Landscape Research*, 38, 799–817.

Redpath, S. M., Young, J., Evely, A., Whitehouse, A., Sutherland, W. J., Arjun, A., Lambert, R. A., Linnell, J. D., Watt, A., & Gutiérrez, R. J. (2013). "Understanding and managing conservation conflicts". *Trends in Ecology & Evolution*, 28(2), 100–109, https://doi.org/10.1016/j.tree.2012.08.021.

Sandole, D., Byrne, S., Sandole-Staroste, I., & Senehi, J. (eds.). (2009). *Handbook of Conflict Analysis and Resolution.* Routledge. London and New York.

Scheffran, J., Brzoska, M., Kominek, J., Link, M., & Schilling, J. (2012). "Climate change and violent conflict". *Science*, 336, 869.

Sidaway, R. (2005). *Resolving Environmental Disputes: From Conflict to Consensus*, EarthScan. London.

Sluiter, R., & de Jong, S. M. (2007). "Spatial patterns of Mediterranean land abandonment and related land cover transitions". *Landscape Ecology*, 22, 559–576.

Stohl, M., Lichbach, M. I., & Grabosky, P. M. (2017). *States and Peoples in Conflict. Transformations of Conflict Studies.* Routledge. London and New York.

Woodroffe, R., Thirgood, S. J., & Rabinowitz, A. (eds.). (2005). *People and Wildlife: Conflict or Coexistence?*, Cambridge University Press, Cambridge.

Young, J., Watt, A., Nowicki, P., Alard, A., Clitherow, J., Henle, K., Johnson, R., Laczko, E., Mccracken, D-., Matouch, S., Niemela, J., & Richards, C. (2005). "Towards sustainable land use: Identifying and managing the conflicts between human activities and biodiversity conservation in Europe". *Biodiversity and Conservation*, 14, 1641–1661.

12 The governance of collective actions in agro-silvopastoral systems in Tunisia

A historical institutional analysis

Aymen Frija, Mariem Sghaier, Mondher Fetoui,
Boubaker Dhehibi and Mongi Sghaier

12.1 Introduction

In North African countries, public authorities have been aware of the extent of the challenges to be met by silvopastoral systems in terms of ecosystem management, food security, territorial governance and solidarity-based economic development. This has resulted in many institutional reforms during recent decades (Nefzaoui, 2004; Banque Mondiale, 2010; DGF & Banque Mondiale, 2015; Bourbouze, 2018). However, a lack of proper understanding of the processes of natural resource degradation has led to poorly informed interventions and policies that have sometimes exacerbated degradation. Agro-silvopastoral areas in these countries are subject to conversion to crop cultivation, over-exploitation by livestock, over-extraction of woody biomass by uprooting woody species for charcoal production and increased aridity due to both climate change and extraction of water (Fuhlendorf et al., 2012; IPBES, 2018; Prince et al., 2018). In Tunisia, population growth during the last century has exerted high pressure on the environment (Abaab & Guillaume, 2004). This is particularly true in the agro-silvopastoral areas of southern Tunisia where overgrazing, biodiversity losses and a reduction of pastoral productivity and the socio-economic fragility of the local population have been observed. In Tunisia, forest ecosystems and rangelands occupy about 5.6 million ha (35% of the whole country's surface). Rangelands represent 80% of this area and occupy 4.5 million ha, mostly in arid zones (DGF & Banque Mondiale, 2015). About 14% of the country's population lives in forest or rangeland areas. Most of these households are considered to be poor and are heavily dependent on silvopastoral activities, which provide 30 to 40% of their income, 15–25% of the feed needs for their livestock (this value was 65% in the 1960s, according to Nefzaoui, 2004) and 14% of household energy needs. These systems generated around 14% of agricultural GDP and 1.3% of the country's GDP in 2012 (DGF & Banque Mondiale, 2015).

In fact, about 3.7 million ha of this total rangeland area are located in the six arid governorates of South Tunisia and receive less than 200 mm of rain annually, on average. This southern part of the country also contains about

DOI: 10.4324/9781003028437-13

800,000 ha of arable land and about 17,000 ha of forest-shrub plantations (mostly integrated within rangelands or around arable lands and oases to protect them from wind erosion). Almost all the collective land in the arid regions of Tunisia is classified as rangeland. Collective lands are defined in Tunisia as elusive property, imprescriptible and owned in common, under the administrative control of a group, each head of family having the right only to a share of legal enjoyment (Ben Othman, 2014).

Awareness about the sustainability threats to these collective pastoral resources is gradually increasing at the national level. This is partly influenced by a global awareness of the international community about the importance of silvopastoral ecosystems and the wide range of ecosystem services that they provide to local communities (Ash et al., 2012; Fuhlendorf et al., 2012; Herrera et al., 2014; UNEP, 2016; Liao & Fei, 2017; IPBES, 2018; Prince et al., 2018). At the same time, we are witnessing a quiet and consistent evolution of the governance modes of these spaces in Tunisia, with increasing public recognition of the need to involve land right-holders' and communities in decision making regarding the management of the resources. Traditional governance modes and customary laws regulating access, use and management have been disappearing and replaced by formal rules, with the expressed aim of increasing the sustainability of forestry and rangeland resources. However, only one-third of the rangelands are currently "governed" by the Forest Code (law), involving strong and direct state control and adequate pastoral planning and rational management (e.g. fallowing and rotation) (DGF & Banque Mondiale, 2015; Gamoun et al., 2018). The remaining collective rangelands are governed by farmers' organizations and generally show poor performances in terms of both collective action and sustainability.

Tunisia adopted a new constitution in January 2014, which is committed to decentralization as a new form of territorial governance, with local authorities mandated to decide about local urban and rural planning and resource management issues. The implementation of this new form of decentralization should enable local communities to become more involved in strategic territorial development, which has clear relevance for pastoral and rangeland areas. But this also means that other stakeholders could become involved in rangeland management, adding more challenges for effective governance mechanisms.

The objective of this chapter is to describe the historical and current institutional dynamics and changes that have occurred in south Tunisia during the last century. These changes have led to the loss of customary land governance modes, as well as the emergence of new governance models. The chapter will also present and discuss current governance failures and their causes and report on policies that could potentially enhance the role and sustainability of rangelands in Tunisia. The lessons learned from this analysis may also apply to other pastoral areas in North Africa, especially Algeria and Morocco.

12.2 The characteristics of silvopastoral systems in south Tunisia

Several types of agro-silvopastoral ecosystems exist in Tunisia. (A map showing the main agro-silvopastoral ecosystems in the country can be found in DGF & Banque Mondiale (2015). The northern systems are dominated by forests and pastures integrated within larger areas of arable crop and benefit from a relative abundance of rainfall (above 400 mm/year on average and reaching up to 1000 mm in certain mountainous areas in the northwest regions), which allows for a relatively high annual production of biomass. The second type of system is the steppe areas of central Tunisia which receive an average of 200 to 400 mm of rain annually. This type is characterized by the abundance of scrubland and shrubs well adapted to drought and forms a large steppe area grazed by livestock, in addition to some niches of irrigated agriculture and forests.

The third type is the southern "agro–pastoral systems" (mostly in the six governorates of the South: Gafsa, Gabes, Kebili, Tozeur, Medenine and Tataouine) which occupy around half of Tunisia's territory, with a mean annual rainfall of between 50 mm and 250 mm. These are mostly rangelands with sparse tree and shrub cover (see Figures 12.1 and 12.2). In these areas, most of the forest trees have been planted to protect arable land and oases from desertification (Figure 12.2), or to provide a source of animal feed

Figure 12.1 Silvopastoral landscape in central Tunisia (Haddej, in Sidi Bouzid governorate).

Source: Illustrative photos of agro-sylvo-pastoral systems from central Tunisia (IRA-ICARDA credit).

Figure 12.2 Extensive olive plantations in the southern rangelands of Tataouine in south Tunisia.

Source: Illustrative photos of agro-sylvo-pastoral systems from southern Tunisia (IRA-ICARDA credit).

(in the case of shrubs). This chapter focuses on this type of agroecosystem in which a few niches of silvopastoral systems continue to exist. A first sub-type is characterized by the existence of the Carob (*Ceratonia siliqua*), Eucalyptus and Argan (*Argania spinosa*) (both introduced species) Rosemary (*Rosmarinus*) and Esparto grass (*Stipa tenacissima*) with agriculture behind *Jessours*[1] (typical of Matmata in Gabes governorate). Another system (notably in the area surrounding the national park of Bouhedma) is characterized by the existence of *Acacia tortilis*, steppe and cereals and legumes. The last type can be found in Gafsa governorate (around the national park of Orbata) with forest species such as *Pinus halepensis and Juniperus phoenicea* alongside rangelands and small olive groves with other crops.

Figures 12.1 and 12.2 show the distribution of agricultural land in southern Tunisia across rangelands, forests and arable cultivated areas. They show a strong dominance of rangelands and tree crops (mostly olive plantations).

12.3 The dynamics of rangeland socio–ecological systems

12.3.1 Transition from pastoral to agro–pastoral systems

Traditionally, pastoral production systems in southern Tunisia were based on the use of large rangeland areas including transhumance to central regions and even neighbouring countries (Algeria and Libya).

The most significant land-use change recorded since the 1950s in this area has been from rangeland to cropland and from a completely autonomous pastoral system based on mobility to the more sedentary system dependent on external feed resources. This was coupled with the transformation of agricultural systems, mainly from pastoral breeding to intensive forms of irrigated and rain-fed agriculture (Nefzaoui & El Mourid, 2008). These changes were mainly induced by agricultural development policies, the privatization of collective land, the reduction of pastoral resources and the limitations of transhumance and the mobility of herds and humans. With the improvement of rural infrastructure and increased settlement, agro-pastoral activities started to concentrate in areas close to urban agglomerations that were using poor grazing management of neighbouring areas, which accelerated degradation. This was aggravated by the lack of proper access (in time and space) to biomass resources in the vast areas of collective rangelands.

This transformation process accelerated in the 1960s–1970s and up to the 1990s due to demographic growth and policy changes. One of the most significant policy changes was related to the "land privatization policy" adopted in 1964, which was accelerated at the beginning of the 1970s. The land privatization process aimed at allocating collective land to individual tribe members with the objective of endowing people with valuable land titles that could be used as security with banks in order to get better access to credit (Ben Saad, 2011) and increase their investments.

The partial privatization of the collective rangelands reduced their size and led to the introduction of new agricultural activities that limited the traditional mobility patterns of some pastoralists (see Selmi & Elloumi, 2007 for an assessment of the impact of privatization of collective lands on resource usage and production systems). At the same time local agricultural systems were increasingly integrating cereals and tree plantations (mostly olives) with livestock rearing (Figure 12.3).

Overall, the sheep and goat populations more than doubled between 1964 and 2010, while the rangelands area has declined by more than 25% in the same period (Banque Mondiale, 1995).

The emergence of this new "agro-silvopastoral system" was accompanied by the conversion of the most fertile grazing lands to cereal, olive, or almond production and establishing (almost) year-long grazing of the remaining rangelands. These changes contributed to the disappearance of traditional governance institutions and affected the way rangelands were accessed and used, including more overgrazing, cultivation and the cutting of natural vegetation and deforestation (IRA-IRD, 2003; Sghaier & Fetoui, 2006), all of which exacerbated desertification and consequently reduced biotic resources.

12.3.2 Drivers of rangeland socio-ecological degradation

It is widely recognized that rangeland degradation is the consequence of many factors, with the most important being excessive grazing (Prince et al., 2018).

Figure 12.3 Resting rangelands using fencing.
Source: Mouldi Gamoun, ICARDA.

Rangeland degradation may result in reduced livestock carrying capacity, which will have an aggregated effect on communities' livelihoods if alternative livelihood options are not available. The Intergovernmental Science-Policy Platform on Biodiversity and Ecosystem Services (IPBES) report stated that urgent and concerted actions need to be taken to cope with this dramatic situation (IPBES, 2018). In Tunisia, as in other regions in the Mediterranean basin, biophysical and socio-economic factors interacted and reinforced each other and induced events that triggered a transition of rangelands.

Policy failures in Tunisia have also led to weak resource rights and governance, the weak influence of rangeland stakeholders (users such as pastoralists and agro-pastoral communities, in addition to organizations and administrations dealing with rangeland management who were largely marginalized) and insufficient or inaccurate data, information and knowledge sharing at different levels (Mortimore et al., 2009; Davies et al., 2017; Tumur et al., 2020). Population growth during the past 40 years has triggered a dramatic increase of livestock in the country.

Changes in land-use patterns had socio-economic impacts that affected sustainable management in the silvopastoral territories. The privatization policy resulted in conflicts among disadvantaged pastoral communities and a lack of spatiotemporal access to natural resources for indigenous and pastoral groups. These changes added more pressure to an already frag-ile system due to the lack of quantity and quality of water. The growing population increased the demand for land for settlement purposes. These changes have led to a fragmentation of spaces and increased pressure on the environment (due to the increased needs and changing lifestyles of pastoral societies).

12.4 Actors and institutions involved in rangeland management: a difficult transition from traditional to new modes of governance

12.4.1 The case of Miâad: a self-governance tribe-based management system

As in other south Mediterranean countries, customary land tenure systems and rangeland governance in Tunisia are very old and have traditionally played an important role in reconciling land use and livelihood patterns (Rignall & Kusunose, 2018). In southern Tunisia, the pastoral economy was domi-nated by insecure environmental and human conditions which obliged these populations to remain united, thus explaining why most of the rangeland was collective. At the beginning of the 20th century, rangelands in Tunisia were strictly managed and controlled by traditional institutions, called *Miâad*, which enjoyed effective power due to their confirmed social status and lead-ership. The *Miâad* was an informal customary institution composed of tribe leaders and ensured the management of collective rangelands and common ownership by making decisions on the opening and closing dates of different rangeland areas for each year. The *Miâad* also intervened in the social life of the tribe, organized movements within the tribal territory, decided on the pastoral transhumance and, in particular, in addition to other technical decisions, ensured the management and respect of the rested area, called *Gdel* in southern Tunisia (see Text Box 12.1 about reviving *Gdel* through pastoral investment projects in Tunisia). In Morocco and Algeria, the term used is *Agdal* which is a community territory whose natural resources are protected and defended. This practice, inherited from a distant past, is omnipresent in *Amazigh* societies in the Maghreb and the Sahara (Auclair, 2012). In agro-nomic terms, the *agdal* can be defined as a collectively agreed prohibition on the extraction of a given (generally vegetative) natural resource within a delimited space, during a certain time frame (Dominguez & Benessaiah, 2017). This technique was also practised in southern Europe where some initiatives to revive it are observed in countries, such as Greece (see Text Box 12.2).

Due to poor and variable rainfall distribution across these rangelands, dif-ferent tribes had different arrangements with other tribes concerning mutual access to their respective rangelands during drought years. The *Miâad* also decided about alliance relationships and intertribal conflict resolution; for

---

TEXT BOX 12.1

Community-based rangeland management in Tataouine, south-eastern Tunisia: Taking successes in land restoration (*Gdel*) to scale

Aymen Frija, ICARDA; Mongi Sghaier, IRA; Mondher Fetoui, IRA; Mariem Sghaier, IRA; Farah Ben Salem, IRA

Key points: There is a revival of the rangeland resting practices (Gdel) in northern Africa.

The rangeland resting technique in southern Tunisia (Tataouine), also locally known as Gdel, is carried out by pastoralists' organizations (local CBOs also called GDA, or Groupement de Développement Collectif, in Tunisia) and coordinated by the technical services of governmental agencies represented by the Regional Administration for Agricultural Development (CRDA), the Livestock and Grazing Office (OEP), the local populations represented by the Land Management Council (LMC) and agriculture development groupings (GDA). The approach is very well known by pastoralists in the region and was inherited and transmitted through customary traditions and local knowledge. Since the 1990s, the approach has been encouraged by technical services of OEP in private rangelands and was extended in the 2000s to collective rangelands. The pastoral population and the local CBOs (GDA and LMC) are strongly involved through a participatory approach (which originally was facilitated by an IFAD investment project called PRDESUD I and then PRODESUD II and PRODEFIL).

The approach consists of fallowing (leaving land without grazing) part of the rangeland for a period of time with the aim of reconstituting the vegetation cover. The resting period can be around 2–5 years depending on the ecosystem's capacity to recover and on climatic conditions. The process usually gives good results in terms of regeneration of vegetation in arid and even desert areas. This technique has been widely practiced in traditional rangeland management systems, evidence of the richness of the indigenous knowledge of pastoralist societies in arid and semi-arid regions. However, this practice was gradually abandoned by pastoral societies due to profound socio-economic changes in nomadism and ancient pastoralism systems. The trend in Tunisia today is towards reviving this technique through appropriate policy and organizational settings, in addition to advanced technical research about partial opening for grazing in rainy years or even seasons instead of fully closing of rangelands for years.

---

example, decisions concerning permission for breeders from one tribe to access the collective rangeland owned by another. The *Miâad's* decisions were strictly respected by all the tribes and, once the decisions were taken, pastoralists could not overturn them and sanctions would be applied to those who over-rode these rules (Nefzaoui et al., 2008). Stakeholders involved in rangeland management at this time were only tribe members, with limited interactions with other tribal representatives in the case of the need to make joint agreements or for conflict resolution.

TEXT BOX 12.2

The collective-democratic governance of pastures in a village in central Greece

Athanasios Ragkos, ELGO – Demeter, Agricultural Economics Research Institute and Stavriani Koutsou, International Hellenic University (Greece)

Key points: A bottom-up that is helping to maintain the tradition of collective rangeland management in Greece based on carrying capacity (per farmer) and grazing period.

Until the beginning of the 20th century, summer rangelands in Greece were managed collectively and democratically by their users (following a bottom-up approach), taking into account the grazing capacity of the rangeland and the number of grazing animals. Later and until now, socio-economic developments and policies have contributed to the change of management from collective to individual, with significant impacts on rangeland quality. However, there are very few cases where rural mountain communities continue to manage their pastures based on traditional knowledge on their own initiative.

A typical example is rangeland allocation in Neochori Ipatis (central Greece). The management system consists of the following steps: The President of the community calls sheep farmers to move their flocks to the village on the first Sunday of June (not earlier) for the distribution of rangelands. Farmers gather in the central square of the village, after Sunday prayers. The President invites each farmer to sign a statement regarding the size of his flock while placing his hand on the gospel. Then the President presents the list of available community rangelands, which are not mapped, but which have specific boundaries and the farmers are well aware of their location and grazing capacity. The President then calls on the farmers to declare one by one in which pasture they wish to graze their animals. During the distribution, the basic principle that larger flocks receive larger rangelands is followed. Consequently, if one farmer has more animals than another who preceded and has already chosen particular rangeland, he displaces the latter and takes the rangeland. This distribution is valid until 15th August. After that date, all flocks are free to graze everywhere. This traditional system ensures the sustainability of rangelands, the balance between the number of animals and the food available and at the same time satisfies the farmers' sense of justice.

12.4.2 From Miâad to formal community-based organizations, passing through councils of notables and elected "land management councils"

The main historical institutional milestones of rangeland governance in Tunisia are illustrated and characterized in Figure 12.4 and Table 12.1. The *Miâad* was replaced, for security reasons, in 1918 by strictly controlled councils of notables, and then by elected collective land management councils

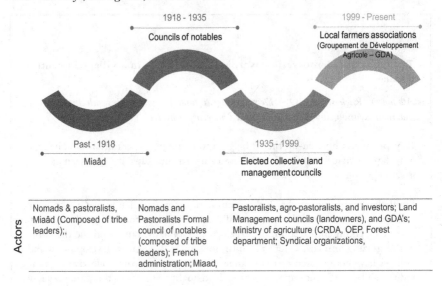

Figure 12.4 The most significant rangeland governance institutions that operated during the 20th Century in southern Tunisia.

(CLMCs) in 1935 (see Table 12.1 for more description). Under the law of 14 January 1971 (Ben Saad, 2011) the Tunisian Government revised the missions and roles of the CLMCs giving them more responsibility, which included all the "collective" (community) acts. The members of these local institutions were elected by the community for a five-year period. The CLMCs had several missions, the most important being privatization of collective lands. They also included promoting rangeland development and improving the social and economic conditions of community members. The councils became responsible for organizing and planning the use of rangeland areas, managing rest periods for rangeland sand other restoration techniques and dealing with land tenure conflicts (Mares & Lahmayer, 2019). However, several factors (e.g. administrative supervision, lack of resources and skills, poor governance, the non-participation of local populations, conflicts between rich and poor and between farmers and pastoralists) combined to limit the role and the achievements of these management councils.

The end of the 1990s was marked by another very important institutional reform related to the creation of local farmers' associations (Law No. 99-43 of May 1999) currently called GDAs (Groupement de Developpement Agricole, essentially community-based organizations (CBOs)). The GDAs were created in addition to the already existing CLMCs, which continued to operate as landowners' bodies and continued to be consulted for issues related to land tenure and ownership. The GDAs were set up to manage collective and private rangeland resources (under collective land tenure systems), which are mostly owned by members of the same tribe or neighbouring tribes. This reform was part of the policy of disengagement of the state in order to give a

Table 12.1 List of main stakeholders involved in rangeland restoration under different land tenure systems in South Tunisia

Stakeholder	Category/level of intervention	Role	Land tenure where the actor is intervening
Collective Land Management Council (CLMC)	Development/ local	Facilitating dialogue between the technical services, authorities and local communities. An important role in rangeland management.	Collective rangelands
Regional Administration for Agriculture Development (CRDA)	Development/ regional	Planning, implementation, monitoring and evaluation of agriculture, rangeland management and restoration.	Collective & private rangelands
Local delegate administrations	Political, decision-maker/ regional and local	Represented by the governor at the regional level and the delegate at the local level. They play a crucial role in collective land management and rangeland restoration.	Collective rangelands
Rangeland Research Institution (IRA)	Research/ regional	Research institution giving its scientific and technical support to all stakeholders with an interest in rangeland restoration.	Collective & private rangelands
Tunisian Union of Agriculture and Fishing (UTAP)	Syndical/local	Plays a syndical role and gives support to the CBOs in the region. It plays an important role in rangeland restoration.	Collective rangelands
Livestock and Grazing Office (OEP)	Development/ regional	Non-administrative public enterprise responsible for the development and promotion of the livestock and pasture sector also acting as adviser and technical reference point for public authorities.	Private rangelands
Agricultural Development Grouping (GDA)	Collective resource management/ local	A formal structure involved in agricultural and pastoral areas that plays a specific and central role in the participatory management of collective rangelands.	Collective & private rangelands

Source: Own elaboration based on FGD results (2019).

greater role to local populations in rangeland governance and in investing in local resource management and restoration. The GDAs are involved in agricultural and pastoral areas and therefore play a central role in the participatory management and mobilization of collective actions in their respective rangelands. The bundle of rights and responsibilities assigned to the GDAs resulted in a gradual marginalization of the CLMCs by both public administration and rangeland users. In some cases, there is overlap between the missions of GDAs and CLMCs resulting in governance failures in terms of GDA–CLMC coordination. For example, the development and implementation of the community development plan is the responsibility of the GDA, while decisions relating to land use and access are with the CLMC. This decisional dichotomy in the same territory has caused conflicts between the two institutions and led to the blocking of pastoral development activities, particularly in collective rangelands. This type of failure is less pronounced when there is an overlap in the membership of the boards of both organizations.

The GDAs are usually led by breeders' representatives. They provide services to member pastoralists who use rangelands, such as the development of relevant grazing infrastructure and the monitoring and maintenance of watering points (Ben Saad, 2011). These structures are supported by a set of other public, private and collective institutions, which intervene directly or indirectly in the governance of collective rangeland (Table 12.1). Some of these institutions are public technical and research agencies, and others are private entities dealing with financing agricultural activities or investing in public goods and infrastructure.

12.4.3 An analysis of changes in rangeland actors and governance attributes

Rangeland governance in Tunisia has been characterized by a shift from being based on a few users and unique decision-making bodies (under the *Miâad* and CLMC) to a more complicated scheme involving a wide diversity of actors/stakeholders, whose tasks and mandates sometimes overlap yet which have relatively few interactions or coordination mechanisms (Table 12.2). There is also a lack of sanctions and their level of enforcement (Table 12.2 and Figure 12.5). More facets of governance failures are shown in Figure 12.5.

12.5 Recommendations for enhancing rangeland governance

This section provides some guidance about the pathways and actions needed to improve rangeland management in Tunisia, based on the assessments conducted in the previous sections. The following recommendations are related to each of the governance attributes analyzed in Table 12.1.

12.5.1 Intervening actors and interests

There is a clear need to review who should be intervening in rangeland management in Tunisia. This would probably lead to a reduction in the number of public and collective participants, some of them being merged, or even the creation of new institutions that could help with coordination at different levels and play a lead role in local decision-making. More thought and effort should be given to ensuring broader consultation of stakeholders and to harmonizing the institutional landscape so as to reduce conflicts of interest. The regulatory powers and the functioning and efficiency of CLMC and GDAs, need to become more organisationally effective ensuring: (i) more

Table 12.2 Rangeland governance characteristics in the different historical periods

Type of management	Land tenure system	Actors involved in use and management	Regulatory agent	Conflict resolution mechanisms (level of enforcement)	Management performance indicators (who is assessing them)
Miâad (Past–1918)	Collective rangelands	Tribal leaders and members. Other users from different tribes	Tribe notables are the core governing body that decides about all rangeland use and management	Exist (strongly enforced)	State of vegetation cover (assessment based on local traditional knowledge)
Councils of Notables (1918–1935)	Collective rangelands	Tribal leaders, colonial administration	Under the supervision of colonial administration	Exist (strongly enforced)	State of vegetation cover (assessment based on local traditional knowledge)
Elected Collective Land Management Council (CLMC) (1935– Present)	Collective rangelands	Elected tribal members, national government	CLMC under the supervision (in the framework) of rules settled by the national administration	Exist (moderately enforced)	State of vegetation cover (assessment based on local traditional knowledge and recently done by the Livestock and Pasture Agency - OEP)
Local farmers' associations (GDAs) (1999– present)	Collective and private rangelands	CLMC, farmers' associations, breeders' representatives, individual users, a few governmental agricultural (livestock and forest) departments	GDA, under the supervision of CLMC and with clearance from different public agricultural administrations	Ambiguous and sometimes lacking (weakly enforced)	Organizational and financial performances of the GDA (public administration); state of vegetation (public livestock department) and still some existing assessment based on local traditional knowledge

empowerment of local institutions, particularly regarding their respective mandates, status, structures and operational rules/tools; and (ii) strengthening the involvement of local actors and community organizations in the processes of planning, monitoring and managing pastoral resources and ecosystem services offered by rangelands.

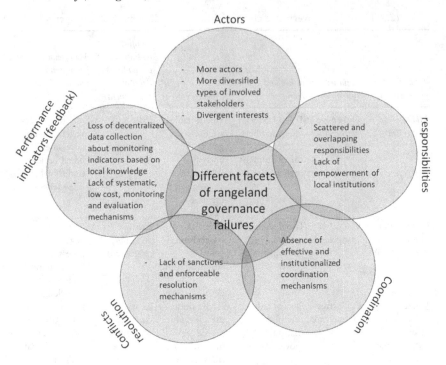

Figure 12.5 Different facets of current rangeland governance failures.

12.5.2 *Re-defining and clarifying responsibilities*

It is important to avoid the confusion created by the overlapping rights and responsibilities of some of the existing institutions currently involved in rangeland management, such as the GDAs and CLMC. It is also important to further empower local organizations (with technical and management aspects) and public administrations (with additional technical and financial resources) on specific aspects that can help them achieve their objectives. It is not enough to give someone a responsibility – it is equally important to provide the necessary resources to be able to handle these responsibilities. Capacity building of local institutions in the field of good rangeland governance is necessary to increase the effectiveness of these institutions for real and effective participation in decision making.

12.5.3 *Enhancing coordination mechanisms*

The creation of effective coordination mechanisms (or even additional coordination entities if needed) is necessary to enhance the implementation of activities, leverage the efforts of different stakeholders and reduce their divergences. Such an institution could be established by merging some of the existing administrations and agencies. It is also important to note that rangeland governance should be tackled from a wider perspective on pastoral development, and this can only happen through better coordination with relevant actors involved in the economic development policy of pastoral territories. The evolution of pastoral areas, with the transition to agropastoralism and the emergence of new economic activities, requires new forms of development based on enhanced social capital (communication and coordination) with the aim to reduce pressure on rangeland resources.

12.5.4 Conflict resolution mechanisms

This aspect is very important especially in the context of collective rangelands, where free-riding of local rules is a frequent phenomenon (Fuhlendorf et al., 2012). In Tunisia, the formalization of local farmers' associations led to the disappearance of traditional social sanctions and was not accompanied by formal sanction mechanisms that could be effectively applied to free-riders.

12.5.5 Monitoring and performance indicators

The Tunisian government needs to develop and implement an effective monitoring and evaluation tool of the technical, managerial and financial aspects related to local rangeland management performance. This will help to better monitor and assess degradation levels, and also identify information and capacity development gaps that need to be plugged in to further empower those actors engaged in rangeland management.

Some of the above recommendations can be achieved by improving the legal and regulatory framework. The legal and regulatory framework that organizes institutional, economic, social, technical and environmental aspects of rangelands in Tunisia is weak, and it is imperative to start developing and implementing a pastoral/rangeland code (a set of laws and regulations). This rangeland code should be accompanied by the design and implementation of proper rangeland strategies and related investment actions. In addition to the regulatory framework, the economic situation and livelihoods of pastoral communities need to be strengthened as this will contribute to reducing pressure on rangelands.

12.6 Conclusion

This chapter summarizes the historical institutional dynamics concerning rangeland governance in Tunisia with a summary of their drivers and implications. Agro-silvopastoral dynamics and the trend towards agricultural exploitation of privatized rangelands have created new alternatives for local populations to improve their livelihoods and reduce their vulnerability to climate variability and change. Nevertheless, this has also caused new challenges for sustainable development which requires better governance of natural and mainly pastoral resources.

The chapter also provides an overview of the major governance failures currently observed and depicts their causes. More work is required to further empower rangeland governance institutions in Tunisia. Governance systems should particularly be more inclusive in terms of partnerships, better attributions of responsibilities and related accountability, enhanced local coordination skills of leader stakeholders, explicit and enforceable conflict resolution mechanisms and well-designed performance indicators for enhanced monitoring, evaluation and accountability in the pastoral areas. The case of Tunisia is very illustrative of other similar pastoral cases in North Africa and so the results and recommendations of this study can also be applied to other pastoral areas in the region.

Acknowledgement

This work was undertaken as part of the CGIAR Research Program (CRP) on Institutions, and Markets (PIM) (under Flagship 5 on Governance of Natural Resources) led by the International Food Policy Research Institute (IFPRI). The work was also partly supported by CRP Livestock (led by ILRI).

Note

1. Small dyke structures within water catchment areas. They are built from soil and allow for water harvesting and rainfed agricultural activities

References

Abaab A. & Guillaume H. 2004. "Entre local et global: pluralité d'acteurs, complexité d'intervention dans la gestion des ressources et le développement rural". *IRD Editions, Collection Latitudes*, 23, 261–290.

Ash A., Thornton P., Stokes C. & Togtohyn C. 2012. "Is proactive adaptation to climate change necessary in grazed rangelands?", *Rangeland Ecology Management*, 65: 563–568.

Auclair L., 2012. "Introduction: Un patrimoine socio-écologique à l'épreuve des transformations du monde rural". In Auclair L. & Alifiqui M. (coord.). *Agdal, Patrimoine Socio-Ecologique de l'Atlas Marocain*, Ed. IRCAM-IRD, Rabat, 23–71 pp.

Banque Mondiale. 1995. *Une Stratégie pour le Développement des Parcours en Zones Arides et Semi-Arides*. Annexe III Rapport technique –Tunisie, 162 pp.

Banque Mondiale. 2010. *La Génération des Bénéfices Environnementaux pour Améliorer la Gestion des Bassins Versants en Tunisie*. Preparé par Croitoru L., Daly-Hassen H., Cherni A., Sterk G., Bird N., Zanchi G., Frieden D. & Oka A., Tran L. Rapport no. 50192-TN.

Ben Othman H. 2014. *Pour une Nouvelle Stratégie de l'Habitat, l'Accès au Foncier. Diagnostics et Recommandations*. Ministère de l'Equipement, de l'Aménagement du Territoire et du Développement Durable, 109 pp.

Ben Saad A. 2011. "Les Conseils de gestion des terres collectives en Tunisie entre mauvaise gouvernance et marginalisation. Cas de la région de Tataouine, Sud Tunisien". *Institut des Régions Arides de Mednine, Tunisie. CIHEAM, Option Méditerranéennes. Série B. Etude et Recherche*, 66, 73–84.

Bourbouze A. 2018. "Les grandes transformations du pastoralisme Méditerranéen et l'émergence de nouveaux modes de production". *Watch Letter - Lettre de Veille Du CIHEAM, Animal Health and Livestock, Mediterranean Perspectives*, 39: 7–12.

Davies J., Herrera P.H., Ruiz-Mirazo J., Mohamed-Katerere J., Hannam I & Nuesri E. 2017. *Améliorer la Gouvernance des Terres Pastorales, Guide Technique pour la Gouvernance des Régimes Fonciers n°6*. Sous la direction de Caterina Batello, Rome.

DGF & Banque Mondiale, 2015. *Vers une Gestion Durable des Ecosystèmes Forestiers et Pastoraux en Tunisie. Analyse des bénéfices et des coûts de la dégradation des forêts et parcours*. Etude a été réalisée par Lelia et Hamed Daly sous la supervision de Taoufiq Bennouna et Youssef Saadani. 68 p. Ministère de l'Agriculture, Des Ressources Hydrauliques, et de la Pêche. Tunis, Tunisie.

Dominguez P. & Benessaiah N., 2017. "Multi-agentive transformations of rural livelihoods in mountain ICCAs: The case of the decline of community-based management of natural resources in the Mesioui agdals (Morocco)". *Quaternary InternationalI*, 437, 165–175.

Selmi S. & Elloumi M. 2007. Tenure foncière, mode de gestion et stratégies des acteurs-le cas des parcours du Centre et du Sud tunisien. Vertigo, Hors Série 4, Novembre.

Fuhlendorf S.D., Engle D.M., Elmore R.G., Limb R.F & Bidwell T.G. 2012. "Conservation of pattern and process: developing an alternative paradigm of rangeland management". *Rangeland Ecology & Management*, 65(6), 579–589.

Gamoun, M., Louhaichi M. & Ouled Belgacem A. 2018. "Diversity of desert rangelands of Tunisia", *Plant Diversity*, 40(5), 217–225.

Herrera P.H., Davies J. & Baena P.M. 2014. "Governance of rangelands in a changing world". In Herrera P.H., Davies J. & Baena P.M. (eds.). *The Governance of Rangelands Collective Action for Sustainable Pastoralism*, Routledge, Oxon, 32–44 pp.

IRA-IRD. 2003. *La Désertification dans la Jeffara, Sud-est Tunisien, Pratiques et Usages des Ressources, Techniques de Lutte et Devenir des Populations Rurales, Rapport Scientifique de Synthèse*. Elaboré par IRA/IRD/CRDA Gabès et Médenine, Médenine, Tunisia, 148 p.

Liao Chuan & Fei Ding. 2017. "Pastoralist adaptation practices under non-governmental development interventions in Southern Ethiopia". *The Rangeland Journal*, 39(2), 189–200.

Mares H. & Lahmayer I. 2019. *Analyse de la Situation Foncière en Vue de la Réparation de la Stratégie REDD+ en Tunisie*. FAO, Roma.

Mortimore, M. with contributions from Anderson S., Cotula L., Davies J., Faccer K., Hesse C., Morton J., Nyangena W., Skinner J. & Wolfangel C. (2009). *Dryland Opportunities: A New Paradigm for People, Ecosystems and Development*, IUCN, Gland, Switzerland; IIED, London and UNDP/DDC, Nairobi, Kenya.

Nefzaoui A. 2004. "Rangeland improvement and management options in the arid environment of central and south Tunisia". In Ben Salem H., Nefzaoui A. & Morand-Fehr P. (eds). *Nutrition and Feeding Strategies of Sheep and Goats under Harsh Climates. Options Méditerranéennes;* Série A. Séminaires Méditerranéens Zaragoza, Spain. 59, 15–25.

Nefzaoui A. & El Mourid M. 2008. *Rangeland Improvement and Management in Arid and Semi-arid Environments of West Asia and North Africa*. IDRC-IFAD, Ontario, Canada.

Prince S., Von Maltitz G., Zhang F., Byrne K., Driscoll C., Eshel G., Kust G., Martínez-Garza C., Metzger J.P., Midgley G., Moreno-Mateos D., Sghaier M. & Thwin S. (2018). "Status and trends of land degradation and restoration and associated changes in biodiversity and ecosystem functions". In Montanarella L., Scholes R. & Brainich A. (eds.). *The IPBES Assessment Report on Land Degradation and Restoration. Secretariat of the Intergovernmental Science-Policy Platform on Biodiversity and Ecosystem services (IPBES)*, Bonn, Germany, 315–495 pp.

Rignall K. & Kusunose Y. 2018. "Governing livelihood and land use transitions: the role of customary tenure in south-eastern Morocco". *Land Use Policy*, 78, 91–103.

Scholes R., Montanarella L., Brainich A., Barger N., ten Brink B., Cantele M., Erasmus B., Fisher J., Gardner T., Holland T.G., Kohler F., Kotiaho J.S., Von Maltitz G., Nangendo G., Pandit R., Parrotta J., Potts M.D. Prince S., Sankaran M. & Willemen L. (eds.). 2018. *Summary for Policymakers of the Assessment Report on Land Degradation and Restoration of the Intergovernmental Science-Policy Platform on Biodiversity and Ecosystem Services*. IPBES secretariat, Bonn, Germany 1–44.

Sghaier M. & Fetoui M. 2006. "Le statut foncier des terres: un facteur déterminant des évolutions socio-environnementales". In Genin D., Guillaume H., Ouessar M., Ouled Belgacem A., Romagny B., Sghaier M. & Taamallah H. (eds.). *Entre Désertification et Développement. La Jeffara Tunisienne*. IRD-IRA-Céres Ed, Médenine, Tunisia, 303–313 pp.

Tumur, E., Heijman, W.J.M., Heerink, N. & Agipar B. 2020. Critical factors enabling sustainable rangeland management in Mongolia. *China Economic Review*, 60, 101237.

UNEP. 2016. Combating Desertification, Land Degradation and Drought and *Promoting Sustainable Pastoralism and Rangelands*. A post second session of the United Nations Environment Assembly (UNEA-2) consultation workshop for stakeholders in East and Southern Africa region. Terrestrial Ecosystems Unit UN-REDD & landscapes. UNEA2, 23–27 May, Nairobi, Kenya.

13 Multi-actor platforms as a mechanism for actively bringing together actors and their interests

Maria Helena Guimarães and Pedro M. Herrera

13.1 Introduction

Silvopastoral systems have a great diversity of governance mechanisms, which have run these systems since historical times and still do so. This is due to the different social, economic and ecological conditions that affect them and the variety of people involved. However, as shown in this book's introduction, there are global economic and market trends that are pressing to homogenize primary production, which deeply affects all extensive farming production. The tensions and conflicts (see Chapter 11) generated between the complexity demanded by silvopastoralism and the simplification favoured by industrial agriculture are deeply affecting the governance of silvopastroral systems, with undesirable outcomes in terms of their performance, adaptation, resilience and survival. The conditions driven by current trends (such as globalization and climate change) and simplified management models are not appropriate for the subtleties of flexible rights of use and access needed to keep silvopastoral systems viable (Herrera, 2014).

Governance implies the interaction between individuals, groups, structures, traditions and institutions (Lockwood et al., 2010). Interactions occur when some sort of common interests exist, providing a kind of glue that motivates the interaction. When such glue disappears so does the interaction (as detailed in Chapter 9). There are many ways in which people and institutions relate around the governance of silvopastoral systems. In this chapter we describe and discuss platforms that aim to promote interactions in order to influence individual and collective decision-making at farm, community and policy levels. The focus is upon characterizing multi-actor platforms that aim to influence the governance of silvopastoral systems. These platforms promote interactions between the different types of actors described in the first section of this book, thereby changing the nature and format of existing relations. Through this characterization, we aim to distil the lessons learned that can be useful for future initiatives. We believe that such multi-actor platforms can be important instruments to rebuild the governance of silvopastoral systems, and set in motion citizen-driven processes to restore the functionality of these systems.

DOI: 10.4324/9781003028437-14

Why do multi-actor platforms have a potential role to play in the governance of silvopastoral systems? Work by Elinor Ostrom (1998) and many others show that communication, learning and dialogue are essential in initiating and supporting collective actions. When channels of communication do not exist, people and institutions tend to decrease interactions to those that are easier, more direct and more comfortable.

> I discuss two major empirical findings that begin to show how individuals achieve results that are 'better than rational' by building conditions where reciprocity, reputation, and trust can help to overcome the strong temptations of short-run self-interest.
>
> ([Ostrom 1998])

In this line of thought, humans might be better thought of as *Homo cooperaticus*, who are naturally willing to work together but may quit if they are suckered, rather than *Homo economicus*, who are rational, narrowly self-interested and never willing to cooperate (Anderies and Janssen, 2012). Cooperation is needed in silvopastoral systems, specifically in the dehesa and montado, in order to sustain their multifunctionality ensure by complex management (Barroso et al., 2013). Multi-actor platforms that encourage interactions and promote cooperation therefore, they can play an important role. One of the main foci of this chapter is how to establish such platforms

13.2 The importance and challenges of multi-actor platforms in silvopastoral systems

In the dehesa and montado there are several challenges in developing collective actions to promote their sustainability. These silvopastoral systems are managed by a multiplicity of actors, pursuing a variety of economic activities with several types of business models and distinct impacts on their resilience. Such diversity implies that we should not look for linear and straightforward solutions to the challenges of sustainability. Even if they do exist, their incorporation into decision-making is not immediate or, at least, is not happening at a pace that counteracts the decline of these silvopastoral systems.

Several possible reasons for the lack of collective actions towards the sustainability of silvopastoral systems can be identified (Guimarães et al., 2018). One of the reasons is the definition of silvopastoral systems.. For example, in the case of the montado, there is no formal institution that gathers all farmers and landowners who own the land or work the system. In many cases, a montado area may be included in a larger management unit that can include vineyards or olive production (Fragoso et al., 2011). This leads the actors to align with institutions that are more relevant to their core business model (i.e. cork or livestock). It was only in 2014 that a federation focused on the dehesa was created in Spain (see Text Box 13.1).

TEXT BOX 13.1

FEDEHESA – Federation for the Dehesas

Maria Pia Sanchez Fernandez, FEDEHESA, Spain

Key points: The integration of multiple actors and interests in one single organization

Fedehesa, is important because it has bought together primary producers, processing, agri-food, forestry, engineering and consulting companies, researchers and universities with specialized departments. This has allowed it to develop a global vision of the problems and to connect on the ground realities with research, politics and the market sector. In this way, the actual impact of problems experienced by dehesa managers or owners is discussed with the organization, which passes them on to those member institutions, research centres and other entities best able to deal with the problem. The results are, on many occasions, the creation of strategies that allow problems to be addressed from a multidisciplinary perspective. As an example of this collaboration, we can address the problem of bovine tuberculosis, which debate initiated the creation of the GOSTU[1] operational group in which hunting organizations, livestock organizations and research centres work together to provide viable solutions. Other important milestones of this collaboration have been the Interreg Prodehesa Montado Project, jointly developed by public and private actors in Spain and Portugal. This project, currently in its third year, has improved the public value of the dehesa and montado in many aspects including the social perspective. The latest collaboration project is Life Live-Adapt, promoted by Fedehesa and the University of Córdoba to adapt extensive livestock farming to climate change. The achievement of these objectives is possible thanks to the organization's own dynamics, whose strategic lines are defined in the General Assembly, which meets every four years, as well as the Board of Directors, which is in charge of implementing those actions. The Federation's work follows two main lines: lobbying on behalf of dehesa in all relevant forums, private and public institutions and the development of national and international programmes or activities for raising awareness, dissemination and direct actions that benefit the dehesa, from an environmental, social and economic perspective.

Another reason that might explain the lack of collective actions is the spatial distribution of key actors who are physically distant from each other. The tradition of the *latifundi* systems is still very present in the governance structures, as explained in Chapters 3, 8 and 11. In many cases, the heirs have left the rural areas and live and work elsewhere, or just use the estate for leisure (Santos, 2004). Frequently, when the time comes to take over the management responsibilities they feel ill-equipped to do so and, with little support, just manage to run a complex and low-profit farm business.

Collective actions might also be a challenge due to the difficulty of including all types of actors. Ideally, multi-actor platforms should include active representatives of each typology of actors but that does not always occur. Even though actors are invited to participate, sometimes they choose not to, since they do not consider themselves to have the mandate to take part or there are not any real opportunities to do so. At other times, the ways to participate are not adequate. Overcoming this challenge is a difficult task, but the benefits are clear. Users and workers in silvopastoral systems (often outside the platforms) hold specific forms of knowledge that other actors do not have. That fact can be decisive in uncovering some of the issues that explain these systems' decline. Self-exclusion is a challenge not only felt by those promoting multi-actor platforms, but also by landowners and land managers. Very often they referred to the need for establishing a network of workers and users they can trust that can contribute to safeguarding the land. For example, in the montado, one of the biggest fears of land managers in the season of cork extraction is robberies, which are increasingly frequent due to the absence of people on the farm. Another concern often expressed is the lack of shepherds who can develop the grazing activities that these systems depend upon.

Despite these challenges, multi-actor platforms focused on silvopastoral governance might well play an important role in the development of governance models that contribute to the maintenance of multifunctional silvopastoral systems. These platforms can be physical or virtual spaces designed using network systems and collective approaches (Cristovão et al., 2012; Guimarães et al., 2017) to engage different actors involved in the issue at hand. Their implementation does not follow the same processes or logic that underlie associations or cooperatives. As we will see with the two cases described in this chapter, multi-actor platforms can have distinct ambitions and scales, but their design principles are transversal. They are not about technological transfer or science dissemination, but about creating an arena where knowledge coming from different societal sectors can be presented, discussed and mingled with other perspectives, attitudes and actions. These platforms experiment with new methods and practices related to co-construction, facilitation and brokerage (i.e. interference) within networks so that ingrained thought-styles[2] can be changed by the incorporation of different knowledge presented or created by the actors' interactions. This kind of initiatives seek to develop a process of interactive learning, empowerment and improved governance that aims to enable actors to be collectively innovative and resilient. These initiatives are still being implemented, so our goal is to discuss what can we learn about them, how can they be replicated and the impacts that they are having.

13.3 Overview of two actual initiatives

13.3.1 The Spanish platform on extensive livestock farming and pastoralism

While not only focused on dehesa, the Spanish Platform on extensive livestock farming and pastoralism[3] is an example of the most representative multi-actor network in Spain that addresses silvopastoral systems. Running since

2013, the Platform is a bottom–up, nationwide initiative, supporting the role of extensive livestock productions in social-ecological systems in Spain. The original idea was developed by Entretantos Foundation,[4] which currently runs the Platform's Technical Secretariat. As stated in its vision and mission, the platform aspires to be a large open and participatory organization, bringing together people and entities from all over the country to build capacity, deliver a voice and advocate for pastoralism and the extensive farming sector. The platform advocates for the whole set of extensive livestock production systems, arguing that they efficiently use local, land-based resources with production models that are compatible with sustainability and maintaining ecosystem and social services.

The platform's actions include promoting best management practices, such as the use of native breeds, the mobility of livestock, the reduction of external inputs, improving animal welfare and scheduling livestock densities so they are optimally adapted to the resources available in each territory at any time. Other activities focus on the social and ecological side, encouraging public recognition for pastoralism and extensive farming, stimulating a supportive legal framework for pastoralism and its proper consideration in policies (especially CAP), amplifying the voices, image and visibility of pastoralists and extensive farmers and, finally, improving sustainability related to pastoralism: land management, employment, rural development, profitability, economics, adaptation and mitigation of climate change, ecosystem services, biodiversity, social relationships, risk management, governance, etc.

The main roles of the platform are coordinating the different actors, building capacity among the sector and developing a common voice. There are several points in the platform's composition that contribute to it being a functional vehicle for dialogue. First of all, shepherds and farmers, both women and men, are considered to be central stakeholders, key decision-making, who are considered in all the platform's actions. This ensures that the platform's proposals are always practical and do not place excessive burdens on the shoulders of primary producers. Second, the platform supports grassroot organizations that directly represent pastoralists but avoids interfering in their dynamics while encouraging them to deploy their own voices. Last, the platform actively searches for balance and exchange among the different groups linked to extensive farming: 1) farmers and landowners; 2) academics and researchers; 3) government agencies, consultants and related professionals; and 4) civic society (conservationists, NGOs, rural development groups, etc.). Gender issues are also deeply embedded in the platform, with specific work with women farmers. The organization and construction of the platform are also innovative. The structure is kept to the minimum, it being a light, flexible and almost virtual organization, based on the exchange of experiences and knowledge. To be part of the platform interested groups and individual people only have to sign a one-sheet agreement of the main aims of the platform and to commit to respect both the people and their opinions.

The platform has two main decision-making mechanisms: a mailing list and the "Botanic Meetings", face-to-face, professionally facilitated rendez-vous's that are celebrated yearly in the Botanic Garden of Madrid. All decisions run through the mailing list and, when controversial, they are diverted to the meetings. The non-controversial proposals are approved by assent and those who propose them become responsible for developing them in proper terms. This truly multi-actor perspective is one of the fundamentals of the platform's work, and largely responsible for its success. While there has been a long history of pastoralism-support movements in Spain, the platform has managed, for the first time, to integrate all actors with their different perspectives in a single structure of debate and sharing, with benefits in terms of generating a stronger image.

The mail list, with around 400 participants, is the core of the platform, and also the main communication tool, where all information flows and specific activities, campaigns and events are designed and shared with the network (Figure 13.1). These actions are complemented with the communication plan including a website,[5] publications, multimedia editions, presence on social networks and participation in events. Additionally, an annual workshop "Territorios Pastoreados" (*Grazed Lands*) is currently in its 5th edition and highlights the most successful experiences and dialogues around the key issues in the field.

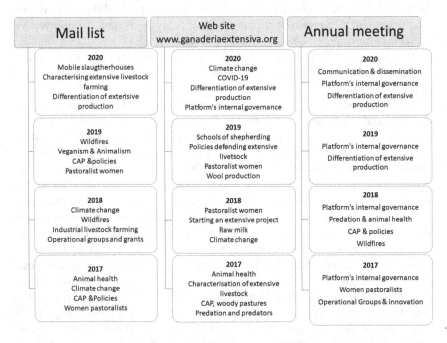

Mail list	Web site www.ganaderiaextensiva.org	Annual meeting
2020 Mobile slaugtherhouses Characterising extensive livestock farming Differentiation of extensive production	**2020** Climate change COVID-19 Differentiation of extensive production Platform's internal governance	**2020** Communication & dissemination Platform's internal governance Differentiation of extensive production
2019 Wildfires Veganism & Animalism CAP &policies Pastoralist women	**2019** Schools of shepherding Policies defending extensive livetsock Pastoralist women Wool production	**2019** Platform's internal governance Differentiation of extensive production
2018 Climate change Wildfires Industrial livestock farming Operational groups and grants	**2018** Pastoralist women Starting an extensive project Raw milk Climate change	**2018** Platform's internal governance Predation & animal health CAP & policies Wildfires
2017 Animal health Climate change CAP &Policies Women pastoralists	**2017** Animal health Characterisation of extensive livestock CAP, woody pastures Predation and predators	**2017** Platform's internal governance Women pastoralists Operational Groups & innovation

Figure 13.1 Main topics addressed by the platform on extensive livestock farming and pastoralism since 2017, including most active mail list conversations, most visited web posts and key themes in the annual meetings.

The network of the platform exchanges information and experiences about fields of interest: management, mobility, policies, animal health, land and pasture management, wildlife and the environment being some examples. It feeds all the interests of the platform's members, from details of the latest scientific papers on the sector to calls for assistance. In addition, multidisciplinary teams have published two technical reports, one on woody pastures and their consideration in the Common Agricultural Policies (Herrera, 2015), and the other a first attempt to define and characterize extensive livestock farming in Spain (Ruiz-Mirazo et al., 2017).

The platform has become the main supporting and dialogue tool for extensive farming in Spain. It has gained momentum and performed a great deal of lobbying for the safeguard of grazing systems, giving voice to the sector in some critical spaces, such as the Common Agricultural Policies (CAP) and the climate change debate. The last public campaign was on the UN Climate Summit COP 25, held in Spain in 2019, that demanded proper consideration of the role of pastoralism in fighting and adapting to climate change.

13.3.2 *The Tertúlias do Montado, Alentejo Portugal*

In Portugal, there is a similar initiative named Tertúlias do Montado. The word *Tertúlia* means an informal gathering of people, sharing a passion for the same subject, who meet to talk about, say, current affairs and to share knowledge, opinions and experiences. The initiative started in May 2016 and by November 2019, twenty-two structured interactions had occurred between researchers, students, landowners, land managers, public administrators, private companies and NGOs. The Tertulias do Montado aims to create a structured and ongoing dialogue between all the actors interested in the sustainability of the montado.

The underlying philosophy of the initiative is the hypothesis that co-construction of knowledge is a long-term process that requires a larger and broader dialogue than financed research projects enable. During each session, considering the specific topic being dealt with, the objective is that the complexity of the issue is explored, taking into account the diverse perspectives; abstract and case-specific knowledge are linked, and different typologies of knowledge are co-produced for what is perceived to be the common good. All of this is done, under the guidance of skilled facilitators, by representatives of different disciplines and perspectives, from both private and public sectors, and civil society. Such goals follow the definition for transdisciplinarity put forward by Christian Pohl (2011). The initiative was started by academics from the Mediterranean Institute for Agriculture, Environment and Development (MED).[6] There is no specific funding for the initiative, and it has mainly been developed through the internal resources of MED. This gave the initiative freedom to create its own agenda and rhythm. There are no established deadlines and no specific outcome or result that is aimed at. Participation in each meeting session is open to anybody interested, up to a limit of 40 participants per session. The openness of the initiative is purposeful. If the initiative was restricted to predefined participants it would not have the potential to expand and to allow the participation of actors who are the most difficult to involve. In line with this openness principle, all information on the *Tertúlias*, as well as, the reports of each session, are publicly available. Online visibility is provided by a blog,[7] where all the meeting sessions are publicized (the press and other dissemination channels are also used) and the reports of the sessions can be consulted.

At present, the contact list includes 300 individuals including researchers, students, landowners, land managers, public administrators, private companies and NGOs. Each session has a planned structure that usually involves one or more specific people who are invited to give an opening talk and ensure that the requisite knowledge is available during the discussions. Each topic is explored by guiding questions and participants are often asked to work in smaller groups to enhance active participation. Each session is planned, developed and summarized by the facilitator. Indoor and outdoor sessions have been organized and most of the sessions last for 3 hours. In the first session (in May 2016), the participants collectively developed a common agenda that resulted in a list of 17 issues considered important to discuss together (Table 13.1).

The common agenda continues to be open for the inclusion of new topics. The decision regarding which issue to discuss next is done by the participants before the end of each session. Some issues are discussed in just one session, others may take more than one. The participants are also invited to suggest how to organize the following session, although the structure of the

Table 13.1 The common agenda of Tertúlias do montado

1	Climate change scenarios and adapting strategies	10	Application for montado as a UNESCO cultural landscape heritage site
2	Payment for non-market ecosystem goods and services	11	Increasing societal awareness of the montado
3	Accessibility	12	Tree ageing and death
4	New ways to capitalize the montado	13	Reforestation
5	Impact of the drop in cork prices	14	Policies that would recognize montado specificities and align conflicting goals
6	Muck spreading	15	Soil and its vital importance
7	Monitoring system and activities	16	Grazing management and impact
8	Legislation and the lack of specialized training for cork harvesters	17	The montado's survival in the context of the economic crisis and demographic changes.
9	Specific valorization of the holm oak montado	...	...

session is defined by the facilitator following the inputs received. Feedback is often received in person but also by answering an evaluation questionnaire delivered at the end of each session in an anonymous format. Each session is tape-recorded so that a follow-up report can be developed. The report of each meeting summarizes the main points discussed, questions answered and those that still remain. So far, the average participation has been 29 participants per session.

In 2019, 100 participants from Tertúlias do Montado replied to an evaluation questionnaire. One of the objectives of this survey was to collect information on the impacts of the initiative. Figure 13.2 shows that in 90% of the interviewees reported that the main outcomes of this platform were acquiring knowledge and new contacts. Enhanced trust was the third most frequent reply (60%), followed by concrete actions (45% of respondents). Such concrete actions were reported mostly by landowners and land managers (60%) followed by researchers (29%).

13.4 A comparison of the two platforms

The use of multi-actor platforms as coordination mechanisms to strengthen landscape governance is growing (Kusters et al., 2018). This concept is still not consolidated or widely discussed in the scientific literature. Yet, we found a few recent projects that aim to implement multi-actor platforms in several domains, in different countries and at different scales (e.g. REMIX,[8] FAIRWAYIS,[9] SHERPA[10]), and some guidelines and references about the challenges faced by these initiatives (Heiner et al., 2017). Hopefully, in the coming years, more examples of such platforms can enrich the comparison developed here. The following box presents some of the emerging multi-actor platforms focused on rural topics (Text Box 13.2).

The Spanish platform on extensive livestock farming and pastoralism is a participatory and flexible organization, with a clear mission of supporting silvopastoral systems. It tries to influence policy design towards the development of collective and inclusive activities, delivering a clear message, which is put forward by a diverse group of participants. It has nationwide coverage and is managed by a non-academic institution. Tertúlias do Montado is more a dialogue platform than a participative one (i.e. related to deliberation) since it is not explicitly oriented towards advocating for specific policy changes. The Tertúlias do Montado is a regional initiative, initiated by academia and based on knowledge sharing. It has maintained its distance from political discourses and lobbies. Some participants have developed such activities, but outside the Tertúlias framework.

Both initiatives share a common ground as multi-level, multi-actor and multi-disciplinary, working through a collaborative approach with the actual stakeholders linked to silvopastoral systems. The experiences of the two platforms allow us to identify transversal features that can be translated into lessons learned.

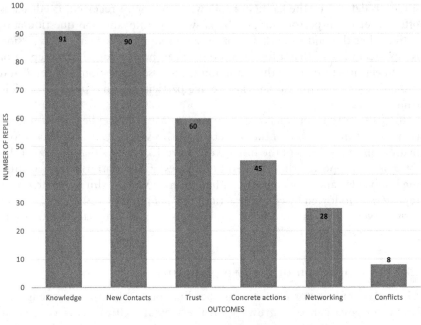

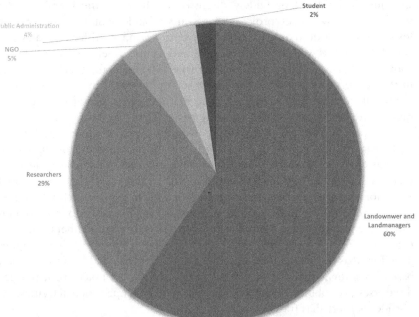

Figure 13.2 (a) main benefits of participating in Tertúlias do Montado. (b) Typology of participants who indicated concrete actions as an outcome of participating in Tertúlias do Montado.

TEXT BOX 13.2

Multi-actor platforms – forums in which actors from science, policy and society meet, share and co-create knowledge

Pedro Miguel Santos, Consulai,[11] Portugal

Key points: A platform driven by the H2020 funds and the explicit goal in this program, to strengthen research-practice interaction: to continue in the coming Horizon Europe

Multi-actor platforms (MAPs) are discussion forums within the SHERPA project (Sustainable Hub to Engage into Rural Policies with Actors); a research project funded by the EU through Horizon2020 and carried out by a consortium of 17 partner organizations (Figure 13.3). The overall objective of SHERPA is to gather relevant knowledge and opinions that can contribute to the formulation of recommendations for future policies relevant to EU rural areas and support actors in rural development to take emerging opportunities and protect their areas from long-term threats of decline.

The interactions between research, policy and citizens take place at a local level in up to 40 MAPs with a MAP at the EU level. Each platform includes a balanced representation of active members representing three communities: 1) local citizens and businesses, 2) scientists and 3) policymakers. Each MAP involves at least 12 active members, a facilitator and a monitor from SHERPA.

The SHERPA project will produce position papers based on topics relevant to rural areas. The position papers will be included in future discussions and processes concerning European development policies and the research agenda for rural areas. The MAPs are crucial for basing the position papers on knowledge from actors at the local level.

In Portugal, CONSULAI (the national SHERPA partner) is in the process of developing 4 MAPs, which will discuss several topics launched by the SHERPA project. To date two MAPs have been established in Alqueva (MAP Alqueva) and the central region (MAP rural.pt).

Figure 13.3 The SHERPA project logo.

The first topic to be discussed in MAP Alqueva is related to rural policies to protect and enhance biodiversity through the preservation, creation and management of landscape features. In MAP rural.pt the long-term vision for rural areas is being discussed.

The aim is to boost demand-driven innovation and the implementation of research, creating synergies between actors and increase impacts through a process of genuine co-creation of knowledge, focusing on real problems and opportunities.

13.5 Lessons learned

13.5.1 The need to understand the context – the responsibility of the promoters

The issues that need to be dealt with in regards to the sustainability of silvopastoral systems are complex and require the production of system-based knowledge instead of generalist, decontextualized and reductionist knowledge (Guimarães et al., 2018). Therefore, the context in which interaction platforms will be implemented needs to be recognized and taken into account. We might assume that the promoters of such platforms are well-rooted in this context and therefore would no need for such a contextualization process. Yet, even locally based promoters are often anchored in a specific identity, looking at the problems from one thought style or a specific perspective that may only cover one subsystem (Pohl, 2011). This implies that they may lack a holistic view of the system (Seifferta and Loch, 2005; Gaziulusoy and Boyle, 2013). This leads us to suggest the use of tools that allow a holistic overview, that promote reflection and allow the planning needed for the implementation and progress of these platforms. It is important to understand and dissect the components, and interlinks that form the dynamic issues to be dealt with in these platforms, including the values and worldviews that affect decisions (Seifferta and Loch, 2005; Gaziulusoy and Boyle, 2013; Williams et al., 2017; Guimarães et al., 2018). In the case of the montado it was important to understand and take into account the traditional value system of the region, and the established power relations, to adapt the language used and how stakeholders were approached at the beginning. The start of the initiative benefited from the social capital already accumulated by the research institute that organizes the Tertúlias (Mckee et al., 2015). In the Spanish case, contextualization come out from field of work. Extensive farmers working in a wildfire prevention project felt a need to develop their own voice, so Entretantos provided support that was crucial, first to the development of some farmers' associations, and eventually to the meetings that gave birth to the platform.

13.5.2 The logic of being of value – what is in it for stakeholders?

People are reluctant to devote their time to collective activities with no visible practical outcomes unless they understand that the benefits could compensate the effort. However, the benefits that stakeholders should receive are often neglected or not addressed properly when initiating participatory processes (Reed, 2008; Roloff, 2008). The lack of benefits for participants is a known source of failure in participatory processes and even harms the local dynamics, creating a profile of stakeholders 'deceived' into participation and no longer willing to take part. Therefore, promoters should start by thinking about the possible benefits from the first stages of planning and include them in the calls and early dissemination activities. In the Spanish example, the fact that the issues being discussed were defined by the participants, was one of the main reasons why engagement continues and interest increased. Another reason mentioned by participants in Portugal is the horizontal structure of the platform and the focus on landowners and managers who play a central role in the future of the montado. They are the ultimate decision-makers and are perceived as such in the Tertúlias do Montado. In Spain, most participants find that the flow of information from different sources and addressing many interests is one of the greatest assets of the initiative. Position and scientific papers, calls for grants and events, policy claims and even job offers are widely shared on the platform, creating a rich sense of practical value. Moreover, the generation of a rationale that supports pastoralism, aims for better policies as well as the simplicity and horizontality of the platform have allowed very different people to work together in a collaborative and respectful way.

13.5.3 The difference between leading and facilitating

One of the key issues in multi-actor platforms is the fine border between leading and facilitating the process. In implementing such platforms there has to be someone who starts it; but, who has the 'authority'? Perhaps those who provide resources? Or have good intentions? Or the most active followers? The reply to these questions can vary from case to case and there is no right or wrong answer. Nonetheless, initiatives might fail because of bad leadership and failed expectations (Sarkis et al., 2010; Haucka et al., 2015; Reypens et al., 2016). Sometimes, the outcomes are not planned, or they interfere with others, or there are hidden dynamics, or the process does not follow the expected path, so promoters abandon it. Other initiatives fail because of lack of assistance or because participants are no longer interested, or the topics have become meaningless. Finally, there are processes that fall because their promoters are unable to keep a balance between being interested parties and leaders and the platform is perceived as partial. Platforms need to be managed in a flexible way, adapted to the capacities of the participants, with the major principle being that the promoters facilitate while the participants lead. This implies a shift from standard presentations, which have a speaker and an audience that passively hears the discourse, formulates questions and individually creates meaning from the interaction. In multi-actor platforms, participants are asked to actively contribute to the topic, to share their own experiences and, by guided questions, collectively build up a meaning that the group can support. Such mindset is not directly understood and incorporated by many participants of such platforms.

The responsibility of the promoters is to set a safe environment for exchange that is transparent, inclusive and equitable. The responsibility of the participants is to take decisions collectively on what is discussed, when to move forward, change the subject or develop parallel actions to the platform. The creation of the common agenda of issues to be discussed in Tertúlias do Montado is an example of a way to show the group that the promoters are merely providing the space while the content is collectively defined by the participants. In the Tertúlias do Montado case such a mindset was progressively achieved, including the acceptance of uncertainty regarding best practices, recognition of knowledge gaps and of the diversity of the group. During the *tertulias*, landowners were motivated to share their experiences and to explain their settings and conditions. Such sharing helped the group understand the diversity of cases and the impossibility of applying the same solution to all cases. In the Spanish MPA, the platform tries to maintain a balance in terms of territorial representation, gender equality and the presence of farmers. Granting a key role to actual farmers is crucial. Some initiatives designed to support silvopastoral systems have been perceived by farmers as biased towards nature conservation or cultural issues, leading them to disengage. The Spanish platform prevents that from happening by granting a specific voice to farmers (e.g. assigning the public representation role to a farmer, including farmers in all public events and mixing farmers with researchers and other actors at every possible opportunity).

13.5.4 Skilled facilitation is key for success

Proper facilitation is a key issue in developing multi-actor platforms, especially when dealing with people from conflicting backgrounds. At least one skilled facilitator is essential and, depending on the size and scope of the project, it is desirable to have a skilled facilitation team. The facilitator's role can be likened to scaffolding (Jordan, 2014) which means providing a support structure like the one needed when erecting the walls of a new building.. Metaphorically, the verb 'to scaffold' refers to the provision of the external support that a person or a group may need to build skills, learn new things, construct a solution to a complex problem or develop a strategy for attaining the desired goal (Jordan, 2014). Skilled facilitation is a competence that can be learned by those interested; there are several training courses available in conflict resolution and mediation.

This role is recognized and understood by participants in the Tertúlias do Montado, in which there are several power relations between older and younger farmers, landowners and land managers, users and researchers. Knowledge about such power interactions is used when forming smaller discussion groups and defining methodologies. Nonetheless, such competencies and skills are not infallible and there are records of strong personalities becoming disinterested by Tertúlias due to two main reasons: its horizontal and informal nature, and the exposure and discussion of knowledge gaps. In the Spanish case, facilitation is at the core of the extensive farming platform initiative. A 4-person highly qualified facilitation team with previous experience helps the platform to plan and carry out its main actions. They do most of the technical work, drawing support, experience and knowledge from the most active participants and sectors represented, and they then ease the way towards agreeing on collective actions and mobilizing resources.

13.5.5 *The crossroads of mingling with different types of knowledge, information and misinformation*

The sharing of knowledge, experiences and information is a cornerstone of these platforms. Although misinformation cannot be fully controlled, recognizing this risk is the first step to overcoming it. Secondly, it is important to consider that the production of knowledge in science and the production of knowledge by experience is distinct. While the former aims at generalization, the second cannot be directly generalized. It is important that this difference is made explicit to all participants and that both ways of knowledge are treated with equal respect. This implies a departure from an action-oriented discourse (often unfamiliar for researchers and scientists); but the goal here is not to identify 'the truth' but to make explicit the limitations of the knowledge that is being shared, including that produced by science.

To take an example that emerged in the montado case: the structure of the root system of *Quercus* trees and the negative impact of disturbing the soil are facts, scientifically tested. Nonetheless, the experience of many farmers over 5 years indicates that soil disking is important for shrub control and does not affect the tree, since the effect of cutting the roots of the trees only has a visible impact 10–20 years later. Making this mismatch in temporal scale explicit is important. From the scientific side, a good example is the results attained in recent research projects that are testing new approaches for water conservation in the montado. While preliminary results show promising results at a farm level; the necessary replications are still to be developed and thus generalizations are not yet possible. The role of the facilitator during the discussion of these techniques was to tune down the discourse of enthusiastic researchers who tended to present preliminary results as scientifically robust solutions to specific problems. It is also important to highlight the importance of the design phase. Inviting the holders of other forms of specific knowledge to assure its presence during discussions is one example. In this respect, it is important that promoters only allow the platform to move forward after both (or more) sides of the coin have been discussed. In the end, it boils down

Figure 13.4 Presentation of the Spanish platform in last COP 25, Chile, celebrated in Madrid (Spain). People on stage include (seated, from the left) four top researchers (Marta Rivera, Gerardo Moreno, Elisa Oteros and Pablo Manzano), three farmers (Joan Alibés, Monte Orodea, from Ganaderas en Red, and Pia Sánchez from FEDEHESA), an activist (at the podium, Concha Salguero from Trashumancia y Naturaleza) and a member of the facilitation team (Pedro Herrera), jointly presenting a discourse advocating the benefits of silvopastoral systems in the context of climate change.

to making explicit the limitations of the different dimensions of knowledge and its sources, as well as, acknowledging ignorance and blind spots.

The work that the Spanish platform is doing on redesigning the way decision-makers, scientists and other actors see and assess the behaviour of extensive livestock farming systems in relation to climate change is another good example of how knowledge-sharing can deliver innovation and new perspectives to deal with real-life problems. Figure 13.4 shows this Platform making a presentation at COP 25, with the presence of researchers, farmers, activists and facilitators, all of them simultaneously advocating for silvopastoral systems in the context of climate change but with different slants. Researchers clarified the systemic approach to greenhouse gas fluxes in silvopastoral systems and how they are managed by farmers. Farmers explained how they deal with these problems through better management and finding practical solutions, while supporters described the need for better policies and markets, showing a common approach with people on the front line (Herrera, 2020)

13.5.6 Keeping a safe and structured platform

Multi-actor platforms should embrace the basic principles of participatory research: active democracy, a safe place for participation, empathy and respect for all participants, community involvement in decision-making and adaptation of participation to different scales and degrees (Bergold and Thomas, 2012). When thinking of the role of a facilitator and of the scaffolding we are recognizing the need to organize face-to-face interactions as structured and safe places for dialogue. The importance of this feature merits further details. The first necessary step is to establish a simple set of rules that everybody agrees upon and is willing to respect. These rules are basic for any useful dialogue, but they should be known, adjusted and accepted from the outset. Although a facilitator, using the art of scaffolding, will gradually educate the group to achieve the desired level of communication, taking care of all participants also involves organizing the physical space (examples in Figure 13.5). A safe space that leads participants into the appropriate mindset implies organizing the physical space appropriately. This means that, in a session, participants are in equal physical positions, a semicircle or circle chairs indicates that no one is in a higher hierarchical position, even if they might have distinct roles within a session, for example, someone who has been invited to bring specialized knowledge into the discussion (Figures 13.5a and 13.5d). Identification tags and talking

(a)　　　　　　　　　　(b)

(c)　　　　　　　　　　(d)

Figure 13.5 Examples of safe and structured dialogue spaces in the Tertúlias do Montado: (a) a plenary moment in a circle format so that each participant can see the others; (b) organization in small groups so that participants have sufficient time to express their understanding of the topic, listen to others and build a shared perspective; (c) the conclusions of each working group are summarized in plenary by a group spokesperson; (d) some topics are introduced by a specialist with an oral presentation followed by a clarification moment.

without the use of academic titles gives a sense of informality and equality. Having a coffee break can stimulate the mingling of participants and improve their levels of attention, as well as showing consideration for the well-being of those in the room. Ideally, rooms should allow a rapid reorganization of chairs and tables so that participants can easily shift in the different phases of a session, e.g. to smaller groups (Figure 13.5b) and later gather in plenary (Figures 13.5a and 13.5c). Good time management is one important capacity, but difficult to deliver. Assigning equitable time to each speaker, balancing the time assigned to each question and intervention, limiting the number of times any individual is allowed to speak, encouraging reticent people to talk and finishing the interventions on time in order to be able to recapitulate and summarize is an ability that facilitators need to learn and practice.

13.6 Conclusions

The governance of silvopastoral systems is changing and the question of how to steer it towards resilience and sustainability is important. Formal and top-down institutions have not been able to create the necessary conditions to prevent the decline of these systems, so new forms of governance need to be put in place. Multi-actor platforms can provide a safe place for dialogue and interaction between stakeholders and can contribute to collective actions that strengthen the governance of silvopastoral systems. Opening communication channels, sharing knowledge, experience and capacity, are demonstrable benefits of these platforms. Multi-actor platforms have different aims, ambitions and scales, yet they share similar design principles. They rely on inter-active learning, capacity-building, empowerment and the co-construction of alternatives, and achieve similar outcomes in terms of innovation, resilience and adaptation.

The wider development of such platforms can be seen as part of a larger strategy to improve the governance of silvopastoral systems. These platforms might feed or substitute the progressively outdated institutions which do not overcome inertia, co-produce innovation or preserve the economy. In such a strategy a framework should be put in place to guarantee the quality of the processes, the allocation of enough time and resources, the existence of sound and skilled facilitation, the inclusion of gender balance, equality and the inclusion of deep democratic values in each platform.

Notes

1. http://gostu.es/
2. A thought-style is characterized by a specific way of looking at the world and distinguishing relevant and irrelevant aspects, i.e. looking at things from a specific perspective (more details in Pohl, 2011).
3. http://www.ganaderiaextensiva.org/
4. http://www.entretantos.org/

5. http://www.ganaderiaextensiva.org/
6. https://www.med.uevora.pt/
7. https://tertuliasdomontado.blogspot.com/
8. https://www.remix-intercrops.eu/
9. https://www.fairway-is.eu/
10. https://rural-interfaces.eu/
11. https://consulai.com/

References

Anderies, J.M. & Janssen, M.A. (2012). "Elinor Ostrom (1933–2012): Pioneer in the interdisciplinary science of coupled social-ecological systems". *PLoS Biol*, 10(10), e1001405.

Barroso, F., Menezes, H. & Pinto-Correia, T. (2013). "How can the land managers and his multi-stakeholder network at the farm level influence the multifunctional transitions pathways?". *Spanish Journal of Rural Development*, 4 (4), 35–48.

Bergold, J. & Thomas, S. (2012). "Participatory research methods: A methodological approach in motion". *Forum Qualitative Sozialforschung/Forum: Qualitative Social Research*, 13 (1). Art. 30.

Cristovão, A., Koutsouris, A. & Kügler, M. (2012). "Extension systems and change facilitation for agricultural and rural development", in I. Darnhofer, D. Gibbon, & B. Dedieu B (eds.). *Farming systems research into the 21st century: The new dynamic* (pp. 201–227). Dordrecht, The Netherlands: Springer.

Fragoso, R., Marques, E., Luca, M.R., Martins, M.B. & Jorge, R. (2011). "The economic effects of Common Agricultural Policy on Mediterranean montado/dehesa ecosystem". *Journal of Policy Modeling*, 33, 311–327

Gaziulusoy, A.I. & Boyle, C. (2013). "Proposing a heuristic reflective tool for reviewing literature in transdisciplinary research for sustainability". *Journal of Cleaner Production*, 48, 139e147.

Guimarães, H., Fonseca, C., Gonzalez, C. & Pinto Correia, T. (2017). "Reflecting on collaborative research into the sustainability of Mediterranean agriculture: A case study using a systematization of experiences approach". *Journal of Research Practice*, *13*(1), Article M1.

Guimarães M.H., Guiomar N., Surová D., Godinho S., Pinto-Correia, Sandberg A., Ravera F. & Varanda M. (2018). "Structuring wicked problems in transdisciplinary research using the social–ecological systems framework: An application to the montado system, Alentejo, Portugal". *Journal of Cleaner Production*, 191, 417–428.

Haucka, J., Steinc, C., Schiffere, E. & Vandewallef, M. (2015). "Seeing the forest and the trees: Facilitating participatory network planning in environmental governance". *Global Environmental Change*, 35, 400–410

Heiner, K., Buck, L., Gross, L., Hart, A. & Stam, N. (2017). *Public-private-cvic partnerships for sustainable landscapes: A practical guide for conveners—Ecoagriculture partners and the Sustainable trade initiative*. EcoAgriculture Partners. Fairfax.

Herrera, P.M. (2014). "Searching for extensive livestock governance in inland northwest of Spain", in Herrera, P.M., Davies J. and Manzano P. (2014). *The Governance of Rangelands. Towards collective action for sustainable pastoralism*. Routledge. Earthscan. London

Herrera, P.M. (Coord). (2015). *Informe sobre la elegibilidad para pagos directos de la PAC de los pastos leñosos españoles*. Plataforma por la ganadería extensiva y el pastoralismo. Valladolid.

Herrera P.M. (Ed.). (2020). *Ganadería y Cambio Climático: Un acercamiento en profundidad.* Fundación Entretantos y Plataforma por la Ganadería Extensiva y el Pastoralismo. Valladolid.

Jordan, T. (2014). "Deliberative methods for complex issues: A typology of functions that may need scaffolding". *Group Facilitation: A Research and Applications Journal*, 13, 50–71.

Kusters, K., Buck, L., de Graaf, M., Minang, P., van Oosten, C. & Zagt, R. (2018). "Participatory planning, monitoring and evaluation of multi-stakeholder platforms in integrated landscape initiatives". *Environmental Management*, 62(1), 170–181.

Lockwood, M., Davidson, J., Curtis, A., Stratford, E. & Griffith, R. (2010). "Governance principles for natural resource management". *Society & Natural Resources*, 23(10), 986–1001.

McKee, A., Guimarães, M.H. & Pinto-Correia, T. (2015). "Social capital accumulation and the role of the researcher: An example of a transdisciplinary visioning process for the future of agriculture in Europe". *Environmental Science & Policy*, 50, 88e99.

Ostrom, E. (1998). "A behavioral approach to the rational choice theory of collective action: Presidential Address, American Political Science Association, 1997". *The American Political Science Review*, 92(1), 1–22.

Pohl, C. (2011). "What is progress in transdisciplinary research?" *Futures*, 43(6), 618–626.

Reed M.S. (2008). "Stakeholder participation for environmental management: A literature review". *Biological Conservation*, 141, 2417–2431.

Reypens, C., Lievens, A. & Blazevic, V. (2016). "Leveraging value in multi-stakeholder innovation networks: A process framework for value co-creation and capture". *Industrial Marketing Management*, 56, 40–50.

Roloff, J. (2008). "Learning from multi-stakeholder networks: Issue-focussed stakeholder management". *Journal of Business Ethics*, 82, 233–250.

Ruiz-Mirazo, J., Herrera, P.M., Barba, R. & Busqué, J. (2017). *Definición y Caracterización de la Ganadería Extensiva en España.* Ministerio de Agricultura y Pesca, Alimentación y Medio Ambiente. Madrid 2017. NIPO: 013-17-199-2.

Santos, R. (2004). Economic sociology of the modern latifundium: Economic institutions and social change in southern Portugal, 17th–19th centuries. Sociologia, Problemas e Práticas. 2004, n.45, pp. 23–52. Lisboa.

Sarkis, J., Cordeiro, J.J. & Vazquez, D. (2010). *Facilitating Sustainable Innovation through Collaboration. A multi-stakeholder perspective.* Springer. New York.

Seifferta, M.E.B. & Loch, C. (2005). "Systemic thinking in environmental management: Support for sustainable development". *Journal of Cleaner Production*, 13, 1197–1202.

Williams, A., Kennedy, S., Philipp, F. & Whiteman, G. (2017). "Systems thinking: A review of sustainability management research". *Journal of Cleaner Production*, 148, 866–881.

14 Time for collective actions

Innovations for silvopastoral territories

Gerardo Moreno, Pedro M. Herrera,
Maria Helena Guimarães and
Isabel Ferraz-de-Oliveira

14.1 Collective actions to face territorial challenges

Silvopastoral territories host specific features that lead to an assemblage of multiple actors working in a common territory. These features include the multiple uses and products of silvopastoral territories, the complex property owner-tenant networks and the communal uses and the public ecosystem services provided. Together, these characteristics offer a lot of room for different collective actions, which are understood here as collaborative work for a common good (Ostrom, 2000).

Silvopastoralism, usually practiced at large territorial scales in Mediterranean countries, has always been dependent on interactions between different actors (Barroso et al., 2013). Yet, as discussed in Chapter 3, a decrease in labour needs and depopulation, among other factors, reduced the number of actors in the territory and increased the distance between them. This hinders the chances of developing stable interactions and, at a higher level of collaboration, of collective actions. Although farmers tend to act as single independent agents with respect to their land (Dolinska and d'Aquino, 2016), they also have a deep-rooted capacity to work collectively if they perceive it to be in their interests (Markelova et al., 2009). Indeed, traditions of rural and peasant cultures weaving formal or informal interactions have been, and still are, the base of landscape sustainability across the world (Ostrom, 2000). Working collectively is one of the pillars for the current vigorous development of agroecology as a response to the increasing dependency of farmers on external agents (Ploeg, 2020). Farmers belonging to communities of practice are more empowered to innovate than those working individually or just with expert support (Dolinska and D'Aquino, 2016).

In this chapter, we look at several collective initiatives that address territorial governance structures implemented either within silvopastoral systems or similar territories. These initiatives are organized in four innovation types, following Beaufoy's (2017) proposal.

- Social and institutional (SI), which covers the social aspects of actors, such as the capacity to collaborate and self-organize, as well as the functioning of public institutions that can facilitate innovation-generating processes.

DOI: 10.4324/9781003028437-15

Table 14.1 Overview of the innovative initiatives presented in this chapter

Typology	Initiatives	Status of application in silvopastoral systems	Broadness
Social and institutional innovations (SI)	Local consumer groups	Implemented	Niches
	Collectivizing services		Widespread
Regulation and policy innovations (RP)	Improve the adaptation of regulatory frameworks	Implemented	Niches
	Result-based payments (RBP)	In development but not yet implemented	Niches
Products and market innovations (PM)	Certification and branding	Implemented	Becoming increasingly widespread
	Recovery of consumption of silvopastoral products	Still poorly implemented	Niches
Farming techniques and management (FM)	Regenerative grazing	Implemented	Niches
	Conservation agriculture for healthy soils	Implemented	Niches

- Regulations and policy (RP), which cover the design of norms that can support or favour innovation rather than impose barriers for its progress.
- Products and markets (PM), which includes the development of new products and commercialization processes that allow value to be added.
- Farming techniques and management (FTM), which include innovations that reduce cost, increase efficiency and are able to promote ecological and socio-economic objectives.

Table 14.1 provides an overview of the initiatives detailed in the following sections, the scale of each one and the degree of implementation (i.e. planned but not implemented up to fully functioning). Our goal is to provide an overview of the possibilities rather than to advocate the benefits of any specific pathway since the appropriateness and success or failure of different initiatives are highly context-dependent.

14.2 Social and institutional innovations (SI)

In recent years, there has been a growing interest in making agricultural and other natural resource projects and programmes more effective, reinforcing collaboration among individuals, groups and institutions. One of the more popular examples is the rapid growth of local consumer groups (LCG), which are an expression of new forms of relationship between producers and consumers. In addition, the collectivization of services regarding, for example, herding, feeding, night-penning, milking, health care, shearing, transhumance, as well as haymaking, pruning and debarking are examples of SI that are of relevance for silvopastoral territories.

14.2.1 Local consumer groups (LCGs)

LCGs have grown in popularity around the world in recent years. They often emphasize the re-connection of producers and consumers against faceless and placeless industrial agriculture (Papaoikonomou and Ginieis, 2017). These groups are motivated by concerns over the standards of food quality, animal welfare and nature conservation, qualities which are all high in silvopastoral systems, as well as wishing to support local primary producers and develop 'short circuits'

Although there are a huge variety of models, two of them are particularly relevant: Community Supported Agriculture (CSA) and Responsible Consumption Communities (RCC). Others rooted in self-sufficiency, such as social markets, and barter, in the sense of non-monetized markets also occur (Della Porta et al., 2015). Sharing risk between local farmers and consumers is central to the CSA model. Consumers 'buy' shares in the local farm before planting/rearing starts, and later receive a weekly box with available products. If for any justified reason the products cannot be delivered, the CSA consumers are not usually reimbursed. On the other hand, RCC is mainly local, neighbourhood-based group whose main purpose is the collective purchase of products according to ethical criteria decided by the group. These criteria are usually (1) a preference for small, local producers; (2) an avoidance of intermediaries, as members want a direct relationship with the producer and (3) purchasing products produced under fair labour conditions. In Andalusia, González et al. (2012) found more than 100 short supply chain initiatives in territories dominated by dehesas, including 45 LCG and 15 consumer cooperatives. Some of them were initiated by consumers, others by producers, and others were the result of shared efforts and actions. Some projects were supported by public bodies, local action groups or agrarian associations, but in all cases, the role of 'network facilitators', who contribute to the development and continuity of these experiences, creating links between them and facilitating the exchange of knowledge and resources, was essential. The presence of products from dehesa farms is appreciable (sausage, meat, cheese, honey ...), although not the majority and shows a great capacity for growth. Nevertheless, the success of these initiatives is more likely where there are few producers with close access to large urban communities, which rarely is the case for depopulated silvopastoral territories.

14.2.2 Collectivizing services

The collectivization of farm activities aims to increase farmers' efficiency, profitability and free time. Examples of collective herding can be found among reindeer breeders in northern Scandinavia. During the warm season, the individual herds belonging to each of the units in the district are managed collectively, as a loose but cohesive herd (Valinger et al., 2018).

Another example is the organization of Animal Health Defence Groups, which are common in Spain (ADSG: Agrupación de Defensa Sanitaria Ganadera), France (GDS: Groupements de Defense Sanitaire) and Portugal (OPP: Organização de Produtores Pecuários). In France, the public animal health policies rely on regional veterinary public services, but rural veterinary practitioners back up the veterinary services' personnel by carrying out 'health mandates' such as vaccination. Breeders participate in GDS through the management of regulated prophylaxis, training in preventive measures and continuous information updates. Through mutualization, GDSs manage the financial compensation for farmers, complementing government aid. Similarly, the Spanish ADSG and Portuguese OPP are widely found in rural areas, mostly organized by small groups of farmers (there are more than 2000 groups in Spain and 104 in Portugal), frequently associated with breeder cooperatives. These groups are considered essential for the eradication and control of the main livestock diseases, and for the design of nutrition and reproduction programs, but also enable solidarity in a traditionally individualistic sector, creating an associative fabric that brings together practically all farms at a local/county scale (Amat-Montesinos et al., 2019).

14.3 Regulatory and policy innovations (RP)

As described in Chapter 9, silvopastoral systems are affected by multiple regulations that concern agriculture cross-compliance and payments, livestock welfare, forest management and nature conservation, alongside food security rules for multiple products. Accordingly, some collective actions are working to develop tailored measures for specific sectors and for the system as a whole. We have selected two examples encompassing farmers' associations that aim to establish 1) rules that are best suited to small producers; 2) multi-sectoral networks that aim to develop agreed-on result-based payment schemes. We refer to these collective actions as RP initiatives, as they work on designing new norms to favour the maintenance of silvopastoral systems, but they can also be viewed as SI initiatives since they involve a strong capacity for collaboration. Many other RP-SI initiatives can be found in the frame of the 'new commons', in which local communities search to convert territories to contemporary agro-environmental uses (Woestenburg, 2018).

14.3.1 Improve the relevance of regulatory frameworks

Nowadays, the main handicap facing silvopastoral production is that standards and regulations are not adapted to their reality and production systems. Regulatory problems often prevent their further, and better, development. One example that aims to meet this challenge is the Spanish Network of Field and Artisanal Dairies (QueRed)[1] which supports small livestock producers who are deeply rooted in their territory, and frequently hold an important element of the cultural collective consciousness. QueRed aims to be the interlocutor with governments to improve and adapt the regulatory framework for artisanal cheese and other livestock-based products, working on legislative changes from local to European levels. QueRed is a member of the Farmhouse and Artisan Cheese and Dairy Producers European Network (FACEnetwork),[2] an association with members in 15 countries. Besides farmers and artisanal producers, QueRed includes animators who promote the consumption of artisanal cheeses, by offering the members and associated cheese producers the opportunity to participate in fairs and festivals. They also develop training and dissemination activities upon demand of producers, in addition to advising and tutoring associates with personalized information and training. Despite their efforts, progress is slow as relaxing the strict sanitary regulations designed for, and by, large industries is challenging.

14.3.2 Result-based payments (RBP)

RBP aims to provide higher payments to farmers who produce higher quality environmental services, (Byrne et al., 2017). RBP scheme measures appear to be a promising initiative for collective actions in silvopastoral systems (Guimarães et al., 2019). Schemes of this nature have focussed on, for instance, biodiversity conservation (Byrne et al., 2017), high nature value faming systems (Muñoz-Rojas et al., 2019) and the conservation of grassland habitats (Sainte Marie, 2014).

RBPS in Spain is still in an early stage of development, though some regions have been involved in similar schemes for a long time. The most well-known is the RAPCA, the network of grazed areas for wildfire prevention in Andalusia (see Chapter 16 in this volume), while the most advanced has been developed by GAN-NIK in Navarra under the umbrella of an European Results Based Agri-environmental Payment project (RBAPS Schemes).[3] This project relies on the assessment of indicators that are agreed among farmers, administrations and an environmental NGO as the third party, and defined for quick monitoring that can be performed by the farmers themselves.

The project Sembrando Dehesa,[4] is an example of how private funds and foundations can work together with conservationist NGOs for nature conservation, getting experts and those who work day to day managing silvopastures and grazing flocks to work hand in hand. Project actors have collaborated in the definition and implementation of good management practices, integrating traditional knowledge into new management schemes to respond to new societal demands. They have produced guidelines of user-friendly indicators of good management practices for the Iberian dehesa and montado that can be used in future RBP schemes. This collaboration also builds support for tailored public policies and consumers' awareness about the importance of these systems. Its goal is a model that is socially and economically sustainable, that restores the natural and cultural values of the dehesa and montado. However, the life of these initiatives beyond the funding period is always uncertain, and only the creation of stable cooperation structures can sustain project-derived outcomes.

In this regard, The Spanish Platform on Extensive Livestock Farming and Pastoralism (see Chapter 13), is advocating for the implementation of practice-based eco-schemes focused on pastoralism and extensive livestock farming. In the Montado, there is a focus on a structured agri-environmental scheme, a 'result-based payment' in the CAP post 2020, which aims to reward farmers for achievements on the condition of the montado as a multifunctional system (Ferraz-de-Oliveira et al., 2019). The innovative co-construction of this scheme involves farmers, public officers and researchers in the different steps of development of a pilot case, in a Natura 2000 site and its surroundings, in Alentejo, south Portugal.

14.4 Products and market innovations (PM)

Current concerns about the quality of life, including food quality and nature conservation, have created new opportunities for the economies of silvopastoral territories. Differentiation and different value chains could contribute to the maintenance of silvopastoral territories and the livelihoods of producers within them. In this section, we present examples of two collective initiatives aiming to revitalize traditional silvopastoral products through exploring new marketing strategies and creating fashionable and healthy products based on local food resources.

14.4.1 Certification and branding

In Spain, the De Yerba Association is promoting the consumption of grass-fed meat by facilitating contact between producers and consumers. They organize collective purchases of featured products to reach a minimum order quantity. Their webpage[5] promotes the restaurants and consumer groups that purchase grass-fed meat and other grass-feed initiatives. All producers have to comply with a transparent internal breeding protocol. The association also provides detailed information on grass-fed meat, including its benefits, recipes, news and even organizes visits to member farms. This, and other initiatives, aim to add value to silvopastoral products by promoting differentiation and awareness. They are rooted in direct trust between producers and consumers; although this trust becomes more difficult to build when the distance between producers and consumers is great (especially for urban citizens) it is often provided by independent third-party certification bodies.

In a similar vein, Patagonia (Argentina) also has an initiative for silvopastoral territories, GRASS certification (their Grassland Regeneration and Sustainability Standard; Borrelli et al., 2013), which encourages sustainable grazing throughout Patagonia and creates a value chain that aims to maintain demand for wool sourced in a sustainable manner from certified farms. Today the initiative includes more than 160 producers with over one million sheep grazing more than 1.3 million ha. Elements of this protocol are being evaluated and applied in many other countries including the USA, Mongolia and Spain.

Certification is an example of a PM innovation insofar as it seeks to collectively identify a common message and establish a clear and simple message for consumers. Yet, the growing number of eco-labels, brands and differentiated productions has generated confusion among consumers and to some extent damaged the image of environmentally friendly productions. Current recommendations suggest the unification and simplification of messages (Moon et al., 2017). Following this recommendation, a group formed by the Entretantos Foundation and the Spanish Platform on Extensive Livestock is working to establish an umbrella branding for all extensive livestock production that can unite and coordinate the myriad of small artisanal brands (while allowing them to keep their own identity) seeking to differentiate their products as coming from silvopastoral systems.

14.4.2 Recovering the use of silvopastoral products

Acorns have a long history both as food for humans and feed for animals. Although in the last quarter of the 20th Century acorns were mostly used to feed animals (particularly for fattening free-ranging pigs in Iberian Peninsula), there has been, in the last decade or so, a renewed interest in the human consumption of acorns and related products. This is part of a present consumer-driven trend for sustainable and functional foods and is resulting in a new dynamic involving various acorn-related sectors such as production, processing, commercialization, cooking and consumption.

In Portugal, this acorn movement was initially mostly driven by one farm that started to work on the recovery of old gastronomic uses of acorn (e.g. acorn flour for bread), but also pushed forward research for new products such acorn based drinks. This dynamic was embraced by a municipality in Central Alentejo, Montemor-o-Novo, that since 2016 has been organizing, yearly, the annual acorn week. During the acorn week, restaurants in Montemor-o-Novo serve acorn-based meals and food products. The week is used as a time for creating awareness among locals and visitors of the value of acorns as food for humans, rather than just feed for animals. Another initiative both resulting from, and driving, this dynamic is the organization for the last 3 years (2017–2019) of the Iberian Conference on Acorns that gathers a wide range of delegates interested in the use of acorns as human food. Besides these collective initiatives, a number of new start-up small companies are transforming increasing quantities of acorns for human consumption.

In 2020, the *Confraria Ibérica da Bolota* association was created, gathering members from Portugal and Spain. Its objective is to promote acorns, and the diversity of products that can be made from them, as well as to protect their ecosystem of origin. This association brings together silvopastoral farmers, researchers, small-scale processing industries, chefs and consumers. Although there has been a renewed interest in the consumption acorns in recent years, its potential as food or additive remains largely unexplored and private and communal initiatives should be supported by a public research agenda.

14.5 Farming techniques and management (FM)

Given new consumer preferences and public regulations, farmers have had, in many cases, to adopt new management practices to meet new standards of food security, livestock welfare and nature conservation. While many changes in the traditional practices come with technological advances or are imposed by administrations, there are also voluntary changes that, in some cases, are adopted collectively. Here we present two examples that have been started a few years ago, one concerning regenerative livestock farming in the dehesa and the other conservation agriculture for healthier soils at the core of the montado. We also include a Text Box 14.1 describing land stewardship initiatives in Spain, an example of silvopastoral practices that enhance nature conservation.

BOX 14.1

Land stewardship for enhancing nature value inn conservation-based silvopastoral systems

Gerardo Moreno, Universidad de Extremadura, Spain

Key point: Land stewardship initiatives are becoming common in Spanish dehesas

Land stewardship involves the conservation and enhancement of a property's most important natural resources and involves long-term planning. Land stewardship is developing strongly in Spain, and there is a dense network of collective actions (see map below) that is integrated in the state-wide Network of Land Stewardship Entities. The network includes a multitude of stakeholders, divided into actors (who manage landscapes on the ground, such as landowners, farmers, gardeners and foresters), enablers (e.g. administrations, researchers and funding bodies that provide supportive frameworks), facilitators (bridging organizations, such as land care groups and environmental NGOs) and civil society.

The Fundación Biodiversidad (a government body on nature conservation) provides technical support to the Land Stewardship Platform[9] and is a meeting point for all stakeholders around land stewardship. The collaboration has helped the success of land stewardship in Spain, with 2487 land stewardship agreements, covering 370 272 ha, agreed with 166 entities (2017 data). As shown in the Figure 14.1, many of the actions are based in central-western Spain, where more than 50,000 ha of dehesa are managed under some form of land stewardship agreement. For instance, the NGO Fundación Naturaleza y Hombre (Nature and Human Foundation[10]; manages a territory of about 132,600 hectares of Natura 2000 sites, which occupy a large proportion of the Spanish dehesa.

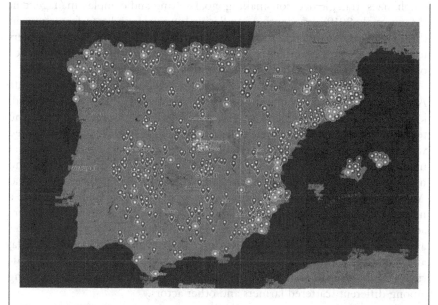

Figure 14.1 Map of Land Stewardship initiatives in Spain in 2017. The goals mostly concerned the conservation of habitats (23%), conservation of fauna species (21%), maintenance of traditional management practices (12%).

The Network and the Platform have worked together in the White Book of Land Stewardship[11] advocating for collective and participatory management in nature conservation and work actively to include extensive livestock within the regulatory frameworks for nature conservation and agriculture. This is reflected by the creation, in 2019, of the Por Otra PAC,[12] a coalition which is lobbying for the creation of a new payment measure (agro-eco-scheme) for silvopastoral territories managed under land stewardship agreements.

14.5.1 Regenerative grazing

There is a growing movement to develop certification for agricultural systems that are deemed to be not just sustainable, but regenerative (Elevitch et al., 2018). Regenerative farming systems focus on improving soil health through increasing soil carbon and life in the soil. Within the wide umbrella of regenerative agriculture, adaptive grazing (under any of its numerous terms, such as Prescribed Grazing, Holistic Management, or *Voisin* Grazing Management) is being progressively adopted by more and more dehesa farmers, despite the scientific evidence on the productive and ecological advantages of these practices still being controversial (Hawkins, 2017).

Adopters have to take decisions on the grazing system, which combines high pressure grazing livestock with frequent rotations, together with many other organizational adaptive and holistic decisions. The adaptive grazing method emphasizes strategic decision-making, goal-setting and complex management (Mann et al., 2019). Its adoption implies making profound transformations in the daily management of the farm and the farmers' life. This motivates adopters to be eager to share experiences and information and has led to the emergence of two nationwide associations in the Spanish dehesas over the last decade.

The Iberian Regenerative Agriculture Network (Red Ibérica de Agricultura Regenerativa)[6] was formed by individuals and entities of all kinds that share the regenerative vision and seek to promote the tools and techniques to pursue regenerative agriculture. This network is connected with Holistic Management International,[7] a non-profit organization that maintains an international network of educators and 'land stewards' who use holistic management strategies. Similarly, ALEJAB (Asociación Juntas Arreglamos la Biosfera)[8] is a nationwide association that brings together a network of farmers, advisors and land stewards working towards regenerative livestock farming. Both associations facilitate collaborative initiatives among farmers across the main silvopastoral areas of the Iberian Peninsula, and specifically, promote pioneer farms that act as living labs holding multiple training events that facilitate knowledge sharing among networking among different scattered farmers and other actors.

14.5.2 Conservation agriculture for healthy soils

A group of Alentejo farmers, concerned about the economic and environmental sustainability of their farms, and in particular the risk of soil degradation, have abandoned conventional soil tillage and moved to conservation agriculture. This informal group gathers around a field researcher, who is respected both among the research and farming communities, and functions as the 'leader' of the group. The group has been meeting for about 15 years to share experiences, knowledge, technical and scientific support. A significant part of the group manages the montado system applying the principles of conservation agriculture mainly through improving soil fertility by using mineral corrections with no-tillage, direct seeding and grazing management to increase soil organic matter, soil fertility and pasture productivity. The group has a fluctuating number of farmers (about 20) and is informally constructed. There is no real institutional support and no funding to support meetings or even travelling expenses. Replication of such an experience requires funding for specific training in order to be able to train field technicians capable of providing technical assistance in similar structures.

14.6 Lessons learnt and future perspectives

In this chapter, we have outlined examples of collective actions implemented at different territorial scales that involve different governance schemes, based on the adaptation of traditional practices to new labour conditions, lifestyles, market conditions and societal demands. These examples show that these new social organizations focus not only on the production of deliverables but also on three key strategic processes: (1) fostering the drivers of local mobilization; (2) improving the quality of local productive elements; and (3) having a coherent learning strategy (Divay, 2016).

Given the low productivity of extensive silvopastoral farms, it is essential to derive economic advantages from the provision of the natural and cultural values of these systems to guarantee their economic survival. However, the pace at which new market demands and environmental changes arise exceeds the capacities of individual managers to react accordingly, generating a need for joint participatory actions. To encourage the success of many of the innovations described in this chapter, social and institutional arrangements need to change and be supported by reinforcing regulations and policies that influence products and markets, as well as by the adoption of new FM practices. In this regard, large collectives of consumers have become very important buyers of the produce of silvopastoral territories, opening up new opportunities for the livelihoods of silvopastoral farmers. Working together with local consumer groups, different initiatives for the certification and/or intermediation among producers and urban consumers are branding silvopastoral products, enlarging their scope, visibility and economic viability in the marketplace.

The examples described here also show that environmental groups, mostly conservationist NGOs funded by national and international foundations and public programmes (notably the LIFE programme of the European Commission), are becoming key actors in silvopastoral areas. These organizations are boosting collective actions for the conservation of nature values, and are also responsible for different schemes that reward collective approaches that aim to conserve biodiversity and grassland habitats, support high nature value silvopastoral systems, improve connectivity, water flows and water quality and prevent wildfires. In this regard, public result-based payments and landscape stewardship schemes have the potential to provide good results and pilots need to be established that can be properly monitored and assessed.

Bureaucratic burdens and the unwillingness of traditional famers to change are two barriers that slow down the spread of these collective actions. A hybrid governance approach might be one way of overcoming these obstacles. This approach means that decisions on objectives take place in a top-down, centralized manner (society), while decisions on actions (to achieve the objectives) are taken in a bottom-up territorial manner (by farmers and/or the local population). The communal use of silvopastoral territories, widely practiced across the world in the past, requires new governance schemes that, even if founded on traditional structures and rules, incorporate new approaches and actors.

What the examples presented here show is that spontaneous large-scale collective actions are difficult to organize and the problems of up-scaling local initiatives may demand the involvement of an external third party to guide and coordinate such actions (Jagers et al., 2019). This implies the need for farmers, consumers, environmental experts and organizations, government agencies, and the local population, to work together, sensing 'what should be', watching 'what is' and finally doing 'what can be done' to increase the economic, social and ecological resilience of silvopastoral farms and territories. Collective actions can play an important role here. The COVID-19 pandemic situation was a chance to add new arguments to this discussion (see Text Box 14.2).

BOX 14.2

Collective action during COVID-19 state of alarm

Pedro M. Herrera, Fundación Entretantos, Spain

Key points– The COVID-19 crisis forced collective action in Spain in order to cope with the devastating effect on small-scale livestock production and generated joint actions, creativity and collaboration.

The coronavirus crisis in Spain led to numerous initiatives to deal with the enormous toll that the pandemic took on small producers. The consequences of the confinement were especially harmful to dehesa and other land-based production and were fuelled by the co-occurrence of several factors. First, there was the closure of producer markets, (although big distributors kept their businesses intact). Farmers also suffered from the closure of slaughterhouses and local shops, followed by harsh speculation that depleted the prices and distribution of their products. Last but not least, it coincided with the Easter season, a sales peak for dehesa produce (lamb, goat kids, cold meats, etc.). This situation quickly became an emergency that generated collective action from producers, conservationists and other NGOs. More than 600 organizations sent a letter to the two ministries in charge, defending local production, local food, producer markets and short circuits during the COVID-19 pandemic. In addition, a constellation of collective initiatives emerged during March and April 2019, of which this box shows three actions aiming at supporting the areas and production of the dehesa.

The first one, led by QueRed, proposed to authorities ten measures for small dairy and livestock farming, including authorizing on-site slaughter of lambs and kids, shared workrooms for producers, collaboration with the food chain and the promotion of public purchasing of agroecological food supplies. The second initiative was a campaign of collective donations and purchases of goat products, developed by Cabrandalucía (Andalusian Goat Breeders Association). This #YoNoRompoLaCadena campaign (I don't break the chain) coordinated purchases and donations (made by producers and supporters' organizations and individuals) of goat products (milk, cheese, meat...) directed to hospitals, elderly residences and other key institutions (Figure 14.2). Finally, stakeholders in lamb production joined together under the campaign "At Easter we eat lamb", establishing an online shopping and door-to-door delivery system to reach individual homes wishing to buy a lamb for Easter. The actors of this campaign were the EA Group (a second-degree institution encompassing Extremadura and Andalusia sheep cooperatives), along with the PGI (Protected Geographical Indication) and the inter-professional organization for sheep and goat producers.

#YoNoRompoLaCadena

#ConsumeProductosCaprinos

Figure 14.2 Poster of the YoNoRompoLaCadena campaign (I don't break the chain), promoted in Spain in support of small farmers during the months of the curfew due to COVID.

Notes

1. https://www.redqueserias.org/actividades-que-estamos-desarrollando/
2. https://www.face-network.eu/
3. https://www.rbaps.eu/
4. https://www.wwf.es/nuestro_trabajo/alimentos/sembrando_dehesas/
5. https://www.lacarnedepasto.com/
6. https://www.agriculturaregenerativa.es/
7. https://holisticmanagement.org/
8. http://www.manejoholistico.net/
9. https://custodia-territorio.es/la-plataforma
10. https://fnyh.org/
11. https://custodia-territorio.es/libro-blanco-de-la-custodia-del-territorio
12. https://www.custodia-territorio.es/novedades/politica-agraria-comun-coalicion-por-otra-pac

References

Amat-Montesinos, X., Martínez, A. & Larrosa, J. (2019). "La ganadería extensiva en el desarrollo territorial valenciano. Reconocimiento público y experiencias sociales". *TERRA. Revista de Desarrollo Local*, 5, 32–54.

Barroso, F., Menezes, H. & Pinto-Correia, T. (2013). "How can the land managers and his multi-stakeholder network at the farm level influence the multifunctional transitions pathways?" *Spanish Journal of Rural Development*, 4 (4), 35e48.

Beaufoy G. (ed). (2017). Comparative Collection of High Nature Value Innovations, Experiences, Needs and Lessons, from 10 European "Learning Areas". THE HNV-LINK COMPENDIUM – HNV-Link WP2, deliverable 2.6.1. Cuacos (Spain).

Borrelli, P., Boggio, F., Sturzenbaum, P., Paramidani, M., Heinken, R., Pague, C., Stevens, M. & Nogués, A. (2013). Grassland regeneration and sustainability standard (GRASS). The Nature Conservancy-Ovis XXI SA, Buenos Aires.

Byrne, D., Maher, C., Alzaga, V., Astrain, C., Beaufoy, G., Berastegi, A., Bleasdale, A., Clavería, V., Donaghy, A., Finney, K., Kelly, S.B.A., Jones, G., Moran, J., O'Donoghue, B. & Torres J. (2017). Developing result-based agri-environmental payment schemes (RBAPS) for biodiversity conservation in Ireland and Spain. Grassland resources for extensive farming systems in marginal lands: major drivers and future scenarios. *Grassland Science in Europe*, 22, 299–301.

Della Porta, D., Andretta, M., Calle, A., Combes, H., Eggert, N., Giugni, M.G., Hadden, J., Jiménez, M. & Marchetti, R. (2015). *Global Justice Movement: Cross-national and trans-national perspectives*. Routledge. Oxon.

Divay, G. (2016). "Public performance and the challenge of local collective action strategies: Quebec's experience with an Integrated Territorial Approach". *International Review of Administrative Sciences*, 82(3), 472–489.

Dolinska, A. & d'Aquino, P. (2016). "Farmers as agents in innovation systems. Empowering farmers for innovation through communities of practice". *Agricultural Systems*, 142, 122–130.

Elevitch, C. R., Mazaroli, D. N. & Ragone, D. (2018). "Agroforestry standards for regenerative agriculture". *Sustainability*, 10(9), 3337.

Ferraz-de-Oliveira, M.I., Guimarães, M.H. & Pinto-Correia, T. (2019). "The Montado case study. Co-construction of locally-led innovative solutions". *La Cañada*, 31: 16–17. ISSN 1027-2070.

González, I., de Haro, T., Ramos, E. & Renting H. (2012). "Circuitos cortos de comercialización en Andalucía: un análisis exploratorio". *Revista Española de Estudios Agrosociales y Pesqueros*, 232, 193–227.

Guimarães, M.H., Esgalhado, C., Ferraz-de-Oliveira, I. & Pinto-Correia, T. (2019). "When does innovation become custom? A case study of the Montado, southern Portugal". *Open Agriculture*, 4, 144–158.

Hawkins, H. J., Short, A. & Kirkman, K. P. (2017). "Does Holistic Planned Grazing™ work on native rangelands?" *African Journal of Range and Forage Science*, 34(2), 59–63.

Jagers, S. C., Harring, N., Löfgren, Å., Sjöstedt, M., Alpizar, F., Brülde, B., Langet, D., Nilsson, A., Almroth, B.C., Dupont, S. & Steffen, W. (2019). "On the preconditions for large-scale collective action". *Ambio*, 49, 1282–1296.

Mann, C., Parkins, J. R., Isaac, M. E. & Sherren, K. (2019). "Do practitioners of holistic management exhibit systems thinking?" *Ecology and Society*, 24(3), 19.

Markelova, H., Meinzen-Dick, R., Hellin, J. & Dohrn, S. (2009). "Collective action for smallholder market access". *Food Policy*, 34(1), 1–7.

Moon, S. J., Costello, J. P. & Koo, D. M. (2017). "The impact of consumer confusion from eco-labels on negative WOM, distrust, and dissatisfaction". *International Journal of Advertising*, 36(2), 246–271.

Muñoz-Rojas, J., Pinto-Correia, T., Thorsoe, M. H. & Noe, E. (2019). "The Portuguese montado: A complex system under tension between different land use management paradigms". In *Silvicultures-Management and Conservation*. IntechOpen. London. doi: 10.5772/intechopen.86102.

Ostrom, E. (2000). "Collective action and the evolution of social norms". *Journal of Economic Perspectives*, 14(3), 137–158.

Papaoikonomou, E. & Ginieis, M. (2017). "Putting the farmer's face on food: governance and the producer–consumer relationship in local food systems". *Agriculture and Human Values*, 34, 53–67.

Ploeg, J. D. Van der (2020). "The political economy of agroecology". *The Journal of Peasant Studies*, Ahead-of-print, 1–24. doi.org/10.1080/03066150.2020.1725489.

Sainte Marie, C. de (2014). "Rethinking agri-environmental schemes. A result-oriented approach to the management of species-rich grasslands in France". *Journal of Environmental Planning and Management*, 57(5), 704–719.

Valinger, E., Berg, S. & Lind, T. (2018). "Reindeer husbandry in a mountain Sami village in boreal Sweden: the social and economic effect of introducing GPS collars and adaptive forest management". *Agroforestry Systems*, 92(4), 933–943.

Woestenburg, M. (2018). "Heathland farm as a new commons?" *Landscape Research*, 43(8), 1045–1055.

15 Collective and individual approaches to pastoral land governance in Greek silvopastoral systems

The case of sheep and goat transhumance

Athanasios Ragkos and Stavriani Koutsou

15.1 Introduction

The modernization of agriculture has caused significant changes in the quantity and diversity of farm outputs as well as in agricultural production systems and the institutions associated with them. Rural areas have witnessed depopulation and the new production systems that have been adopted have had a significant environmental impact. This has prompted the engagement of non-rural groups which increasingly demand to have an input into decision making regarding farming practices and the use of natural resources. These changes were encouraged by national and supranational (Common Agricultural Policy (CAP) policies, which favoured and accelerated a transition from traditional production to new models, one of the key features of which has been the individualization of the producer.

The effectiveness of the governance of common resources is an issue of particular importance for Greece when it comes to the use of its pastures since these are owned by the State and can be considered to be public goods. As will be shown in this chapter, the transition towards more intensive and market-oriented livestock production, combined with policy changes, has altered a long-standing system that had previously largely ensured the sustainable use of pastoral land. A gradual disconnection of livestock production from the land is being witnessed and the sector's economic performance is now driven by capital investments and a more efficient organization of labour. The latter pursuit became more important with the economic crisis, that hit Greece harder than most other places and principally involves more efficient use of unpaid family labour (Ragkos et al., 2018). The former, on the other hand, has been a driving force for decades and involves modernization through using improved sheep and goat breeds, new machinery and buildings and variable capital for feedstuff, veterinary drugs and other purchased inputs. This pattern of intensification largely ignores the vital role of grazing in the economic performance

DOI: 10.4324/9781003028437-16

of livestock farms. For instance, a recent comparison showed that, with proper grazing management, extensive farms can achieve cost savings of up to 47% compared to intensive ones (Ragkos et al., 2014a).

Extensive livestock systems are constantly shrinking, partly due to the strategy of Greek governments of supporting intensification in primary production in general and partly due to a disabling legislative framework over access to pastures. The existing framework has not adopted a long-term sustainable perspective and pushes farmers to intensify and to reduce their use of pastoral resources. Operating under such a bureaucratic system, farmers are frequently overwhelmed by their private interests and strive to meet their own needs while neglecting the long-established collective management of common resources.

Silvopastoral systems are particularly affected by these trends since they make efficient use of lands with a high level of woody biomass which is of low quality in terms of forage production and generally not suitable for other uses. These lands play a vital role for small ruminant nutrition during the prolonged Mediterranean dry-period (June to October) (Papachristou and Papanastassis, 1994) and allow extensive farms to operate without having to buy feedstuff from markets. They have traditionally been used by sheep and goats in Greece's mountainous and less-favoured areas. However, inefficient governance can jeopardize the viability of these vulnerable extensive systems and lead to the loss of their important socio–economic functions.

The purpose of this chapter is to demonstrate how the evolution of pasture management from traditional/collective to bureaucratic/individualistic patterns has led to diverse governance schemes across the country and how these developments threaten the sustainability of silvopastoral transhumant production. The main argument is that governance schemes can incorporate significant traditional knowledge rather than 'impose' uniform solutions for all livestock systems and areas and thereby become more sustainable. This is also very relevant for reforging the disrupted relationships between land use and livestock production.

We have chosen to focus on transhumance in this chapter as it has a traditional and multifunctional character and is practiced in almost every part of the country. Transhumance is characterized by a unique and formidable connection to land, which makes it of specific importance in understanding how the development of land use systems affects the socio–economic and environmental performance of livestock production. The particularity of transhumant silvopastoral systems also relates to the use of low-productivity pastoral land, which is typically not used by other livestock systems.

The chapter provides an assessment of various governance schemes for pastoral land in several different areas of Greece. After providing a critical description of the actual framework for the allocation of pastoral land, five governance case studies are presented, each of which faces and poses different challenges. Based on this analysis, the chapter seeks to pinpoint key aspects of the dynamics of silvopastoral transhumant systems that could be translated into tools to support all pillars of sustainability both now and in the future.

15.2 The current situation in the Greek transhumance sector

Transhumance involves a seasonal and circular movement of flocks towards mountainous pastures in the summer and their return to their specific winter domiciles (Nyssen et al., 2009). This system of livestock production is typical of numerous Mediterranean and Balkan territories (Pardini and Nori, 2011; Vallerand, 2014). According to Bunce et al. (2004), the term transhumance encompasses all extensive livestock systems based on displacement. However, not all forms of extensive animal production are transhumant nor all systems involving livestock movements can be categorized as such. Indeed, extensive systems are also referred to as 'pastoral', describing livestock production systems that are based predominantly on grazing of natural vegetation (Bourbouze and Chassany, 2008; Farinella et al., 2017). On the other hand, the latter also includes 'nomadism', in which livestock move across areas depending on the availability of resources without specific settlements (Vallerand, 2014). Therefore, transhumance constitutes a type of pastoralism that differs from nomadism. It is interesting to mention that transhumant farmers, following this circular movement, have created establishments (e.g. houses, barns, sheds) in both the lowlands and the highlands.

Transhumance is highly dependent on the availability of pastoral land in the highlands – where they are the mainland users (Karatassiou et al., 2015) – and the lowlands. Although the number of flocks is decreasing, the total number of animals remains constant resulting in increases in grazing pressures in specific areas. This means that pastoral resources are simultaneously facing problems of over-and under-grazing: over-grazing entails a reduction in the diversity of broad-leaved forbs and legume species (Papanastasis, 2002), while under-grazing is connected to encroachment and desertification (Sidiropoulou et al., 2015) and favours the expansion of some grass species over the ones more preferred by animals grazing in pseudo-alpine areas.

Changes in grazing pressures and structural developments in Greek sheep and goat transhumance are closely related to the modernization of the sector in recent decades. In general, the sector has maintained its traditional character, when it comes – for instance – to milking by hand (Lagka et al., 2014), the rearing of local breeds (Loukovitis et al., 2016), low dependence on fixed capital (Ragkos et al., 2014a) and the low use of antibiotics and purchased feedstuff. Modernization is mainly sought through the introduction of imported dairy breeds and uncontrolled crossbreeding to increase productivity. In addition, many farmers make investments in machinery and buildings, mainly aiming to cope with the increased requirements of keeping their animals confined in the lowlands in winter.

Transhumance fulfils important social, cultural, economic and environmental functions both at the local scale and nationwide (Hadjigeorgiou, 2011; Ragkos and Lagka, 2014; Koutsou et al., 2019). It is the main source of employment in many mountain communities across the country, supporting rural livelihoods and averting depopulation trends. It is also a profitable activity, partly as the use of natural pastures in summer generates cost savings in feeding expenses and partly due to the provision of EU payments (Ragkos et al., 2014a). Transhumance also plays a cultural role, safeguarding norms, traditions, customs and practices, which led to its inscription as an Intangible Cultural Heritage of Humanity in late 2019 (ich.unesco.org). The environmental role of transhumance is two-fold, involving the valorization of local sheep and goat breeds and the maintenance of high nature value (HNV) habitats.

In Greece, there are two main systems of transhumance: small ruminants (sheep and goats) and cattle, with notable differences between them. Cattle transhumance is steadily developing across the country because it generates high economic returns due to policy support, lower labour requirements, and satisfactory meat prices due to the shortage in beef meat in the Greek market (the country is a net importer, as only around 30% of demand is met by national production) (Ragkos et al., 2013; Koutsou et al., 2019). It is estimated that there are about 45,000 transhumant bovines throughout the country, mostly in the central and north-eastern parts (Ragkos et al., 2013). While bovine transhumance is not of direct relevance to silvopastoral systems, its expansion is of importance for the governance of pastoral land in specific areas (Koutsou et al., 2019), as will be shown in Section 15.4.

Despite this growing trend, sheep and goat transhumance remains the prevailing form of transhumance in Greece. After a serious decline in the number of farms during the 20th century, processed data from the Greek Payment and Control Agency for Guidance and Guarantee Community Aid (OPEKEPE) show that there are currently 3,051 transhumant farms rearing more than one million sheep and goats (about 7.5% of the national flock). This does not include unregistered flocks on Greece's many islands or flocks that do not move every year. The distances moved can vary greatly from 10-20 km (local movements from the valley plains to the mountains named 'trasterminance') to 300 km and more. The flocks leave the lowlands in spring and usually remain in the mountains for 4 to 8 months, depending on the area.

Transhumant farms can be found in all Greek regions, although there are notable differences amongst them, even within the same region. The typological study of transhumant farms by Ragkos et al. (2014b) discerned five types.

1 **Goat farms** (11.7%) predominantly engage in local movements and are found mainly in the southern Peloponnese (Lakonia) and in northern Greece (Central Macedonia), where the climate and soil conditions favour goat farming. It is evident that this type prevails in silvopastoral systems, as goats are the main users of woody pastures.

2 **Mixed farms** (15.5%) that rear sheep and goats. They engage in local movements either on foot or using trucks. This type is particularly relevant for silvopastoralism, as it is common for areas with woody

vegetation, shrublands and olive groves, and is particularly prevalent on the islands of Thassos, Chios and parts of Crete, as well as in the southern Peloponnese.

3 **Farms engaging in remote movements** (25.4%). Flocks of this type move at least 100 km in spring or summer to reach their summer domiciles and graze mostly grasslands at relatively high altitudes and – more rarely –shrublands.

4 **Farms performing small local movements** (30.1%) are the most abundant. They are scattered throughout the country, mainly in lowland regions near mountains. Flocks are moved, on foot, 5-15 km away from their winter base in order to take advantage of natural vegetation in neighbouring highland areas, some of which are of relevance for silvo-pastoral systems, especially in the southern part of the country.

5 **Small regional farms** (17.4%) rear sheep and goats and engage in local movements. This type is most common in the western part of the country but also on the island of Evia, where these transhumant flocks have developed a silvopastoral system of importance for local livestock production, as sheep graze mainly on phryganic lands and pastures with significant woody compounds.

Another aspect that needs to be taken into account in this typological study is the heterogeneity in the ways that livestock movements are performed, with long-distance transhumance on foot being almost completely substituted by the use of trucks. Local movements, as well as 'trasterminance' are still done on foot in most parts of the country, but only a very small proportion of long-distance transhumance (more than 100 km) occurs without the use of trucks. The abandonment of transhumance on foot has led to encroachments across the paths/routes traditionally used by transhumant flocks. However, the silvopastoral landscapes are not severely affected by this development, as most goat and mixed flocks are moved by foot.

15.3 Transhumance and land uses in Greece

15.3.1 The current rangeland allocation system in Greece

The importance of sheep and goat transhumance contrasts with the increasing trend of intensification and disconnection from land currently occurring in the Greek ruminant sector. Pursuit of a productivist development model in previous decades, combined with that of an easier lifestyle, led many extensive livestock farms to choose to become sedentary or to confine their animals indoors throughout the year. Although the available official statistical data cannot capture this transition, its effects are evident in the steady increase of ungrazed pastures, with the subsequent adverse effects of encroachment and the likelihood of wildfires. This is particularly prevalent among traditional silvopastoral systems throughout Greece.

Rangeland governance evolves and adapts to external conditions and policies over time. Until the first decades of the 20th century, transhumance operated in the form of *tseligato*, an informal cooperative form in which farmers united their flocks and their families to collectively face dangers under the leadership of the owner of the largest flock (Karavidas 1931; Kavadias, 1996; Koutsou et al., 2019). Transhumant farmers had developed a system for assessing the grazing capacities of pastoral land based on traditional ecological knowledge (empirical observations of nature, vegetation and weather conditions). Each *tseligato* would graze its animals on a particular part of the mountain each year, based on these criteria. Nonetheless, transformations in transhumant communities after the 1930s – and mainly during the second half of the 20th century – brought the prevalence of separate family farms and competition in land use and led to the abandonment of much collective rangeland management.

Today, traditional ecological knowledge is rarely taken into account in the allocation framework for pastoral land (Koutsou et al., 2019). The Greek system of rangeland use is communal (Ragkos et al., 2020), which is not unusual for Europe (Oosterhuizen, 2011). The communal element of this system is related to most pastures being owned by the State, with farmers who are permanent residents in a specific community allowed to graze their animals within a specified part of the community land for a given period (usually one year or during their stay in the highlands). The management and allocation of common pastures under this system is the responsibility of Decentralized Governments, which also define the grazing fee that farmers need to pay per animal. Farmers who are not permanent residents in the community can either be directly allocated the remaining land by paying the same or a higher fee or gain access through an auction. In this case, the individual(s) who place(s) the highest bid is awarded the use of specified areas for a given period of time (usually one year).

The exact acreage of land allocated to locals depends on their livestock numbers and on a rough estimation of grazing capacity, based on official indicators and literature estimations of the land's elevation (lowland, semi-mountainous, mountainous zones) for four main types of Greek rangelands (grasslands, phryganic lands, shrublands, forested rangelands) (Common Ministerial Decision 117394/2932/30-12-2014). However, the acreage of the area that each farmer is entitled to use is not directly related to the size of the herd/flock and/or the animal species. In some cases, this is the cause of conflicts. In addition, this bureaucratic top-down process does not incorporate local norms, traditional knowledge or specific interventions to maintain productivity (e.g. rotational grazing). Although the use of a specific area is typically awarded for one year, this does not permit users to undertake light interventions to ensure its sustainability (for instance fencing or sowing grass species) and also increases uncertainty about whether a farmer will be allocated land for animals to graze on in the future. The case studies presented in Section 15.4 illustrate some of these issues in practice.

15.3.2 Challenges and the evolution of a new rangeland allocation system

This system is in need of radical reforms in order to effectively reconnect livestock production to land and to achieve higher levels of socioeconomic and environmental sustainability. Nonetheless, this issue has only recently received adequate attention, due to the CAP 2014-2020 introducing changes in the eligibility of pastoral areas for income support payments.

1   Permanent grasslands were characterized as agricultural areas.
2   Grasses should definitely be the dominant type of vegetation (i.e. exceed 50% of the eligible area), although previously shrublands and other types of grazing land were also equally eligible, even if the rock coverage was significant. Exceptions were allowed when the basic grazing material (forage) was traditionally other than grasses (EC/2393/2017).
3   Permanent grasslands can only be eligible if they are appropriate for grazing, without preparatory activities or additional interventions, including the condition that they have not been ploughed in the last five years.

By combining these amendments, Greece introduced five types of eligible areas, according to the percentage of woody vegetation and rocks. Pastures with a high percentage of such cover were those of most importance for silvopastoral systems, and the reduction in their eligibility had a severe impact on these systems, causing additional pressure to silvopastoral farmers. Although a subsequent Regulation (EC/2393/2017) provided a partial solution, more action is still required at the national level to properly characterize such pastures and their eligibility.

The imminent threat of significant income losses for silvopastoral farmers and of reduced absorption of CAP funds by the Greek State highlighted the need for a modernized integrated rangeland use system. The Common Ministerial Decision (CMD) 1058/71977/FEK 2331/7.07.2017 introduced the regulatory framework to modernize the system through the deployment of Integrated Grazing Management Plans (IGMP). The key aim of IGMPs is the sustainable use of pastoral lands, according to their grazing capacity. To achieve this goal, the IGMPs will be based on the division of rangeland areas into smaller land parcels (grazing units), each of which will be characterized according to species composition, herbage production, soil properties, topography, etc. The delivery of the IGMPs is therefore dependent on the collection of detailed data from each area. The innovative feature of these IGMPs – for the Greek setting – is the subsequent distribution of grazing units to farmers, so that each farmer will have their own rangeland for a considerable period of time. The allocation will be based on actual flock sizes and on the eligible rights for CAP income payments that each farmer holds so that farmers do not lose payments. Nevertheless, by the middle of 2020, IGMPs had still not been delivered in most areas. For the current transitional period, the previous system still holds.

15.4 Case studies on pastoral land governance in Greece

This section presents five case studies of different pastoral land governance schemes, that seek to be representative of different Greek regions (Figure 15.1) and farm typologies, (as presented in Section 15.2). The main goal is to demonstrate how challenges faced by the transhumance sector, combined with local particularities have led to variations of the system described in Section 15.3 across the country. The case studies differ in terms of the degree of active involvement of actors (mainly local municipalities and farmers).

Case study 1. Northwest Thessaly. This is an important site for Greek transhumance, as in summer it is grazed by more than 100 transhumant flocks, mainly of Types 2 and 3. It mainly consists of grasslands and less of woody pastures – which are mostly grazed by goats – and has not witnessed significant changes in grazing pressures. The traditional way of land allocation still holds, with farmers grazing specific areas according to their flock sizes. Local (municipal) authorities manage the allocation process, which is supervised by the decentralized government. Although traditional knowledge plays an important role in this allocation, on-site research revealed two major challenges. First, the intensification of crop production and the expansion of cropland in the lowlands has reduced the available land for grazing during the 7–8 months per year that they spend there. This is the cause of reduced profitability and uncertainty. Second, in the highlands, sheep and goat are facing increasing competition from cattle farms, whose number has increased in the area, partly as a result of coupled subsidies (see Koutsou et al., 2019, for more details). Due to these changes, a considerable number of local transhumant farmers are considering sedentarization, which will inevitably affect land use in the highland community. The governance system is therefore facing increasing external challenges and locals are looking for options to overcome future barriers.

Case study 2. The Highlands of Grevena. This area is one of the most characteristics of Greek transhumance. According to a recent study by Ragkos et al. (2020), various dynamics have led to three different pathways of vegetation evolution: thinned vegetation due to overgrazing, encroachment due to under grazing and relatively steady vegetation in just a small part of the whole region. These changes demonstrate the effects of the local governance system, which combines the formal framework with traditional knowledge. Still, the Decentralized Government monitors the whole system which is implemented by the municipalities. Farmers graze their animals in the same area – allocated to their families based on customary rights for a considerable period of time – every year from around the 25th May. Traditional knowledge of grazing capacities, climate and ecology used to be vital elements in this allocation, but there are currently no new data to inform potential reallocations. This is very important since, during the years, flock sizes in

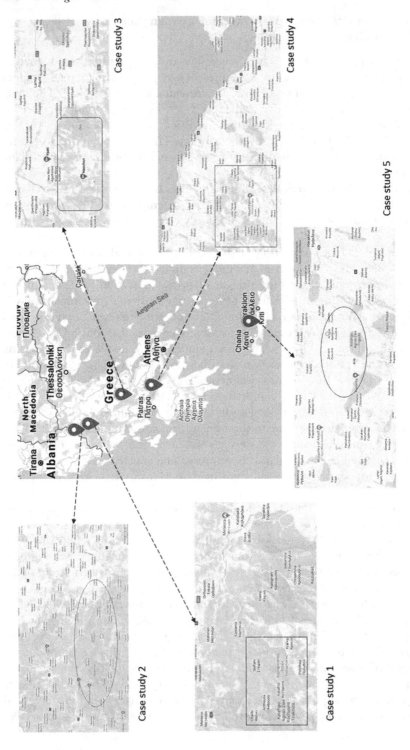

Figure 15.1 Geographical distribution of transhumance case studies.

some parts have increased considerably, while in other areas grazing pressures have been reduced. This explains the complex vegetation dynamics in the area but also demonstrates the failure of traditional knowledge to cope with intensification.

Case study 3. Central Greece. This case study exemplifies the persistence of traditional land allocation in a highland community in Central Greece. As described in detail by Papanastasis et al. (2018), the system is based on mutual understanding and cooperation to ensure the sustainability of local pastures. Pastoral land was divided into smaller parcels – which differ in terms of productivity and acreage – a long time ago and rangeland uses have historically been based on this division. Farmers declare their livestock numbers to a local board and indicate their preferred area, but farmers with larger flocks are generally prioritized for the most productive areas. If parcels are not 'filled' – i.e. there is no one flock to cover their grazing capacities – other flocks are allowed to graze in the same parcel. If the grazing capacity of the whole area is not covered by local flocks, farmers from other communities are invited to participate in the allocation process without auctions. There are no scientific estimations of grazing capacities, rangeland productivity and microclimate, but there is enough practical knowledge to inform the allocation process. It is worth noticing that this knowledge is accepted and respected by all local farmers.

Case study 4. Northern Peloponnese. The case study area is used by almost all types of transhumant farmers. This governance system follows the bureaucratic pattern of the legislative framework without local adjustments and constitutes a typical case of generalization of the auction process. Recently, priority to local producers was reduced and auctions became more frequent in order to increase the economic returns from land leasing. This led to transhumant cattle farmers, with more economic means, starting to take over grasslands in this area. This gradually became a source of conflicts, as sheep and goat farmers – the traditional users of these lands – were sidelined, and competition for silvopastoral areas increased. Largely deprived of their summer rangelands, some small ruminant farmers decided to consider sedentarization and lease private land for grazing in the lowlands. Nonetheless, this option has also considerable drawbacks, as there is also a shortage of pastoral land in the lowlands, due to the expansion of cropland. In this case, reduced access to land is becoming a major threat to the viability of sheep and goat transhumance.

Case study 5. Highlands in Crete (Psiloritis Mountain). The island of Crete still maintains a traditional system of land use, as intensification is not as widespread as in other parts of the country. Small ruminants graze for a considerable period through the year (usually 8–10 months) on rocky and slopey areas throughout the island. Transhumance, in particular, is based on grazing in highland parcels named *mitata* (plural, singular *mitato*), named after the typical makeshift stone buildings, which used to host livestock farmers in summer. Most of these farms are categorized as 'Mixed farms' (Section 15.2). The complex land ownership system of Crete – a result of political changes in the 19th and 20th centuries – has encouraged an individualistic approach towards pastoral land management. The users of the *mitata* are also their informal owners and this entitles them to use the area for years and to be responsible for its maintenance. Cretan pastoralists have a saying: "restful land gives good yields", which demonstrates that they have long realized the importance of sustainable land use, rational grazing pressures and other practices such as rotational grazing. Nonetheless, the recent changes in eligibility have had adverse effects, as many farmers have increased their grazing pressures over time which – combined with the reduced percentages of eligibility of the areas they utilize – threatens them with significant losses of CAP payments.

15.5 Typology and assessment of governance schemes

The findings of the five case studies are systematically presented in Table 15.1. Even though the framework is defined by law and follows a top-down approach, there is flexibility in its implementation leading to local/regional variations. The governance in each case study is evaluated in a qualitative way with respect to four characteristics, which reflect the differences among the five areas and allow to derive basic axes for proposing improvements in governance. The four characteristics are described here, while the most important findings are outlined below.

- **Consciousness/acknowledgement** reflects the degree to which governance incorporates local knowledge and to how end-users and other actors recognize the common character of the land and the need to safeguard productivity and access to it.
- **Cooperation and a collective approach** assess the degree of stakeholder communication in establishing common approaches.
- **Grazing pressure** refers to the intensification of grazing, as a result of increased livestock numbers or the pursuit of increased productivity.
- **Uncertainty** in land allocation and availability relative to its importance for animal nutrition, which reduces economic efficiency and threatens farms' viability.

Three types of governance schemes can be discerned from Table 15.1.

- **Scheme A** involves case studies 4 and 5, which are typical examples of top-down implementation of the existing legislative framework or individualization, with a low degree of cooperation and loose vertical and horizontal relationships. Little or no traditional knowledge is integrated and this leads to environmental issues triggered by the increasing dominance of cattle and to social conflicts between users and the authorities, especially when it comes to the disbursement of CAP payments. These circumstances are threatening the resilience of the silvopastoral system and reducing its sustainability.

Table 15.1 Qualitative assessment of governance in the case study areas

	Case study 1	Case study 2	Case study 3	Case study 4	Case study 5
	Northwest Thessaly	Grevena highlands	Central Greece	Northern Peloponnese	Crete
Consciouness/ acknowledgement	High- traditional knowledge is still important in managing land and maintaining its productivity	High- but decreasing	High- the importance of traditional knowledge and of the maintenance of land productivity is recognized	Low acknowledgement of traditional knowledge combined with the pursuit of more profitable land use	High respect for traditional knowledge
Cooperation and collective approach	Collective management and very few conflicts (mostly between sheep and goat and cattle farmers in the highlands and crop farmers in the lowlands)	Collective management with increasing individualistic behaviour	Collective management	Low (top-down approach) Increasing conflicts between ruminant farmers and between farmers and regional authorities	Individualistic-low, but increasing, distrust in the authorities in charge of allocation
Grazing pressure	Steady across time and area	Encroached, thinned and steady parts	Steady but decreasing over time	Decreasing from small ruminants – increasing from cattle	High and increasing
Uncertainty	Medium-High	High	Low/ Medium	Very high	High

- **Scheme B** includes case studies 1 and 2 and describes collective approaches to pastoral land allocation. The challenge lies mainly in the emergence of adverse external conditions, which, combined with some level of doubt and individualization of end-users, reduces the scheme's ability to cope with challenges. Although it has been proven resilient over time, current socioeconomic circumstances are proving a challenge.
- **Scheme C** involves more integrated approaches, such as case study 3, which combines public and individual participation. This scheme is widely accepted by all involved parties and explicitly acknowledges the need to use traditional knowledge to sustain land productivity.

Even within each of these schemes, there are significant differences in the external conditions and in the efficiency of particular characteristics which affect the effectiveness of their management. For instance, scheme B can be effective in the highlands, but in the lowlands, transhumant flocks are competing with intensive crop production, which makes pastures even more scarce. This leads them to be intensively managed during their stay in the lowlands and this could affect their traditional/extensive character in the highlands during summer. Our personal contacts with local farmers indicate that some of them are intensifying and considering stopping transhumance and remaining in the lowlands all year round. Thus, the discrepancy between conditions in the lowlands and the highlands decreases the importance of this integrated governance scheme and reduces its potential for generalized applicability. On the other hand, the individualistic pattern outlined in case study 5, performs in a satisfactory way in terms of acknowledgement without incorporating many elements of collectiveness. Even the example of Scheme C in central Greece – although it seems to perform well – cannot ensure that the decreasing grazing capacities that are occurring will not bring environmental degradation in the future.

The lack of a scheme, which can take precedence over the other alternatives, partially explains why it is becoming increasingly difficult for silvopastoral transhumant farmers – and other livestock farmers – to access land and to utilize it effectively. Instead of becoming involved in a complex situation, which increases their uncertainty, farmers prefer to shift towards systems that require looser ties with pastoral land. In this sense, the intensification of the small ruminant sector can be considered as an initial effect of the disconnection from pastoral land, but also as a cause, triggering the persistence to the less effective top-down approach of the current system of land allocation.

Sheep and goat transhumance in Greece illustrates how the modernization of livestock production in the country has been affecting and changing silvopastoral systems, which have important social, economic and environmental roles to play. Text Boxes 15.1 and 15.2 demonstrate how this process also has taken place in other Mediterranean regions. Modernization has been ubiquitous - especially under the effects of the CAP (but also elsewhere as we see in the case of Turkey). Rather than looking for novel technological innovations for intensive livestock farms, we might better resort to past experience and adapt it to current conditions, which could provide an alternative that would promote sustainable, efficient and resilient silvopastoral production. In almost all the examples outlined here and particularly in Schemes B and C, traditional knowledge and characteristics have been incorporated in governance, yet have not led to accommodating adaptation to changing conditions. One way of making them more effective could be to combine traditional knowledge with new technological tools, as envisaged by the IGMPs.

TEXT BOX 15.1

The Spanish transhumance and nature association

Jesús Garzón and Concha Salguero (The Spanish enviromental NGO Asociación Trashumancia y Naturaleza)

Key points: The renewal of transhumance in Spain and its role in contributing to the conservation of nature values in the dehesa

Until the start of the 20th Century, more than 5 million livestock grazed the productive winter pastures of south-western Iberia, leaving the dehesas in late April, when the summer dry period begins, to spend several weeks walking to pastures hundreds of kilometres away (Figure 15.2). During their absence, which lasted six months or so until the first autumn rains, the dehesas' vegetation recovered.

This traditional management pattern collapsed in the 20th century with the transport by rail and then by a lorry of the livestock, their products and feed. This has provoked the overgrazing of the dehesas during the critical periods of maximum heat and drought which, combined with the progressive substitution of sheep by cattle of ever-heavier meat breeds, is compacting and over-enriching the soils, polluting and drying out the ponds and streams, and destroying the tree saplings, leading to the degradation of the dehesa.

In 1992 a pilot project was started, to recover long-distance transhumance, and demonstrate its viability and its importance for Iberian ecosystems. The European

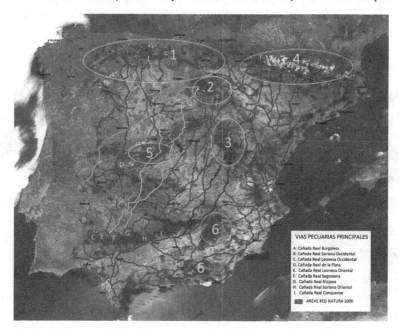

Figure 15.2 The map of transhumance ways in Spain.

Commission supported the effort in 1993–96 with one of the first LIFE projects. During those four years, 6,000 km of drovers' roads, abandoned since mid-century, were walked with more than 14,000 Merino sheep, a feat which had been considered impossible by the leading experts in the subject. The great flocks once again passed through fields, villages, roads and cities such as Zamora, Salamanca, Valladolid Cáceres and even the centre of Madrid (in 1994) and provoked great interest in the revitalization of pastoral culture, a feature of Spanish history for centuries. The support of the municipal and regional authorities, as well as of the police force that helped the shepherds and herders to resolve difficulties along the way, was decisive.

In March 1995, the Spanish Parliament approved a new Drovers' Roads Law, renewing the protection that the network of 125,000 km of livestock routes, covering 420,000 ha since 1273. They were declared to be public assets, with priority use reserved for the movement and grazing of transhumant livestock. This network represents a globally unique heritage that links almost all the parishes of mainland Spain. Since then, transhumance on foot has once again become a habitual practice that links Spain's valleys and mountains, despite having had to overcome many different constraints.

Since the end of the LIFE Project in 1996, the 'Transhumance and Nature' Association has continued to support the families that wish to carry out transhumance on foot, providing them with advice on the paperwork needed and on how to follow the established routes, as well as modern equipment, electric stock fences, solar batteries, quick-folding tents, etc. This involves around 50 families across the country who since 1997 have recovered more than 100,000 km of drovers' roads and use them to drive some 350,000 sheep, goats, cows and horses, to graze around 600,000 ha of high nature value pasture, in the valleys during winter and the mountains during summer (Figure 15.3).

Figure 15.3 Transhumance people at work.

These livestock journeys are hugely popular with the national and international media and have ecological and social importance. A further important contribution has been the advances in research on the dispersion of fertility and seeds by transhumant livestock. One report demonstrated that each 1,000 sheep spread five million seeds and three tonnes of manure between Extremadura and the Cantabrian Mountains, along 600 km of drovers' roads on a daily basis. This long-distance ecological connectivity, similar to what herds of wild herbivores did for millions of years, can only occur through the movements of transhumant livestock. Transhumance is vital for the conservation of Iberia's valuable biodiversity, with many species threatened by the climate crisis unless they are able to adapt to the changing environmental conditions.

TEXT BOX 15.2

Transhumant communities in the western Mediterranean region of Turkey

Ahmet Tolunay, ISUBÜ, Faculty of Forestry, Turkey

Key points: Small ruminants raised by nomadic societies have multiple ecosystem benefits, especially in preventing forest fires. Thousands of kilometres of fire safety lanes are established every year in order to keep forest fires in small areas and to prevent their spread.

Turkey's nomads have, for centuries, seasonally migrated to graze their goats in the basin of the Mediterranean Region. Besides being a production system, raising and breeding angora goats has a cultural and symbolic value. Nomads who follow a transhumance lifestyle even come to Turkey's coastlines from inland mountain areas. The economic life of nomads is based on animal breeding. Geographical mobility and animal breeding are two central factors that affect and shape the daily lives of nomads. During their migrations, their animals have to graze every day and so cannot travel continuously. For this reason, these nomads cannot travel for more than 4-5 hours per day. The roads on Toros Mountain are generally accessible but do not have the capacity to allow their thousands of animals to pass freely.

Although there are legal regulations, the most important factor affecting access to the grassland in winter and the highlands and wetlands in the forest in summer, thus determining the migration routes, are the ecological conditions of the region. These ecological conditions have influenced the way of life of the nomadic societies who live in harmony with these natural cycles. This factor needs to be taken into account when discussing the sustainable grazing of forest resources.

The areas where transhumants move through and the places that they use as pens and winter quarters they call their 'settlement place and home' and are mostly forest. Village headmen and local shepherds (who do not own these lands) demand money from the nomads under the name of a 'land settlement price', which can amount to large sums of money. Nomads, who agree to pay these high prices get the right to grazing, a situation that can cause various conflicts.

The presence of agricultural areas along the migration routes or projects by the forestry administration for reforestation, regeneration, rehabilitation and pasture improvements can create difficulties during migration. Often these areas are surrounded by wire fencing with the drinking water resources obstructed. Increasingly transhumants cannot use their usual routes and are migrating greater distances which leads them to use highways as an alternative. This leads to conflicts with careless drivers who cause accidents and the loss of life and property.

15.6 Conclusion: targeted proposals to boost the governance of silvopastoral resources

Although there is a specific legal framework for the allocation of pastoral land in Greece, there are several variations, a reflection of the attempts of end-users looking for ways to adapt it to local conditions and particularities. This analysis has shed light on several aspects of such variations and discerned three different schemes, without clearly identifying one that is inherently superior to the others. As a general conclusion, the more participatory Schemes B and C reflect society's increasing interest in sustainably grazed pastoral land and could also support a transition towards a rediscovery of the relations between livestock production and the land.

The inability of all three schemes outlined here to ensure sustainable land utilization and effective access to users can be partially attributed to a lack of scientific knowledge and data. Modern tools for generating maps and databases with information about the actual characteristics of rangelands (e.g. climatic and soil conditions, acreage, productivity, vegetation composition, etc.) have not yet been widely utilized. Traditional ecological knowledge was enough in the past and constitutes a valuable resource for the present, but the modernization of transhumant silvopastoral systems perhaps calls for more sophisticated technological resources. Modern tools, such as Geographical Information Systems, are actually available to scientists and practitioners who could use them to systematically develop detailed knowledge about rangeland characteristics and grazing capacities thereby informing and enriching such schemes and making them more effective. Such schemes could accommodate new challenges, such as sustainable intensification, economic sustainability and the environmental concerns of a broader range of stakeholders.

It is envisaged that the IGMPs will be based on such data and measurements. The public good character of pastoral land will not be abolished in IGMPs, as the property rights of rangelands will remain with the State. However, the detailed description of rangeland units and the estimation of grazing capacities at that scale, combined with the allocation of their use to specific farmers for longer periods of time, will help resolve issues that have occurred under the unsuccessful implementation of the current scheme. The correct, uniform and integrated application of the IGMPs constitutes a basic prerequisite for developing effective pastoral land governance in Greece.

Departing from the basic goal of providing an overview of the characteristics, quality and contributions of silvopastoral resources in Greece, the evidence we have gathered in this chapter also shows that there is a dire need for novel institutional arrangements to accommodate effective land governance. The State needs to provide a generic framework for effective pastoral land governance, with significant improvements that will be informed by the IGMPs. This framework should also respond to divergent EU characterizations of rangeland eligibility and effectively valorize silvopastoral areas. Current trends compel the emergence of local, flexible, schemes with the active involvement of local actors from the private sector and civil society and arrangements to coordinate them. This pattern prevailed in the past, with local variations, but -with a few exceptions - could not cope with modernization and socio-cultural changes. A return to the past, with the help of modern technologies, would therefore be an effective way to safeguard the future sustainability of silvopastoral and transhumance systems in Greece.

References

Bourbouze, A. & Chassany, J.-P. (2008). "Les enjeux sur le pastoralisme mondial et méditerranéen; vers de nouveaux paysages" in *Réunion Thématique d'Experts: Les Paysages Culturels de l'Agropastoralisme Méditerranéen Actes, 20-22 Septembre 2007, Meyrueis-Lozere*, pp. 41–50.

Bunce, R.G.H., Pérez-Soba, M., Jongman, R.H.G., Gómez Sal, A., Herzog, F. & Austad, I. (2004). *Transhumance and Biodiversity in European Mountains*. Report of the EU-FP5 project TRASHUMOUNT (EVK-CT-2002-80017). IALE publication series nr 1. ALTERRA, Wageningen, The Netherlands.

Farinella, D., Nori, M. & Ragkos, A. (2017). "Change in Euro-Mediterranean pastoralism: which opportunities for rural development and generational renewal?" *Grassland Science in Europe (22) - Grassland Resources for Extensive Farming Systems in Marginal Lands: Major drivers and future scenarios*, 23–36.

Hadjigeorgiou, I. (2011). "Past, present and future of pastoralism in Greece". *Pastoralism: Research, Policy and Practice*, 1(1), 24.

Karatassiou, M., Galidaki, G., Ragkos, A., Stefopoulos, K. & Lagka, V. (2015). "Transhumant sheep and goat farming and the use of rangelands in Greece". *Options Mediterranneens Series A*, 115, 655–659.

Karavidas, K. (1931). *Farming Issues*. Athens, Papazissis (in Greek).

Kavadias, G. (1996). *The Sarakatsans: A Pastoral Greek community*. Athens, Batziotis (in Greek).

Koutsou, S., Ragkos, A. & Karatassiou, M. (2019). "Accès à la terre et transhumance en Grèce: bien commun et conflits sociaux". *Développement Durable et Territoires*, 10(3), 1–19.

Lagka, V., Siasiou, A., Ragkos, A., Mitsopoulos, I., Bampidis, V., Kiritsi, S., Michas, V. & Skapetas, V. 2014. "Milking and reproduction management practices of transhumant sheep and goat farms". *Options Mediterraneennes*, 109,695–699.

Loukovitis, D., Siasiou, A., Mitsopoulos, I., Lymberopoulos, A. G., Laga, V. & Chatziplis, D. (2016). "Genetic diversity of Greek sheep breeds and transhumant populations utilizing microsatellite markers". *Small Ruminant Research*, 136, 238–242.

Nyssen, J., Descheemaeker, K., Zenebe, A., Poesen, J., Deckers, J. & Haile, M. (2009). Transhumance in the Tigray highlands (Ethiopia). *Mountain Research and Development*, 29(3):255–264.

Oosterhuizen, S., 2011. *Archeology, Common Rights and the Origins of Anglo-Saxon Identity*, Oxford, United Kingdom. Blackwell.

Papachristou, T. & Papanastassis, V.P. (1994). "Forage value of Mediterranean deciduous woody fodder species and its implication to management of silvopastoral systems for goats". *Agroforestry Systems*. 27, 269–282.

Papanastasis, V.P., Lyrintzis, G. & Solomou, A. (2018). "Traditional allocation of rangelands in Neochori Ipatis in Fthiotida". *Dimitra*, 21, 9–11 (in Greek).

Papanastasis V.P. (2002). "Ecology and management of pseudo-alpine rangelands". *Proceedings of the 3rd Greek Rangeland Conference, Karpenissi, Greece* pp. 437–445 (in Greek).

Pardini, A. & Nori, M. (2011). "Agro-silvopastoral system in Italy: integration and divestification". *Pastoralism: Research, Policy and Practice*, 1(1), 26.

Ragkos, A., Mitsopoulos, I., Siasiou, A., Skapetas, V., Kiritsi, S., Bambidis, V., Lagka, V. & Abas, Z. 2013."Current trends in the transhumant cattle sector in Greece". *Scientific Papers Animal Science and Biotechnologies*, 46(1), 422–426.

Ragkos, A., Siasiou, A., Galanopoulos, K. & Lagka, V. 2014a. "Mountainous grasslands sustaining traditional livestock systems: The economic performance of sheep and goat transhumance in Greece". *Options Mediterraneennes*, 109, 575–579.

Ragkos, A., Siasiou, A., Galidaki, G. & Lagka, V. (2014b). "A typology of transhumant sheep and goat farms in Greece". *65th Annual Meeting of the European Association for Animal Production*, 25–29 August 2014, Copenhagen, Denmark, pp. 142.

Ragkos, A. & Lagka, V. (2014). "The multifunctional character of sheep and goat transhumance in Greece". *Proceedings of the 8th Pan-Hellenic Rangeland Conference "Rangelands - Livestock farming: Research and Development. Employment prospects for young people"*, 1-3 October 2014, Thessaloniki, pp. 47–52.

Ragkos, A., Koutsou, S., Theodoridis, A., Manousidis, T. & Lagka, V. (2018). "Labor management strategies in facing the economic crisis. Evidence from Greek livestock farms". *New Medit*, 2018 – 1, 59–71.

Ragkos, A., Koutsou, A., Karatassiou, M. & Parissi, Z. (2020). "Scenarios of optimal organization of sheep and goat transhumance". *Regional Environmental Change*, 20(1), 13.

Sidiropoulou, A., Karatassiou, M., Galidaki, G. & Sklavou, P. (2015). "Landscape pattern changes in response to transhumance abandonment on Mountain Vermio (North Greece)". *Sustainability*, 7(11), 15652–15673.

Vallerand. (2014). Seasonal movements and transhumant farms in European Mediterranean. http://www.metakinoumena.gr/https://ich.unesco.org/en/RL/transhumance-the-seasonal-droving-of-livestock-along-migratory-routes-in-the-mediterranean-and-in-the-alps-01470 Retrieved on November 30th 2020.

16 Modernising silvopastoral territories through new governance schemes

Pier Paolo Roggero, Antonello Franca, Claudio Porqueddu, Giovanna Seddaiu and Gerardo Moreno

16.1 Introduction

The traditional mosaic of Mediterranean silvopastoral landscapes is the outcome of the interaction between natural drivers and pastoral, agricultural and forestry activities. The high seasonal and inter-annual variability of Mediterranean grasslands have been important determinants of the abundance and diversity of silvopastoral systems (Rolo and Moreno, 2019). Trees and shrubs provide forage resources that complement the grass shortage at certain times of the year. In addition, the practice of long-distance transhumance, usually between northern and southern regions, and short-distance displacements across local altitude gradients (vertical transhumance) have contributed to extending the seasonal availability of forage resources (Caballero et al., 2011). Finally, livestock has been moved from grasslands to nearby field crops (fodder crops and stubble) and forests to cover the gap between pasture availability and feeding requirements (Mattone and Simbula, 2011). Thus, at the landscape scale, this feeding system involves multi-actor management units that were operated through formal or informal grazing institutions that regulate pastoral societies (Caballero & Fernández-Santos, 2009). Through these practices pastoralism has significantly shaped many Mediterranean landscapes and the bundle of ecosystem services they provide.

With changes in global environmental and trading conditions, the resilience of the Mediterranean silvopastoral systems, with their own flexible, improvised and mobile responses, becomes a strategic resource (Nori and Scoones, 2019) and often the only economic alternative for the less favoured areas of the Mediterranean basin. It is essential to understand the scales at which ecosystem services are produced, managed, consumed, and accessed in order to design management strategies and governance structures that are effective, accurate, and fair (Raudsepp-Hearne and Peterson, 2016). Ecosystem services at different territorial scales are co-produced by the interactions between silvopastoral practices and ecological processes. It is essential to understand these ecological processes and how they influence such relationships and dynamics when seeking to redesign the governance of silvopastoral systems, since agricultural and environmental values and the

DOI: 10.4324/9781003028437-17

socio-economic structures and processes that they create are structurally coupled (Ison et al., 2011).

This chapter describes how changes in the socio-economic role of Mediterranean silvopastoral systems have given rise to new socio-ecological structures at territorial scales that better provide the ecosystem services that are demanded by society. We use case studies from Italy (Sardinia), France and Spain to provide a grounded insight of the territorial organizations that structural couple these ecological and socio-economic processes, which we analyze from a governance-based diagnostic framework (Stayaert and Jiggins, 2007). The objective of this chapter is to help the reader to understand, through an integrated approach and the narrative of concrete case studies, the complex nature of the new roles of silvopastoral territories and how these are driving the creation of new governance structures.

16.2 Silvopastoral stories: lessons from the past and perspectives for the future

The following sections utilize four case studies from across the Mediterranean basin to illustrate issues about the governance of pastoral territories. In each case study, we make a synthesis of the main ecological constraints, stake-holders and stakeholding (i.e. the process through which stakeholders are pro-active (or not) in protecting their interests), institutional frameworks and learning spaces. The case studies focus on four main challenges facing pastoral territories: wildfire prevention (in Andalucia, Spain), biodiversity conservation and territorial marketing (respectively in Monte Pisanu and Bue Rosso, both in Sardinia, Italy) and cultural values and tourism services in the Cevennes (France). Each case study is described using the same structure: the challenges, the main features of the case, the main stakeholders involved and conclusions with the lessons learned, which include the constraints and difficulties. These are not all necessarily success stories, but they showcase the unique characteristics, that emerge from the complexity and structural coupling between biophysical and socio-economic processes.

16.2.1 Grazing for wildfire prevention: a successful programme in southern Spain

Silvopastoral systems are considered to play a potentially important role in fire prevention under Mediterranean conditions even if the application of this effective strategy has been very limited in territorial terms. The response of most Mediterranean countries to the problem of the increasing number of wildfires has generally been to make huge investments in strengthening fire suppression capacity (e.g. for aerial fleets). Despite this, the number of wildfires and the amount of forest lost to them continues to increase, highlighting the need for a shift from short-term suppression measures to long-term preventive policies (Lovreglio et al., 2014; Bagella et al., 2017).

Silvopastoralism can play a role in this context, by exploiting grazing animals to reduce and control fuel biomass (Franca et al., 2012) and promoting the presence of livestock in forest areas in order to limit land abandonment (Rigueiro-Rodríguez et al., 2009). In this case study, we present an example of the advantages of incorporating grazing in the wildfires prevention plan of southern Spain, where the harsh topography, rural land abandonment and natural forest expansion have been behind the worrying increase of large wildfires. The implementation of this publicly supported programme required the participation of multiple actors and a substantial reconfiguration of the governance of silvopastoral territories, leading to stronger participation of local herders in the territorial management plans, decreased wildfire risk and a reduction in public spending.

16.2.1.1 The challenges

Fire prevention is essential for dealing with the wildfire problem in the Mediterranean countries (Marino et al., 2014). New opportunities related to rural development, such as forest biomass extraction for energy purposes or quality recognition for animal products obtained from controlled grazing, may also be important for solving the current problems in fire-prone areas (Marino et al., 2014). It has also frequently been proposed that grazing should be incorporated into fire prevention programmes (Mancilla-Leytón et al., 2013; Lovreglio et al., 2014). This, together with the high costs of mechanical treatments and the limited budgets of public administrations, has triggered different programs of fire prevention based on grazing targeted areas in several Mediterranean countries over the last two decades (Dopazo et al., 2009; Papanastasis, 2009).

16.2.1.2 Implementation and actors' network

A grazed fuel break network was developed in southern Spain after several years of research (Ruiz-Mirazo et al., 2011) conducted in partnership with the Regional Department of the Environment (DoE) and the Andalusian Public Agency of Environment and Water (AMAYA). After pilot work with a handful of farmers from 2005, a fully operational phase, known as RAPCA (Red de Áreas Pasto-Cortafuegos de Andalusia),[1] started in 2007. Initially, there were tens of farmers involved, with numbers increasing to over 200 farmers by 2011, which has allowed the yearly grazing of a total surface area of over 6000 ha of fuel breaks ever since. The programme does not operate through open calls but instead, local shepherds are invited to participate in the programme. They are selected to participate on the basis of their professional record, capacity and availability for grazing on specifically targeted firebreaks. Most of them (94% in 2015) have small-ruminant flocks of traditional sheep and goat breeds that are well adapted to local conditions.

RAPCA rewards shepherds for intensive seasonal grazing to reduce vegetation biomass (fuel) in areas mapped as firebreaks in publicly owned forest-land with a high risk of wildfires. It was introduced as an alternative to the mechanical clearance of firebreaks which involves high costs for the authorities. The payments are calculated in relation to the difficulties in grazing management (for example the steepness of the terrain, the distance from the farm and the type of vegetation). They vary between €42 and €90 per hectare and are subject to compliance with the grazing commitment. Agreements are generally annual, although when linked to a lease they can last up to three or four years. Payments can be reduced or cancelled if the results are not considered adequate in terms of reducing the biomass. Indicators used by inspectors are known as 'utilization rates' and range from 0–5: this involves a visual assessment of how many of the shrubs have been grazed and the overall consumption of the herbaceous layer. The inspector also evaluates the overall vegetation structure in the area.

The DoE and AMAYA manage the programme on behalf of society. As of 2016, there were 5 RAPCA staff members, hired by AMAYA, involved in the programme, preparing the contracts and aligning the different agents (local councils, forest managers and shepherds) involved in the programme. They identify the fuel breaks suitable to be maintained by grazing and verify the contracts are fulfilled through periodic monitoring. They also assess the results leading prior to making the final payments, conducting pre-assessments during the spring and keeping in contact with the shepherds throughout the year. This outcome-based payment scheme was developed by researchers together with RAPCA staff, forest guards and forestry engineers at the DoE. The researchers were seen as a trusted and authoritative source and assumed a bridging role between shepherds and public administration, which reduced mutual distrust in the initial stages and facilitated the implementation of the programme (Varela et al., 2018).

16.2.1.3 Lessons learned

These programs fit well with the current framework of payments for environmental services, the institutional arrangements that compensate producers of positive externalities, channelling financial resources from ecosystem service beneficiaries (i.e. society as a whole in the case of wildfire protection) to service providers (i.e. the shepherds) (Varela et al., 2018). It is estimated that the RAPCA approach saves up to 75% (average 63%) of the costs of managing firebreaks through mechanical clearance with brush cutters, although this calculation does not take into account the costs of administrating and monitoring the different approaches.

The programme has other indirect positive effects, such as forest surveillance, habitat conservation and the increased pastoral value of woodlands. Herders are paid for grazing fuel breaks but they usually graze the whole surrounding woodland, which becomes of higher pastoral value once the whole area becomes more accessible thanks to the fuel break network. The economic value of the whole woodland increases, and it is usually leased to the same herder who is paid for grazing the fuel break network. The reduction in the management (public) costs more than compensates for the programme's costs. Nevertheless, Mena et al. (2016) pointed out that although shepherds are remunerated for grazing fuel breaks, many of their farms are barely viable in the current socio-economic context. They found that best fuel break grazing results were achieved by larger combined flocks of sheep and goats, higher grazing densities in the fuel break area, and longer (year-long) grazing periods.

16.2.2 Biodiversity conservation through silvopastoralism: the case of monte pisanu in Sardinia

Biodiversity conservation in some silvopastoral systems is growing through strengthening the interactions between multi-stakeholders and ecological processes. This case study shows the emerging contradictions in the transition from a traditional mountain pastoral system, still generating pressure on natural resources and the regulatory system of the regional forestry agency, that enforces laws designed to prevent overgrazing and the occurrence of wildfires. In this context, the Natura 2000 framework to protect grazing-dependent habitats of community interest creates opportunities and challenges for shepherds and local administrators involved in managing the local vegetation dynamics.

16.2.2.1 The challenges

In Sardinia, the rural landscape is dominated by a mosaic of agro-silvopastoral systems where livestock, often a mix of sheep, goats and beef cattle graze all year round using different forage resources, (Caballero et al., 2009; Porqueddu et al., 2017) (Figure 16.1). The extensive grazing systems are managed so as to prioritize the best pastures to lactating ewes or beef suckling cows, Short-distance transhumance across gradients of altitude and between grass and woody vegetation extends the availability of forage resources of the lowland farmland during the dry and hot spring and summer months.

Grazing is a major driver of habitat and biodiversity conservation as it generates the necessary open habitats for enhancing the richness of plant, animal and microbial species (Seddaiu et al., 2018). Nevertheless, the regulations of the environmental protection authority (in Sardinia the Regional Forestry Agency Fo.Re.STAS) restrict grazing activities in public protected forests, following prescriptions designed decades ago, when overgrazing was causing land degradation. In protected areas, silvopastoral management plans often arise from top-down decisions based on 'command-control' paradigms (Roggero et al., 2010), often relying on desk studies that do not take into account the socio-economic dynamics, such as land abandonment or the

Figure 16.1 Bue Rosso (Sardo-Modicana breed) suckling cows and calves in the wooded grasslands of Santu Lussurgiu in Sardinia.

depopulation of inland and mountainous areas, resulting in rapid encroachment of grasslands by woody vegetation, which increases wildfire risks (López-Poma et al., 2014), entails the disappearance of valuable grassland habitats (Fracchiolla et al., 2017) and increases drought susceptibility (Rolo and Moreno, 2019).

Forest policies in Sardinia are quite restrictive for grazing. Grazing is allowed from 15th April to 15th July to reduce the biomass fuel and the fire hazard, for a maximum of 3.0 sheep ha^{-1} and 0.5 cattle ha^{-1}. In the remaining periods, grazing is restricted to max 1 sheep ha^{-1} or 0.2 cows ha^{-1}. Grazing is prohibited in newly planted woods, as well as in woods that have recently been fully or partially cut or burnt by fire. Moreover, in the woods and shrublands with a protective function, grazing by goats is prohibited.

16.2.2.2 Implementation and actors network

Monte Pisanu is a state forest, protected since 1886, and managed by Fo.Re. STAS, the forestry agency of the Sardinian Autonomous Region, which deals with forestry and environmental management, multifunctionality, rural landscape protection, research and technology transfer. It is part of a Site of Community Importance (SIC) at an elevation ranging from 600 to 1,259 m a.s.l. The main objective of the management plan is biodiversity conservation: the vegetation is dominated by downy oak (*Quercus pubescens* L.) with rare monumental trees and a biotope of millenary European yews (*Taxus baccata* L.) which are at the southern limit of the species in Europe (Farris and Filigheddu, 2008). The SIC is also a Special Area of Conservation (SAC) and a habitat for a rich wild fauna.

The forest management is designed to achieve multifunctionality by combining conservation goals with wood and animal production. Every year, some 100 tonnes of cork are extracted and sold (production is about 5 kg per tree^{-1}), firewood is produced from coniferous and deciduous trees, other cuts are made on deciduous oaks damaged by biotic and abiotic agents.

Unlike many Sardinian protected forests, extensive sheep and cattle grazing is allowed by Fo.Re.STAS provided that it is compliant with strict rules on the areas to be grazed and the stocking rates. Grazing authorizations allow about 3000 sheep (40 shepherds) and 120 cattle (10 herders) on some 2,500 ha., including 500 ha of private land. Sarda dairy sheep graze where tree density is low. Sarda cattle graze under denser tree canopies. Free-range grazing is continuous in the mountains with beef suckler cattle and seasonal with short distance vertical transhumance for dairy sheep, which graze the most productive areas in the autumn, before lambing, and the lowlands during winter.

Since 2009, intense research activity has been carried out at Monte Pisanu on the effect of grazing exclusion on the agro-ecology of wooded grasslands. It found that grazing exclusion affected plant diversity and pastoral value. Pastoral values (Daget and Poissonet, 1972), based on nutrient content and digestibility, ranged from 8.1 to 44.9 in unfenced areas and were significantly higher, 12.5–52.3 in fenced areas, with high variability between years and sites. The adopted grazing scheme had had a positive effect on the Shannon index of plant diversity and the results from FARSITE's simulations showed that grazing strongly reduced (-77%) the potential burnt area and the rate of fire spread (Franca et al., 2016). The application of site-specific grazing regimes and P fertilization rates contributed to improving the pastures' legume seed bank, thus enhancing the resilience of the grasslands with a low tree density (Franca et al., 2018).

16.2.2.3 Lessons learned

The Monte Pisanu silvopastoral system is an example of how carefully managed grazing and forestry activities can coexist, generating habitats for biodiversity conservation and multifunctionality. However, stakeholders were critical of the top-down approach followed by the regional authority. This critical issue emerged from participatory research carried out during the AGFORWARD Project (Pisanelli et al., 2014). This showed that farmers had positive perceptions on crop and pasture production, plant and animal diseases and weed control, timber/wood/fruit/nut production and quality; while policy-makers had positive perceptions of the scheme's effects on wildfire risks and landscape aesthetics, but negative ones on tree regeneration/survival. Researchers and policymakers gave greater weight to indicators related to biodiversity and wildlife than farmers (Camilli et al., 2018).

The different perceptions between stakeholders are at the roots of emerging controversies in the current management of the Monte Pisanu forest. Farmers claim full recognition of the environmental role of grazing activities through specific agri-environmental policies for silvopastoralism and the payment of ecosystem services.

A sustainable management system could emerge from the coordination between agroforestry and environmental policies, the identification and implementation of suitable financial and normative instruments supporting agroforestry systems at a local scale, including shared grazing management plans, in the framework of the agro-environmental payments of the next CAP, better coordination among stakeholders to generate more synergies and the development of new learning spaces where local actors can share experiences and knowledge.

16.2.3 Bue rosso: territorial marketing in Sardinia

The value added in terms of food and environmental quality of the beef cattle grazing system relying on Sardinian silvopastoral systems (Figure 16.2) is generally lost in the food chain as most calves are sold to feedlots at weaning, thus losing the 'memory' of the first 6-8 months of their lives in the woody grasslands with the sucklers. In this case study, we describe an exception for

Figure 16.2 Image of Les Causses plateau and gorges and the dominant silvopastoral landscape. Taken from: https://www.tourisme-lodevois-larzac.fr/en/the-causses-cevennes.

Sardinia, where the entire beef cattle food chain, based on a Mediterranean rustic breed, relies on mountain silvopasture. However, the added value of the generated food quality and ecosystem services is not yet fully appreciated by consumers and hence subsidies are still vital for the sustainability of this niche production.

16.2.3.1 The challenges

Extensive silvopastoral systems can provide high-quality food products, rich in antioxidants, quality of lipids and fat-soluble vitamins (D'Ottavio et al., 2018). The ability of ruminants to convert plant products that are inedible for humans into high-quality edible proteins in marginal non-arable land contributes to global food security, especially when compared to other animal production systems that rely on human-edible feed such as cereals.

The development of local agri-food chains requires effective market strategies to add value to local products thus making them attractive for retailers and consumers (Bardají et al., 2009). The commercial practices proposed by the Taskforce for Agricultural Markets of the European Commission (Veerman et al., 2016) assume that classical market mechanisms do not provide adequate incentives to support sustainable local production. The market orientation of high-quality agricultural products requires new connections between farmers and consumers, generating added value for local, organic, free-range, GMO-free, or antibiotic-free products, as well as animal welfare standards and quality labels. In an attempt to maintain their competitiveness in the meat production market, some Sardinian cattle breeders have started new pathways to gain new economic strength while maintaining their identity and meeting consumer demand.

This case study shows the potential of EU-funded networking and local development initiatives to trigger virtuous processes that contribute to the revitalization of rural economies, enhance marketing opportunities for rural territories and maintain traditional knowledge and production methods. However, it is evident that the economic fragility of these marginal territories necessitates an ongoing supporting system and marketing promotion which is, at the same time, a major weakness that increases uncertainty and threatens their long-term sustainability.

16.2.3.2 Implementation and actors' network

The case study focuses on the development and implementation of the Montiferru "GAL" (Local Action Group of the Leader II program) partnership, in a Sardinian mountain area. The Montiferru GAL was launched in 2002 to revitalize the socioeconomic system of the area which is based on its livestock and small-scale, craft-based, enterprises. The main goal was to mitigate the gradual abandonment and negative demographic changes which are causing a loss of labour force, knowledge, and the skills required to maintain traditional productions. GAL's actions were focused on supporting the typical local cheese (Casizolu) and meat products of the *Bue Rosso* beef cattle (a local Sardo-Modicana breed), promoting official protocols for meat production, enhancing marketing opportunities and providing capacity building opportunities and networking opportunities with other local specialty producers (e.g. Malvasia wine, honey, and extra virgin olive oil). However, after some initial success, both at local and international scales at the beginning of the 21st century, this livestock system has been losing its competitiveness in relation to imported and more productive cattle breeds, such as Charolais and Limousine.

All the GAL initiatives were facilitated by professional knowledge brokers and involved the Sardo-Modicana breeders, the Montiferru GAL, the Oristano Province Health Department, Slow Food,[2] technicians from the Arezzo Province, a farmers' union (CIA, the Italian Farmers Confederation) and the regional and provincial farmers' associations (ARA and APA). The positive testimonies of external actors highlighted the success story of Chianina meat (Guarino, 2011) elsewhere in Italy, worse regime included the use of finishing feeding and maturing the meat, methods not traditionally used with Bue Rosso meat, and this success story was a great help in getting over the reluctance of local farmers to participate in the scheme. The incentives provided by GAL, which grants up to a non-refundable 80% of the costs of new farming infrastructure, facilitated the adoption of technological innovations.

The Bue Rosso regulations prescribe that calves should be fed by their mothers until weaning and then be left to free-range in the wooded grasslands that include Mediterranean shrubs, holm and cork oaks, myrtle, strawberry tree and chestnuts, over a range of altitudes between 250 and 800 m a.s.l. Before slaughtering, calves are housed for two months and finished with feeds that exclude silage feeds, animal and GMO products.

The communication strategy promoted by the GAL was based on a territorial marketing approach, emphasizing the uniqueness of the products that are derived from natural pastoral resources. The involvement of Slow Food enhanced national and international marketing opportunities and gave the Bue Rosso Consortium the opportunity to participate on several occasions at the Terra Madre – Salone del Gusto in Turin, Italy (www.salonedelgusto. com), one of the most important and prestigious international agro-food/ gastronomy events.

16.2.3.3 Lessons learned

The Bue Rosso experience contributed to a shift in the economy of the Montiferru area not only towards sustainable production but also the integration of recreational activities such as agritourism, typical craft production (cutlery, distillates) and local restaurants, as well as the development of the *albergo diffuso* network (a type of lodging in the private houses in small villages within the Montiferru area).

There are about 70 farms that breed and raise Sardo-Modicana pure breeds, producing around 1000 calves each year. However, the Bue Rosso Consortium has only attracted 25 farmers, a decrease of about 40% since 2007 (Guarino, 2011). This reflects the complex situation which the consortium has experienced, following the initial optimism after GAL's initial interventions and the incentives from the Rural Development Programme for protecting threatened local breeds, which were suspended in 2014 following the new implementation of rural development policies. The problems faced by the Consortium are evident when analyzing the communication strategies that it has in place: a Facebook page with just a few followers and an official website that is almost empty (Sois, 2020). Several factors have contributed to the decline in membership of the consortium, the most important being: (i) the economic crisis which affected demand and led to buyers delaying making payments; (ii) the overly strict farming and rearing prescriptions of the consortium are often not compatible with business-as-usual practices; (iii) a lack of future targeted incentives; (iv) insufficient market opportunities, partly because of the limited product range offered and; (v) the previous incentive system was exposed to fraud on several occasions (personal communication from Celestino Illotto, President of the Bue Rosso Consortium).

The lessons learned by the Bue Rosso case study highlight the need for a continuous support system through incentives and marketing promotion. The future of such niche livestock systems is quite fragile unless the many associated ecosystem services can be fully valued through extra- prices in their products and/or public incentives, and this requires a more conducive normative and institutional system.

16.2.4 Les causses-cévennes, a world heritage site in southern France

In this case study, we present an example of taking advantage of a cultural landscape shaped by centuries of pastoral activities, particularly the transhumance between two well-defined territories. Local actors worked together with local, regional and national administrations to achieve formal acknowledgment of the natural and cultural value of their pastoral territories and practices. Through this, they revitalized the local economy, mostly based on extensive pastoralism and ecotourism. This required new governance structures, with the participation of new actors who operate at very different geographical scales.

16.2.4.1 The challenges

Younger members of livestock farm families in agro-silvopastoral territories are tending to seek alternatives to pastoralism, thus contributing to the depopulation of mountain areas and exposing grasslands to abandonment and socio-economic desertification (Farinella et al., 2017). Nature tourism is an interesting complementary economic activity to the maintenance of their silvopastoral activities and landscapes and can contribute to halting the agroecological and socioeconomic desertification in Mediterranean rural areas (Nori, 2017). Nature tourism and livestock farming also involve complex interactions in the food system (Beudou et al., 2017), with a potential synergy between pastoralism and nature tourism through the production of high-quality typical dairy products, meat, honey and medicinal and aromatic herbs (Nori, 2017).

The landscape of the Cévennes Mountains and the Causses limestone plateau in Languedoc-Roussillon is a mosaic of grasslands, heathland, meadows that are the result of the modification of the natural environment by agro-pastoral systems that have been in use for at least a thousand years (Figure 16.2). Almost all the types of pastoral organization found around the Mediterranean (agro-pastoralism, silvopastoralism, transhumance and sedentary pastoralism) can be found in this 3,000 km² territory. Using specific social structures and distinctive local breeds of sheep, pastoralism has shaped the structure of the landscape, notably the farmhouses, settlements, fields, water management, *drailles* (the drove routes used by cattle and sheep moving to and from the pastures) and common grazing land.

16.2.4.2 Implementation and the actors' network

In 2011 the site was registered on the UNESCO World Heritage List as being of Outstanding Universal Value (OUV), an example of the Mediterranean agro-pastoral cultural landscape, due to the different types of agropastoralism that coexist in the territory today, and their influence on the landscape.

The management of the site requires expertise and a thorough understanding of this particular cultural landscape, which needs to draw on both nature and culture.[3] The Ministry of Ecology is responsible for the enhancement and preservation of the site and is supported by the regional services in charge of the environment and cultural affairs. Following UNESCO guidelines that encourage the participation of a wide variety of actors, a strong governance system was established around three authorities:

  i Territorial Conference: This is the decision-making body that defines the main management guidelines and validates the objectives for the good conservation of the site. It includes site managers, local and regional authorities, local communities, non-governmental organizations (NGOs), representatives of the National and Natural Parks, other stakeholders (e.g. architecture council) and interested partners (e.g. tourism providers).
 ii Interdepartmental Alliance, which ensures the implementation of management guidelines and coordination among existing structures, manages the use of the UNESCO World Heritage label and decides and implements the necessary communication, knowledge and promotional activities.
iii The Steering Committee is the consultative body in charge of providing guidance and recommendations for the management of the site, in collaboration with the territory's local managers. It is assisted by a Scientific Council and five working groups (management; agropastoralism; heritage and culture; landscape; communication and tourism).

16.2.4.3 Lessons learned

UNESCO recognition is helping the region to face the social, economic and environmental problems that are common to this type of landscape throughout the world. This area shows remarkable vitality, due to the strong revival of agro-pastoral systems, and it is an important and viable example of Mediterranean agro-pastoralism. There is a rising demand for the environmental, social and cultural benefits provided by the pastoral rangelands of this UNESCO site (Moreau et al., 2019). The current concerns of people about their quality of life (including food quality and nature conservation) and the development of new economic sectors related to recreational activities (including farm tourism, game hunting, educational services and valorization of local genetic resources) have opened up new opportunities for rural economies within the wider regional economy (Pardini, 2009).

Berriet-Solliet et al., (2018) have analyzed the social-ecological system of the Cévennes National Park, a central part of the UNESCO site (Figure 16.3). They identified the following collective innovations and adaptations to the new market context: market distinction strategies in order to meet a new demand for high-value local and organic food; collective farmers' shops; short supply chains, all of which support local tourism development. They also identified that the newcomers to the area (the neo-rural population) introduced organizational innovations (organic/biodynamic

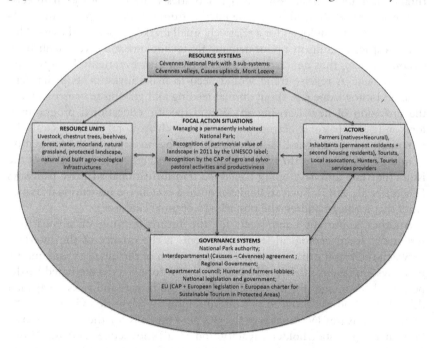

Figure 16.3 Environmentally and socially beneficial outcomes of the social-ecological system in the French Cévennes National Park (Adapted from Berriet-Solliet et al., 2018).

production, short supply chains, innovative transformation processes, etc.) that had positive impacts in the area, despite some problems about accessing land for farming. The arrival of new inhabitants also stimulates demand for new services or supports existing ones.

16.3 Silvopastoral systems as structurally coupled social-ecological systems

The four case studies provide some ground evidence of how ecological and social processes are structurally coupled in large-scale social-ecological Mediterranean silvopastoral systems. The emerging properties such as wild-fire prevention, biodiversity conservation, economic development in mountain areas and the maintenance of cultural landscapes are generated by the dynamic interaction of stakeholders' practices, local institutional frameworks and ecological processes. This implies that the future dynamics will be shaped by investments to enhance the adaptive capacity of pastoral communities and adapt to the local normative system. A recent study made in Sardinia for the development of a climate change adaptation strategy[4] showed that the weak adaptive capacity of rural communities is associated with a chronic lack of material (e.g. technology) and immaterial infrastructures (e.g. education, learning spaces). This emerges from decades of a lack of public investment in remote rural communities where silvopastoral activities still represent the main economic activity, combined with little interest from private companies in developing specific technologies for a relatively small market of potential users. This lack of capacity results in a poor demand for technology and investments from pastoral society, thus generating a vicious circle that eventually leads to land abandonment. A reverse trend would need to make investments in farmers' skills and knowledge e.g. about animal feeding and grazing management and the provision of new learning spaces and specific educational programmes for young silvopastoral entrepreneurs. These silvopastoral systems could be reframed and re-designed as learning systems by investing in the facilitation of multi-stakeholder platforms to share the understanding of the nature of the issues they face and developing concerted transformational adaptive pathways (Toderi et al., 2017). Such strategies require a conducive institutional framework that is able to address the obstacles to optimal land use.

Another important component for the future framing of silvopastoral systems as socio-ecological learning systems is the quality of the interface between them and urban society. The perception of the quantity and quality of ecosystem services delivered by such systems to society is mediated by the perceived advantages for local social groups, the threats and risks for habitat degradation and the provision of interdisciplinary scientific evidence of the required measures (Vrahnakis et al., 2017). This calls for the integration of bottom-up (i.e. stakeholder capacity and local knowledge) and top-down (interdisciplinary science-based evidence provision) approaches combined with strategic communication to potential customers or end-users.

Notes

1. https://ec.europa.eu/environment/nature/rbaps/fiche/rapca-red-de-areas-pasto-cortafuegos-de-andalucia-_en.htm
2. Slow Food is a global, grassroots organization, founded in 1989 to prevent the disappearance of local food cultures and traditions, counteract the rise of fast life and combat people's dwindling interest in the food they eat, where it comes from and how our food choices affect the world around us (www.slowfood.com)
3. http://www.causses-et-cevennes.fr/en/who-are-we/governance/the-governance/
4. http://delibere.regione.sardegna.it/protected/45525/0/def/ref/DBR45368/ (Consulted on 3 May 2020

References

Bagella, S., Sitzia, M., & Roggero, P.P. (2017). "Soil fertilisation contributes to mitigating forest fire hazard associated with *Cistus monspeliensis* L. (rock rose) shrublands". *International Journal of Wildland Fire* 26, 156–166.

Bardají, I., Iráizoz, B., & Rapún M. (2009). "Protected geographical indications and integration into the agribusiness system". *Agribusiness* 25, 198–214.

Berriet-Solliec, M., Lataste, F., Lépicier, D., & Piguet, V. (2018). "Environmentally and socially beneficial outcomes produced by agro-pastoral systems in the Cévennes National Park (France)". *Land Use Policy* 78, 739–747.

Beudou, J., Martin, G., & Ryschawy, J. (2017). "Cultural and territorial vitality services play a key role in livestock agroecological transition in France". *Agronomy for Sustainable Development* 37 (36), 11 pp.

Caballero, R., & Fernández-Santos, X. (2009). "Grazing institutions in Castilla-La Mancha, dynamic or downward trend in the Spanish cereal–sheep system". *Agricultural Systems* 101, 69–79.

Caballero, R., Fernandez-Gonzalez, F., Badia, R.P., Molle, G., Roggero, P.P., Bagella, S., Papanastasis, V.P., Fotiadis, G., Sidiropoulou, A., & Ispikoudis, I. (2011). "Grazing systems and biodiversity in Mediterranean areas: Spain, Italy and Greece". *Pastos* 39, 9–154.

Camilli, F., Pisanelli, A., Seddaiu, G., Franca, A., Bondesan, V., Rosati, A., Moreno, G., Pantera, A., Herrmansen, J.E., & Burgess, P.J. (2018). "How local stakeholders perceive agroforestry systems: an Italian perspective". *Agroforestry Systems* 92, 849–862.

Daget, P., & Poissonet, J. (1972). "Un procédé d'estimation de la valeur pastorale des pâturages". *Fourrages* 49, 31–38.

Dopazo, C., Robles, A.B., Ruiz García, R., & San Miguel, A. (2009). "Efecto del pastoreo en el mantenimiento de cortafuegos en la Comunidad Valenciana". In: *5° Congreso Forestal Español*, SECF, Ávila, Spain.

D'Ottavio, P., Francioni, M., Trozzo, L., Sedic, E., Budimir, K., Avanzolini, P., Trombetta, M.F., Porqueddu, C., Santilocchi, R., & Toderi, M. (2018). "Trends and approaches in the analysis of ecosystem services provided by grazing systems: a review". *Grass and Forage Science* 73, 15–25.

Farinella, D., Nori, M., & Ragkos, A. (2017). "Changes in Euro-Mediterranean pastoralism: which opportunities for rural development and generational renewal?" *Grassland Science in Europe* 22, 23–36.

Farris, E., & Filigheddu, R. (2008). "Effects of browsing in relation to vegetation cover on common yew (*Taxus baccata* L.) recruitment in Mediterranean environments". *Plant Ecology* 199, 309–318.

Fracchiolla, M., Terzi, M., D'Amico, F.S., Tedone, L., & Cazzato, E. (2017). "Conservation and pastoral value of former arable lands in the agro-pastoral system of the Alta Murgia National Park (Southern Italy)". *Italian Journal of Agronomy* 12, 124–132.

Franca, A., Caredda, S., Sanna, F., Fava, F., & Seddaiu, G. (2016). "Early plant community dynamics following overseeding for the rehabilitation of a Mediterranean silvo-pastoral system". *Journal of Grassland Science* 62, 81–91.

Franca, A., Re, G.A., & Sanna, F. (2018). "Effects of grazing exclusion and environmental conditions on the soil seed bank of a Mediterranean grazed oak wood pasture". *Agroforestry Systems* 92, 909–919.

Franca A., Sanna F., Nieddu S., Re G.A., Pintus G.V., Ventura A., Duce P., Salis M., & Arca B. (2012). "Effects of grazing on the traits of a potential fire in a Sardinian wooded pasture". In: Acar Z., López-Francos A. & Porqueddu C. (eds.) *New Approaches for Grassland Research in a Context of Climate and Socio-Economic Changes*. Zaragoza: CIHEAM Options Méditerranéennes: Série A. Séminaires Méditerranéens; n. 1022012, pp. 307–311.

Guarino, A. (2011). "Il dinamismo ai margini: pratiche di sviluppo rurale in aree svantaggiate del Mediterraneo". In Brunori G. (ed.) *Le Reti Della Transizione. Impresa e lavoro in un'agricoltura che cambia*. Felici Editore Srl, Ghezzano (PI). 199 pp.

Ison, R., Collins, K., Colvin, J., Jiggins, J., Roggero, P. P., Seddaiu, G., Steyaert, P., Toderi, M., & Zanolla, C. (2011). "Sustainable catchment managing in a climate changing world: new integrative modalities for connecting policy makers, scientists and other stakeholders". *Water Resources Management*, 25(15), 3977–3992.

López-Poma, R., Orr, B.J., & Bautista, S. (2014). "Successional stage after land abandonment modulates fire severity and post-fire recovery in a Mediterranean mountain landscape". *International Journal of Wildland Fire* 23, 1005–1015.

Lovreglio, R., Meddour-Sahar, O., & Leone, V. (2014). "Goat grazing as a wildfire prevention tool: a basic review". *Forest-Biogeosciences and Forestry* 7, 260–268.

Mancilla-Leytón, J. M., Pino Mejías, R., & Martín Vicente, A. (2013). "Do goats preserve the forest? Evaluating the effects of grazing goats on combustible Mediterranean scrub". *Applied Vegetation Science* 16, 63–73.

Marino, E., Hernando, C., Planelles, R., Madrigal, J., Guijarro, M., & Sebastián, A. (2014). "Forest fuel management for wildfire prevention in Spain: a quantitative SWOT analysis". *International Journal of Wildland Fire* 23, 373–384.

Mattone, A., & Simbula, P. (eds.). (2011). *La Pastorizia Mediterranea Storia e Diritto (Secoli XI-XX)*. Carocci, Rome

Mena, Y., Ruiz-Mirazo, J., Ruiz, F. A., & Castel, J. M. (2016). "Characterization and typification of small ruminant farms providing fuelbreak grazing services for wildfire prevention in Andalusia (Spain)". *Science of the Total Environment* 544, 211–219.

Moreau, C., Barnaud, C., & Mathevet, R. (2019). "Conciliate agriculture with landscape and biodiversity conservation: a role-playing game to explore trade-offs among ecosystem services through social learning". *Sustainability* 11, 310, 20 pp.

Nori, M. (2017). "Migrant shepherds: opportunities and challenges for Mediterranean pastoralism". *Journal of Alpine Research*, 105–4, 19 pp.

Nori, M., & Scoones, I. (2019). Pastoralism, Uncertainty and Resilience: Global Lesson from the Margins. Pastoralism 9, 10. https://doi.org/10.1186/s13570-019-0146-8.

Pardini, A. (2009). "Agroforestry systems in Italy: traditions towards modern management. In *Agroforestry in Europe*. Springer, Dordrecht. pp. 255–267.

Papanastasis, V.P. (2009). "Grazing value of Mediterranean forests". In *Modelling, Valuing and Managing Mediterranean Forest Ecosystems for Non-Timber Goods and Services* 57, 7–15.

Pisanelli, A., Camilli, F., Seddaiu, G., & Franca, A. (2014). Initial Stakeholder Meeting Report: Grazed oak woodlands in Sardinia. 15 October 2014. 9 pp. http://www. agforward. eu/index. php/en/grazed-oak-woodlands-in-sardinia. html.

Porqueddu, C., Melis, R.A.M., Franca, A., Sanna, F., Hadjigeorgiou, I., & Casasús, I. (2017). "The role of grasslands in less favoured areas of Mediterranean Europe". *Grassland Science in Europe* 22, 3–22.

Raudsepp-Hearne, C., & Peterson, G. D. (2016). "Scale and ecosystem services: how do observation, management, and analysis shift with scale—lessons from Québec". *Ecology and Society* 21, 16, 36 pp.

Rigueiro Rodríguez, A., Fernández-Núñez, E., Santiago-Freijanes, J. J., & Mosquera-Losada, M. R. (2009). "Silvopastoral systems for forest fire prevention". In: *Agroforestry Systems as a Technique for Sustainable Land Management*. AECI-Madrid, pp. 335–344.

Roggero, P.P., Desanctis, G., & Sedaiu G. (2010). "Cambiamenti climatici e sistemi agrari: da 'comando e controllo' a 'azioni concertate'". In: Francchia, A. & Occhiena, M. (eds.) *Climate change: la risposta del diritto,,* Editoriale Scientifica, Napoli, Italy. Pp. 359–377.

Rolo, V. and Moreno, G. (2019). "Shrub encroachment and climate change increase the exposure to drought of Mediterranean wood-pastures". *Science of the Total Environment* 660, 550–558.

Ruiz-Mirazo, J. (2011). "Environmental benefits of extensive livestock farming: wildfire prevention and beyond". *Options Méditerranéennes: Série A. Séminaires Méditerranéens* 100, 75–82.

Seddaiu, G., Bagella, S., Pulina, A., Cappai, C., Salis, L., Rossetti, I., Lai, R., & Roggero, P. P. (2018). "Mediterranean cork oak wooded grasslands: synergies and trade-offs between plant diversity, pasture production and soil carbon". *Agroforestry Systems* 92, 893–908.

Sois, E. (2020). "Percorsi agro-turistici nelle aree interne e costiere della Sardegna". In: a cura di Meloni, B. & Pulina P. (ed.) *Turismo Sostenibile e Sistemi Rurali Locali. Multifunzionalità, reti d'impresa e percorsi*. Lexis, Torino, pp. 301–342.

Steyaert, P. & Jiggins, J. (2007). "Governance of complex environmental situations through social learning: a synthesis of SLIM's lessons for research, policy and practice". *Environmental Science & Policy*, 10(6), 575–586.

Toderi, M., Francioni, M., Seddaiu, G., Roggero, P. P., Trozzo, L., & D'Ottavio, P. (2017). "Bottom-up design process of agri-environmental measures at a landscape scale: evidence from case studies on biodiversity conservation and water protection". *Land Use Policy* 68, 295–305.

Varela, E., Górriz-Mifsud, E., Ruiz-Mirazo, J., & López-i-Gelats, F. (2018). "Payment for targeted grazing: integrating local shepherds into wildfire prevention". *Forests*, 9(8), 464.

Veerman, C.P., Valverde Cabrero, E., Babuchowski, A., Bedier, J., Calzolari, G., Dobbin, D., Fresco, L.O., Giesen, H., Iwarson, T., Juhasz, A., Laure Paumier, A., & Šarmír, I. (2016). *Improving Market Outcomes – Enhancing the Position of Farmers in the Supply Chain. Report of the Agricultural Markets Task Force*, Brussels, November 2016.

Vrahnakis Nasiakou, K., & Soutsas, M. S. (2017). "Public perception on measures needed for the ecological restoration of Grecian juniper silvopastoral woodlands". *Agroforestry Systems* doi.org/10.1007/s10457-017-0163-9.

Lessons learnt and ways forward

Gerardo Moreno, Rufino Acosta-Naranjo,
Maria Helena Guimarães and Teresa Pinto-Correia

1 Lessons learnt

Many agricultural practices across Europe have historically supported bio-diversity and multiple ecosystem services, bringing a range of benefits to society. This is particularly the case with silvopastoral systems, which were abundant in different forms across the European continent until the middle of the 20th Century. Since then in, many places, such practices have been replaced by more intensive and specialized farming systems which maximize yields through the unsustainable use of natural resources and at the expense of biodiversity and ecosystem services (Pe'er et al., 2020). At the same time, more marginal farmlands have been abandoned, with a great loss of their mosaic landscapes shaped by humans (Eichhorn et al., 2006; Pinto-Correia, Primdahl, and Pedroli 2018). These processes are driven by socio-economic and technological forces but, in Europe, also supported by public policies (Pe'er et al., 2020).

In the most marginal land, rewilding can be a viable response to ongoing depopulation and the loss of agricultural uses (Navarro and Pereira 2015). However, the transition towards scrublands and forests that follows the abandonment of former agricultural land presents a less favourable balance of ecosystem services, and some disservices, specifically in being prone to large wildfires (Varela et al., 2020). The abandonment of cultivation and other human management of former multifunctional production systems leads to the disappearance of characteristic landscapes and loss of biodiversity (Simonson et al., 2018).

1.1 Multi-production as a strategy for coping with low productivity and strong seasonal fluctuations

Within this scenario, the Iberian dehesas and montados can be viewed as exceptional. They persist under harsh environmental conditions as still living, multifunctional, silvopastoral systems, that provide multiple goods and services, and are acknowledged as farming systems of high nature and cultural value (Moreno et al., 2018). To cope with Mediterranean seasonality, pastoral systems have been shaped as multilayered systems, combining herbaceous and woody forage resources, with flocks moved between lowlands and highlands, either in altitude or latitude (trasterminance and transhumance, respectively). In this way, silvopastoralism is one of the great cultural elaborations of the peoples of southwestern Iberia, creating a diversity of spaces and productions that have been adapted to changing socio-economic contexts for centuries.

While most of these traditional practices have vanished in many developed countries, the populations of the dehesas and montados still make use of many of these traditional practices. As such they represent a considerable contrast to standardized industrial agriculture and the current food regime (McMichael, 2009). These practices are based on the separate mobilization of resources, the productive unilaterality of spaces, the intensive use of the productive advantages of plots of land, mechanization and the massive use of inputs in order to create the optimal conditions for single products. By contrast, Mediterraneansilvopastoral systems have historically had their *raison d'etre* on the adaptation to contexts of marginality (low productivity) and seasonality, through enhancing diversity, seeking not to maximize production of the same product, but of discrete but constant quantities of a variety of produce (Acosta-Naranjo, 2002). The ideal of self-sufficiency and self-employment was based on the mutual services that agriculture, livestock and forestry provide to each other, with livestock offering fertilizer to crops and preventing shrub encroachment; crops offering food to livestock and trees improving pastures and offering food and shelter to livestock, and providing timber and firewood to farmers. A virtuous circle.

1.2 Dehesas and montados, areas with multiple workers, actors and customary use and rules

Dehesas and montados emerged as reserves for pasture used by the local communities. Over time they were shaped into multifunctional and complex agro-silvopastoral systems, and more recently, with the withdrawal of crop cultivation under the trees, converted into silvopastoral systems. Dehesas and montados evolved into large private states (of 100s to 1000s of hectares), frequently owned and managed for generations by wealthy families, who were often poorly integrated into the local community. Today, a single farmer (usually a livestock breeder) may be in charge of different farms, including his own but also those of others, either family members or leased. Other individuals manage single components or commodities, parts of the overall farming system. They may be in charge of forestry products, or specific grazing areas or livestock species, or hunting. Up to a few decades ago, there was a wealth of skilled workers devoted to the management of the multiple components and productions of the system. Some rights of the peasants to use certain resources (certain pastures, mushrooms, wild plants, dead wood, water bodies) have survived for centuries while the local networks went beyond a purely local scale through trasterminance and transhumance. In a few places, local communities have kept their rights over communal lands, whether owned by a group of neighbours or a municipality. These complex networks of actors have co-evolved with the system to regulate the use of multiple resources and the provision of multiple products that have been widely used by the local population, over centuries.

Even today, within one single farm or estate, there may be different individuals who take decisions, which all need to be in tune with each other. Rights of use are defined through both formal (legal regulations) and regionally or locally embedded customary rights, which may exclude other uses. With few exceptions (public paths, stream waters, religious/symbolic patrimony …), owners have the full power to decide who can use what, although they often respect customary rights. But owners differ, and increasingly there is a new breed of the owner, coming from outside, unknown in the community and unaware of community traditions. The specialized workers in each farm unit have been replaced by single multifunctional farm managers, although multiple competencies are still needed to manage all the components of the montado or dehesa, which are now frequently provided by specialized companies or freelance workers. New uses bring new people into the montado and dehesa and different interests come into play. There are multiple governance arrangements, inherited from the past or being re-invented today, as well as tensions and conflicts. Dehesas and montados are thus socio-ecological systems that illustrate the transitions taking place in rural Europe today, and particularly in the Mediterranean (Ortiz-Miranda et al., 2013; Pinto-Correia et al., 2018). They can also show us living examples of close and efficient collaboration among actors in the same territory, and some evidence for pathways towards the sustainable management of other land-use systems across the Mediterranean, and other regions of the world, that face resource scarcity and/or strong seasonality.

1.3 Socio-ecological systems in transition and facing multiple challenges

Mediterranean silvopastoral territories, which are managed as complex socio-ecological systems with strong interdependencies among the actors, and between them and the environment, offer a desirable bundle of ecosystem services (Torralba et al., 2018). Nevertheless, despite the strengths woven by the long-term co-evolution between the system and its actors, dehesas and montados are facing serious challenges that have emerged in recent decades. The growing privatization of economic resources on a global scale has accelerated the privatization of dehesa and montados, and market pressure for farm specialization has dramatically simplified the management models, while the manpower requirements (and supply) on the farms has been drastically reduced (e.g., fences replacing herders). The management of dehesas and montados increasingly relies on a few, instead of many, skilled workers, more on multi-task employees, sporadic workers and hired services and less on skilled on-farm workers: more on purchased fodder and less on transhumance. Women are increasingly making the case for their specific role and positioning in managing silvopastoral systems. The influence of specialized companies hired to manage one or more farms, and hiring professionals, as veterinarians, agronomists, foresters, etc., to support the daily decisions of different farm units is growing. As a result, few people have a continuous relationship with the territory through their work, a primary channel of people's connection with the land.

Due to the combination of multiple actors and inherited ways of using parts of the silvopastoral system by people who are not owners, the rights of use in the montado and dehesa are complex. On most farms, the passage is not legally prohibited, but fences for livestock management have made access more and more difficult, and farms now have fewer people than ever before. The already poor interactions between large landowners and local populations have lately grown weaker. The connectivity between local populations and municipal dehesas has also weakened, and in many cases, the pastures are rented to a few breeders, in some cases outsiders. In this way, physically and symbolically, people's ecological and moral relationship with the territory is being diminished, also leading to a deterioration in local knowledge about the environment (Carolino and Pinto-Correia, 2011; Acosta-Naranjo, 2008; Acosta-Naranjo, Guzmán-Troncoso and Gómez-Melara 2020).

1.4 Enlarging the community for emerging services rooted in local knowledge

While, in the past, dehesas and montados were essential for the subsistence of local communities, the contemporary society still receives multiple socio-cultural, economic and environmental benefits that are far more complex than those resulting from production management. There are now multiple uses that can be framed as 'countryside consumption' (Murdoch and Marsden 2013), such as hikers, rural tourists, naturalists, birdwatchers or nature photographers. Movements to keep publicly open a dense network of drove roads have arisen in recent years (Pinto-Correia et al., 2016). Emerging services such as wildfire prevention, biodiversity conservation or cultural landscapes are generated by the dynamic interaction of ecological processes, stakeholders' practices and institutional frameworks. Global society and its citizens are claiming their right to a healthy environment and its sustainability, against the absolute and legal rights of landowners (Scheidel et al., 2018). And although the State always had the capacity for surveillance, control and sanction, today this acquires a new dimension and becomes more special with the environmental issue. Public involvement in ecosystem management is becoming inserted into local traditions in the communal territories of the dehesas and montadas. Thus, in some way, the farms are not private spaces even if they have an owner; their use affects everyone, even if they are not owners.

1.5 Different farmers' discourses and how they face the future

Farmers have very different reactions to these new scenarios, based on their motivations and aspirations, which can be described as making up coexisting discourses over different modes of farming (see also Pinto-Correia et al., 2019). The main goal of the landowner with a heritage farming discourse is deeply rooted in the conservation of the family heritage but this has been constrained by a long period of socio-economic change that has led to varying degrees of paralysis. These landowners are blind to the possibilities of innovation, in terms of the farms' structure, products, marketing, or the organization of labour. The modern production farming discourse, widespread among the younger generation of 'grand families' and new-entrants from the business world (included companies), looks to specialization and intensification as a strategy. A third discourse relates more closely to land stewardship and is mostly linked to NGOs and foundations, for whom nature conservation and the preservation of natural resources are priorities. This last discourse draws on agroecological principles, which are slowly being taken up by those attracted by the competitive advantages obtained from tools such as ecological and organic certification, which result in higher market prices for their products.

1.6 Institutional arrangements

There are three types of institutional arrangements among actors that contribute towards the production and delivery of multiple goods and services (i) horizontal cooperation amongst farmers in cooperatives and producer organizations; (ii) vertical coordination between actors and institutions in supply chains; and (iii) policy interventions designed to correct market failures and mitigate the effects of key power imbalances. More traditional landowner/ farmers tend to reject innovation, leading to a poor level of horizontal cooperation, with individual trading arrangements clearly outnumbering collective ones. This has led new players and institutions to begin to create new networks within local communities and beyond, to diversify agrarian activities, to enhance economic benefits and promote adaptive and collective resource management in the dehesas and montados the latter of which, actively or passively, represents a form of resistance to modernization.

1.7 New opportunities for collective actions

The pace at which new market demands and environmental changes are emerging exceeds the capacity of individual farm managers to react, making a stronger case for joint participatory actions and decisions. The coordination of actors and their interests can allow the development of collective strategies and actions that can lead to more sustainable, resilient and adaptive multifunctional silvopastoral systems. This topic is the main focus of Section C of this book, which argues that coordination must be a long-term process and must be based on proper facilitation, territorial cohesion and social integration. There are some promising networks working to catalyze collective actions. For instance, the Spanish Association of Dehesa Landowners and Managers (FEDEHESA) integrates owners' and farmers' associations, along with researchers, cooperatives and NGOs. Lately, this initiative has given rise to collective actions implemented at different territorial scales focused not only on the production of goods and services, but also on the three key strategic processes: (i) encouraging the drivers of local mobilization; (ii) improving the quality of locally productive elements and (iii) developing strategic coherence. The search for synergies is enabling new opportunities for individual silvopastoral farmers, but also for silvopastoral territories (such as supporting grazing in silvopastoral territories that are prone to abandonment and, thus, large wildfires). To encourage the success of such collective actions, social and institutional arrangements need to change and be supported by reinforcing regulations and policies that influence products and markets, as well as by the adoption of new farming techniques and management practices. In this regard, the role of women of silvopastoral farms and territories is gaining momentum as they have proven to have more positive attitudes to cooperation and collective actions.

1.8 Conflicts and synergies and the need for finely-tuned governance structures

The list of possible conflicts in such complex social-ecological systems is long, and they require finely-tuned governance structures. Conflicts can be grouped into four categories: political (nature conservation laws, animal sanitary regulations), environmental (wildlife as disease vectors, access to reserved and protected areas), economic (lack of adaptation of market tools, low prices, absence of adequate supply chains) and social (use of infrastructures, interference between users, renewal, ageing, loneliness, lack of social organization). Many of these conflicts can be satisfactorily managed by farmers and other stakeholders as part of their jobs, thereby reducing their impact. Others need instruments for integrative participation in search of consensus and/or 'integrating dissensus' (Anderson et al., 2016). Governance needs to go beyond the resolution of conflicts and should aim at enhancing synergies and cooperation among actors. Two examples illustrate this need, the asymmetrical relation among farmers and large companies and supermarket chains, and the strong dependence of farmers on public bodies (often for EU subsidies). Thus it comes down to these external actors, whether public or private, to ultimately decide what business and/or political goals they wish to pursue (Pinto-Correia & Azeda, 2017; Muñoz-Rojas et al., 2019).

All in all, dehesas and montados are a living arena for many actors, both traditional and recently arrived. The latter are encouraged by a combination of factors: the search for healthier lifestyles bringing wealthy citizens to the countryside increased market delocalization which brings new sellers and new buyers, public policies which create new job profiles in rural areas, more focus on nature conservation and a boost in rural tourism, which create new job niches. Again, the transition of dehesas and montados teaches us about the transition from traditional to modern functionality and the management and governance of Mediterranean agro–silvopastoral farms and territories. In recent decades the montados and dehesas have been subject to many complex regulations in the form of laws, strategic instruments, financial and public funding schemes and action plans, but such complex regulatory frameworks have not done much in helping to tackle the key challenges these areas face. There is even a lack of a common and well-established definition of what a montado or dehesa is. As such, there is still much work to do to update their governance schemes to new times, which brings many different challenges for silvopastoral systems and actors. We look at some of these challenges in the next section.

2 Recommendations on ways to move forward

2.1 Defining owners', stakeholders' and societal expectations for the future of silvopastoral territories

Society expects silvopastoral systems to provide a balance of socio-cultural, economic and environmental benefits. Nevertheless, most citizens are not aware of all the benefits that they receive from these systems or the challenges that they are facing. It is likely that unless we achieve a shift in mind-sets and see discourses that give greater acknowledgement of the cultural and environmental importance of silvopastoral systems and of co-responsibility, the management and governance strategies that are put in place will remain largely aspirational, and ineffective, irrespective of how salient, legitimate or credible the targets may be.

A transition at multiple levels and scales is needed in order to respond to new societal goals and market trends while paying attention to multiple ecological, economic and social challenges that threaten the survival of Mediterranean silvopastoral systems. This transition has to promote an active and innovative use of natural resources. The coexistence of old productive schemes, the still strong hierarchical structures within the traditional community of actors and a certain reluctance to innovate and to accept the arrival of new actors and their ideas are hampering this transition. Instead, landowners and farmers are frequently overwhelmed by excessive administrative burdens, their excessive dependence on untargeted public payments, and the increasing feeling of being mere links in the chains dominated by of large corporations, with little autonomy or sovereignty.

The adaptation of silvopastoral systems to the new market and regulatory conditions seems in many cases *cul de sac*, as changes introduced to fit market requirements and preferences go against the sustainability of the system, and with each change, the real sustainable solutions are further and further away. This means rapid degradation of ecological capital (soil and trees) without achieving either sufficient profitability or sustainability. The evidence of degradation of the system at multiple scales is very evident (Rolo et al., 2020). In order to understand what is at stake today and to safeguard the future of these agro-ecosystems, it is essential to address the management practices and motivations and the ways in which the many individuals and institutions involved influence the management and functioning of these systems. A key gap is the lack of a vision for where the system should be heading, resulting in problems in defining strategies for the future that can satisfy the aspirations of the multiple types of farmers and other actors. This search should focus on coherence (ecological conditions, targets and farmers' capabilities), integration (consideration of all components, processes and actors) and policy coordination (bottom-up approach).

2.2 Gaining a place in the global market for silvopastoral products and services

The production of and demand for high-quality products is more common in small-scale markets, where, silvopastoral systems can develop a momentum through closer interactions with communities of consumers. Equally, silvopastoral systems have a large potential in providing recreational services, which calls for new economic actors and new types of cooperation in which, for example, nature guides could work in close cooperation with landowners, farmers and local tourism operators. Collective actions, engaging new actors, such as consumers, environmentalists and public bodies, are currently building new opportunities for silvopastoral systems, but there are still a lot of barriers between landowners, farmers and new actors, which inhibits the potential of these collective actions. Collective actions require the availability of soft skills, which are hard to obtain but can be developed through training, dedicated public investments and policy tools.

There is a strong need to deepen local knowledge, amassed over centuries of strong dependence on the resources of the territory. This aspect is frequently neglected and needs to be a foundation for the renewal of the dehesa and montados. Future silvopastoral farmers will not necessarily come from local farming families, meaning that dedicated training, provided by skilled local workers with local knowledge will be essential. Extension and other advisory services, which have recently become largely focused on administrative functions, also require a new impulse. As is acknowledged in many other farming systems in Europe (Herrera et al., 2019), these advisory services should be independent of public bodies and commercial interests and available to each farmer. Today's readily available advisory services are overly influenced by the commercial interests of fertilizer, agrochemical and machinery companies. They should be replaced by an independent model financed by farmers themselves through their organizations, which would provide a better guarantee of impartiality. Advisory services should have a strong systems approach, linking all dimensions of the silvopastoral farm system. This will require the integration of bottom-up (i.e. stakeholder capacity and local knowledge) and top-down (interdisciplinary science-based evidence provision) approaches, combined with strategic communication to potential end-users. The concept of Rural Living Labs, much promoted by the European Commission (Zavratnik et al., 2019), as local places for experimentation dedicated to regional farm systems, developing close interactions between research and practice, provides a promising way forward. These LivingLabs can also help develop new constellations of actors and improved collaboration in different regional contexts and work in the context of local socio-cultural and agro-ecological characteristics.

2.3 Building new governance schemes

A hybrid governance approach is probably needed to overcome these barriers. This approach means that decisions on objectives take place in a top-down, centralized manner, while decisions on the actions to achieve the objectives are taken in a bottom-up territorial manner, by farmers and/or the local population). Potential pathways forward include:

- Fostering more receptive attitudes (especially related to trust) and structures that enable horizontal cooperation among farmers;
- Providing improved support for moving towards more efficient vertical coordination strategies along the supply chain;
- Facilitating a shift of mentality of farmers to move from reliance on EU subsidies towards land stewards that are better positioned for providing multiple services to society and
- The inclusion of territorial scales on the targets.

It is essential to correctly identify the relevant users and other actors in each specific context and to understand their interests and skills. The interactions among them should be articulated around their common interests, not just those of the landowners. This is especially problematic in a social universe in which the traces of *latifundia* are present in many territories, and the large landowners and local societies almost inhabit separate universes. Seeking agreement on goals that benefit everyone, despite these differences, is one of the greatest challenges. The lack of strong and independent civil society organizations, including farmers' or environmental organizations, remains a handicap in Mediterranean countries and one that deserves more attention.

2.4 Creating multi-actor platforms to facilitate interactions

Multi-actor platforms, where interactions are promoted in order to influence individual and collective decision-making at the farm and landscape scale, as well as at the community and policy level, are very helpful to rebuild the governance of silvopastoral systems. Through such platforms, new perspectives, new attitudes and new actions can emerge. Spontaneous large-scale collective actions are improbable as, without an external third party to generate such actions, it is difficult to generate a critical mass (Jagers et al., 2020). Platforms need to be created that provide the fair participation of all actors (farmers, consumers, environmental experts and organizations, government agencies and the local population). These should be safe places for participation, where empathy and respect for all participants allow for community involvement in decision-making with participation adapted to different scales and degrees. These platforms need a skilled facilitation team to create the best conditions for joint work and provide adequate tools. Leadership, however, should come from the participants and there should be a focus on landowners and managers as the key actors in these areas.

Another important issue is the asymmetrical interactions between suppliers (farmers) and buyers and processors of dehesa and montado products (cork, livestock, etc.). This is because the two operate on different financial scales and also because of their asymmetrical access to information. Institutions working on this sector need to promote the elaboration of detailed and continuously updated statistics on the production and marketing of dehesa and montado products, included volumes, quality and prices. While the fragmented supply chains could be improved through legislative regulations and institutional tools, greater associationism and cooperation between those on the supply side would be even more helpful.

2.5 Tailoring policies for silvopastoral systems

Public policy needs to recognize the particularities of silvopastoral systems. The highly valued multi-functionality of these silvopastoral systems suggests that policy incentives that promote individual functions (i.e. a sectoral approach that promotes either wood or meat production) are insufficient, or even counter-productive. Decades of top-down policies have resulted in policies that are inappropriate for silvopastoral farms and the landscape. More targeted policies and planning instruments and approaches are urgently needed. One way forward would be to transform direct (EU) payments, (based on acreage) into payments for providing public goods, which would help align the environmental and socio-environmental dimensions of sustainability (Pe'er et al., 2019). Results-based payments seem the most appropriate approach although may need to be phased in on an experimental basis. Funds to support multi-functional farming systems that follow agroecological principles are also much needed. The new Eco-Scheme formula (payment schemes in agriculture aiming at the protection of the environment and climate) can be used to fund high nature value (HNV) farming systems, effective climate change mitigation and adaptation, and instruments that effectively maintain biodiversity and ecosystems (Pe'er et al., 2019). There are other interesting proposals that could be implemented on a larger landscape scale. These include payments for ecosystem and landscape services; the spatial targeting and decentralization of CAP and other public incentives; local landscape strategies that are tightly embedded within the spatial planning framework, and strategic and participatory scenario planning. Payments to compensate for the difficulties to farmers as a result of the introduction of wildlife protection legislation, such as predation and cross-infection of livestock, should also be introduced. This all requires more developed knowledge, grounded on ecological economics, on reliable economic assessments of the value of environmental services (or the cost of their absence) as the basis for fairly rewarding farmers for providing such services. An evaluation of the efficiency of multiple measures already implemented would be an important first step in shaping such context-specific policy instruments.

2.6 A new regulatory framework to halt the degradation of silvopastoral systems in North Africa

The challenges in North Africa are more extreme than those facing the dehesas and montados. There is a dissonance between the regulations imposed by the forest administration and the uses and customs of the local population, who are highly dependent on forest products for their subsistence. This is accelerating the degradation of agroforestry systems in North Africa. There is an urgent need for a modernization of the Forest Codes in order to integrate the local population into decision making and increase the flexibility of the laws with regards to access to forests, whether organized individually or through associations. Recommendations for these countries need to include:

- Empowering local institutions, particularly through clarifying their respective mandates, status, organigram and operational rules/tools;
- Strengthening the involvement of local actors and community organizations in the process of planning, monitoring and managing pastoral resources and the ecosystem services offered by rangelands;
- Creating effective coordination mechanisms and
- Implementing an effective monitoring and evaluation tool for the technical, managerial and financial aspects related to the performance of local rangeland management.

These recommendations can be partially achieved by improving the legal and regulatory framework, but the livelihoods of pastoral communities also need to be strengthened in order to reduce the pressure on rangelands.

3 Research agenda

3.1 Developing a commonly agreed definition to provide the basis for spatial differentiation

Silvopastoral territories are diffuse combinations of agriculture and forest, with greatly differing densities of tree and shrub cover. They also change over time as a result of the interactions between human management and natural dynamics. This means that there is no single unique definition of a dehesa and montado, or other Mediterranean silvopastoral systems, making it impossible to map them with any precision (Moreno and Rolo 2019). This situation has prevented the emergence of appropriate policy instruments and public support measures. To obtain a commonly agreed definition we need to look at the intersections between different definitions. The definition to be used needs to consider the objectives (i.e. what purpose such boundaries would be used for) and the intersections between different layers of the components of these silvopastoral systems. Yet montados and dehesas are not homogeneous, and proper targeting of objectives and tools, both in terms of everyday management as well as in policies, requires differentiation between the different types of combination of tree cover, shrub and pasture and the specific biophysical settings. Such an effort is highly relevant for administrative purposes since policies that relied on simplified definitions of the nature of these ecosystems could well reduce their ecological and productive complexity. A commonly agreed definition could also act as a foundation for branding high-quality products from these areas and developing land stewardship programmes. In these two latter cases, best management practices and sustainability outcomes should be considered to be key criteria.

3.2 Resilience strategies as socio-ecological systems

The multiple ecological, sociocultural and institutional decoupling processes undergone by the dehesa and montado in recent last decades show the need to search for new ecological, social and institutional equilibriums to ensure their survival. The concept of resilience has two complementary perspectives (Ingrisch and Bahn 2018); the capacity to recover from transient changes (engineering resilience) and the capacity to reorganize under a change to reach a new stable state whilst retaining the same essential functions (ecological resilience), The dehesa and montado ecosystems have shown themselves to have a high engineering resilience, as tree regeneration and soil recovery can operate relatively well by decreasing grazing pressure (Pulido et al., 2010). However, as a socio-ecological system, they seem not to have had an opportunity to go back to the pivotal socio-economic role that they historically played for their rural communities (Schröder 2011). The alternative is to find new stable states rooted in territorial capital, the social and natural capital of any given territory, counting on new actors and societal demands (Rolo et al., 2020).

3.3 Innovation mechanisms: technological and organizational

Existing (but endangered) local knowledge should pave the way for the progressive development of innovative proposals. Strategies for the efficient marriage of local knowledge and innovation in the dehesas and montados and other extensive silvopastoral systems, that require both technical and social science contributions, are still poorly explored, although they are much needed. New technologies, such as GPS-based facilities, satellite products and GIS (Geographical Information Systems) are increasingly available to practitioners. The use of these tools is bringing promising proposals to improve pastures and habitat conservation, to optimize the use of forage resources and to track animal welfare, but more research is needed for them to become economically feasible tools, that can be used to avoid the prospect of further losses of employment for the local population.

3.4 The social and institutional drivers of change

There is inadequate understanding of the transitions occurring in the social system and of the social and institutional drivers of these changes. This makes it difficult to recognize possible strategies for launching a sustainability transition. Such knowledge is needed to adopt tailored managerial and governance innovations. In the last four decades, there has been abundant scientific research on the biophysical bases of the dehesas' functioning and dynamics, and the consequences of current management practices for the continued existence of the system. But the solutions that emerged from science have rarely taken into the landowners', farmers' and other actors' aspirations and motivations, or their financial and managerial capabilities. There is a need for efficient mechanisms and languages for the sharing of knowledge between different actors to make science more appropriate and to encourage the uptake of this scientific knowledge. Such integration requires more input from the social sciences to deepen understanding of the current social context in the dehesas and montados. The chapters assembled in this book provide an initial picture of these processes but simultaneously show the lack of broader-scale studies needed to assess the representativity of such a rich picture.

3.5 Linking to the territory

The links of different actors with the territory is another pressing sociological question. We can see a gradient of strong to weak bonds, but also a generalized trend to the weakening of these bonds, together with the progressive loss of local and traditional knowledge. As fewer people work on silvopastoral farms and more people move into silvopastoral territories, from urban or other origins, this decoupling between the rural actors and the silvopastoral system that shapes their territories gets stronger – and recoupling requires new awareness and new mechanisms. Smart solutions for a new governance of the rights of use are required, to support the reconnection of people with their territory and their landscape. Understanding customary rights imply examining the past since many of these rights are embedded in local cultures and have their roots in past practices, power relations and land-use systems. These customary rights are in many cases unique, and made effective collaboration possible between many different actors, with different interests. Strong efforts are needed to keep such forms of collaboration and related skills alive or to renew them. Examples from the Maghreb illustrate how attempts at modernization, through the substitution of local customary rights with alien administrative bodies and regulations have failed. There a need to bridge traditional local customs and the new administrative hierarchies, and responsibilities, in order to create efficient governance structures.

3.6 Acknowledging different patterns of ownership and different motivations

The coexistence of several discourses (e.g. the conservation of family patrimony, modernization and intensification, agroecological farming and land stewardship and nature conservation) offers an ideal scenario of contrasting models for the search for a viable future for dehesas and montados. The objective should not be to seek a single model, the sum of the positive aspects of all these practices, but to establish a fluid exchange of knowledge, ideas and experiences that, together, will define the integrative and sustainable management practices that can provide the goods and services demanded by society. This is a pre-requirement for establishing resilient governance structures that efficiently interweave old and new owners and other actors, such as environmentalist NGOs, tourist agents, large companies and recreational users. Building on links to the territory and ancient customary rights that emphasize resource sharing and collaboration between different interests seems a promising way forward.

3.7 Stronger market integration of food products

The demand for quality and healthy foods produced with environmentally respectful practices, and a high standard of animal welfare in the case of livestock, has increased very rapidly in recent years. To capitalize on this ongoing trend, it is necessary to explore and design mechanisms that allow the creation of trust between producers and consumers, and that adequately remunerate producers with a just share of the added value of such products. The proliferation of local consumer groups, which has accelerated since the COVID 19 pandemic (Ploeg 2020) provides a good opportunity to study and improve the performance of these trust-based networks. However, products and services also need to reach more distant markets and consumers, which requires more formal confidence mechanisms, often involving third-party certification. The potential of quality products from dehesas, montados and other Mediterranean silvopastoral systems has not yet been fully exploited and producers are other supply chain actors need to explore new technology, tools and associative relationships in order to enhance their marketing.

3.8 Identifying and resolving conflicts

Despite the long standing tradition of several and differentiated users of resources in the silvopastoral lands, the new societal demands are creating novel tensions and challenges. These societal expectations are likely to become more pronounced and explicit in the future. The traditional social and economic functions of the land, have been augmented by concerns about its environmental functions which are central to any ecological transition. In short, the agrarian question has been transmuted into a socio-ecological question. For instance, in Spain recent legislation makes it possible for the owner to claim the right of exploitation of products that were traditionally freely gathered by the local population (edible, medicinal and aromatic plants and mushroom) and forbid collection by others on his/her land (Acosta-Naranjo, Guzmán-Troncoso and Gómez-Melara, 2020). Although justified by conservation reasons, the social consequences of this have not been evaluated. Dehesas and montados are also of high value for recreation and provide a connection to nature and culture for visitors. Yet the appearance of strange people on privately owned farms is exacerbating tensions. Another source of conflict is the negative interactions between wildlife and livestock, translated into conflicts between livestock breeders and hunters, environmentalists and administrators in charge of nature conservation programmes. These examples illustrate the need to explore conflict-resolution mechanisms, look for consensus and 'win-win' scenarios as well as the integration of 'dissensus' (Anderson et al., 2016). In this instance, the concept of new commons is gaining momentum (Woestenburg 2018) and deserves to be widely applied in studies of the dehesas and montados, especially where communal lands still exist. The open-access nature of many silvopastoral territories, including but extending beyond the dehesas and montados, makes them an ideal context for studying and implementing strategies to enhance new commons bases.

3.9 Opening up for new actor roles

In the current context of rural aging and masculinization, the opening of social and workspaces for young people and women is essential for generational turnover in natural resource management. The struggle for women's empowerment and visibility in rural areas is long-standing and slow, although some women, whether local or newcomers, have been able to overcome barriers and occupy diverse (and sometimes prominent) roles in different dimensions of silvopastoral systems. Yet there has been little research into how they have achieved this. Insights are also needed into the ways in which women are developing new activities related to the processing and elaboration of products (food, handicrafts, etc.) and becoming involved in the valorization of landscape heritage. It is important to identify the changes that might act as leverage points, in order to promote them through public policies.

In recent decades, different programmes have been implemented by EU and national institutions to attract young people to farming, to counter the aging of farm-heads. There have been mixed accounts of the effects of this policy, with some cases where the funding supported new multifunctional pastoral farms, and others cases it was considered just a way to obtain extra-money by nominally dividing the family farm among siblings (Farinella et al., 2017). It is obvious the sons and sisters of existing landowners/farmers in the dehesas will not be sufficient to maintain these agroecosystems, and young people coming from other social settings will be needed to fill the niche. Some thought needs to be given to strategies to overcome the multiple barriers that these newcomers will find. In the same way, strategies to overcome difficulties emerging from a new composition of, not only, land owners', but also local communities, are strongly needed. To date, newcomers seem to have integrated quite smoothly, but as more arrive and the ancient customary rights and forms of distribution of the rights of use weaken, new forms of integrating the interests of all actors, and preserving a unique multifunctional balance, will be needed.

4 Conclusions

If the combination of science and local stakeholders working together fails to find solutions for the future of silvopastoral systems across the Mediterranean, we risk losing them, either to land abandonment or their replacement by other, more specialized, intensive and homogeneous land uses. While the abandonment of extensive silvopastoral systems offers an opportunity *par excellence* for rewilding programs (Guerrero-Gatica and Root-Bernstein 2019) and for an increase in forest cover in the EU (as envisaged in the EU Biodiversity Strategy for 2030; Varela et al., 2020), the environmental risks are multiple and serious. With the reduction and/or loss of these unique silvopastoral landscapes, multiple, unique, public goods that society currently enjoys will disappear. Furthermore, a key cultural foundation of Iberian identity and that of other Mediterranean regions will be irreparably lost. To avoid this, we need reliable evaluations of the i) social, economic and environmental costs of the abandonment or transformation of silvopastoral territories and ii) of the contribution that silvopastoralism makes to meeting to the Sustainable Development Goals (SDG) defined by the UN (2015) as well as the European Green Deal (European Commission 2019). Such work is essential to advance the social recognition and create the administrative framework needed to support the sustainability of silvopastoral territories' socio-ecological systems.

References

Acosta-Naranjo, R. (2002). Los entramados de la diversidad. Antropología Social de la dehesa. Diputación Provincial, Badajoz.

Acosta-Naranjo, R. (2008). Dehesas de la Sobre Modernidad. La cadencia y el vértigo. Badajoz: Diputación de Badajoz.

Acosta-Naranjo, R., Guzmán-Troncoso, A. J. and Gómez-Melara, J. (2020). The persistence of wild edible plants in agroforestry systems: the case of wild asparagus in southern Extremadura (Spain). *Agroforestry Systems*, 94:2391–2400.

Anderson, M. B., Hall, D. M., McEvoy, J., Gilbertz, S. J., Ward, L., & Rode, A. (2016). Defending dissensus: participatory governance and the politics of water measurement in Montana's Yellowstone River Basin. *Environmental Politics*, 25(6):991–1012.

Carolino, J. and T. Pinto-Correia. 2011. Material landscape, symbolic landscape, and identity in the municipality of Castelo de Vide. *Analise Social,1* 198:89–113.

Eichhorn, M. P., Paris, P., Herzog, F., Incoll, L. D., Liagre, F., Mantzanas, K., Mayus, M., Moreno, G., Papanastais, V.P., Pilbean, D.J., Pisanelli, A. and Dupraz F. (2006). Silvoarable systems in Europe–past, present and future prospects. *Agroforestry Systems*, 67(1):29–50.

European Commission. 2019. The European Green Deal, COM (2019) 640 Final. Brussels.

Farinella, D., Nori, M., & Ragkos, A. (2017). Change in Euro-Mediterranean pastoralism: which opportunities for rural development and generational renewal? Grassland Science in Europe, Vol. 22 – Grassland resources for extensive farming systems in marginal lands, pp 23–36.

Guerrero-Gatica, M., & Root-Bernstein, M. (2019). Challenges and limitations for scaling up to a rewilding project: scientific knowledge, best practice, and risk. *Biodiversity*, 20(2–3):132–138.

Herrera, B., Gerster-Bentaya, M., Tzouramani, I., & Knierim, A. (2019). Advisory services and farm-level sustainability profiles: an exploration in nine European countries. *The Journal of Agricultural Education and Extension*, 25(2):117–137.

Ingrisch, J., & Bahn, M. (2018). Towards a comparable quantification of resilience. *Trends in Ecology & Evolution*, 33(4):251–259.

Jagers, S. C., Harring, N., Löfgren, Å., Sjöstedt, M., Alpizar, F., Brülde, B., … Steffen, W. (2020). On the preconditions for large-scale collective action. *Ambio*, 49(7), 1282–1296.

McMichael, Philip (2009). A food regime genealogy. *Journal of Peasant Studies,* **36** (1): 139–169.

Moreno, G., Aviron, S., Berg, S., Crous-Duran, J., Franca, A., de Jalón, S. G., Hartel, T., Mirck, J., Pantera, A., Palma, J.H.N., Paulo, J.A., Re, G.A., Sanna, F., Thenail, C., Varga, A., Viaud, V. & Burgess, P.J. (2018). Agroforestry systems of high nature and cultural value in Europe: provision of commercial goods and other ecosystem services. *Agroforestry Systems*, 92(4):877–891.

Moreno, G., & Rolo, V. (2019). Agroforestry practices: silvopastorism. Agroforestry for sustainable agriculture, María Rosa Mosquera-Losada and Ravi Prabhu (Eds.), Burleigh Dodds, Series in Agriculural Science. pp 119–164.

Muñoz-Rojas, J., Pinto-Correia, T., Hvarregaard Thorsoe, M., & E. Noe. 2019. The Portuguese montado: A Complex System under Tension between Different Land Use Management Paradigms, F. Allende Álvarez, G. Gómez-Mediavilla, N. López-Estébanez (Eds.), Silvicultures – Management and Conservation. United Kingdom: IntechOpen, pp. 146–164. http://dx.doi.org/10.5772/intechopen.86102.

Murdoch. J and T. Marsden (2013) Reconstituting Rurality. Routledge, London.

Navarro, L. M., & Pereira, H. M. (2015). Rewilding abandoned landscapes in Europe, Henrique M. Pereira and Laetitia Navarro (Eds.), Rewilding European Landscapes. Springer, Cham, pp. 3–23.

Ortiz-Miranda, D., A. Moragues-Faus, and E. Arnalte-Alegra. 2013. *Agriculture in Mediterranean Europe. Between Old and New Paardigms*. 1st ed. Emerald. Location?

Pe'er, G., Bonn, A., Bruelheide, H., Dieker, P., Eisenhauer, N., Feindt, P.H., Hagedorn, G., Hansjürgens, B., Herzon, I., Lomba, A., Marquard, E., Moreira, F., Nitsch, H., Opermann, R., Perino, A., Röder, N., Schleyer, C., Schindler, S., Wolf, C., Zinngrebe, Y. & Lakner, S. (2020). Action needed for the EU Common Agricultural Policy to address sustainability challenges. *People and Nature*, 2(2):305–316.

Pe'Er, G., Zinngrebe, Y., Moreira, F., Sirami, C., Schindler, S., Müller, R., Bontzorlos, V., Clough, D., Bezák, P., Bonn, A., Hansjürgens, B., Lomba, A., Möckel, S., Passoni, G., Schleyer, C., Schmidt, J., & Lakner, S. (2019). A greener path for the EU Common Agricultural Policy. *Science*, 365(6452), 449–451.

Pinto-Correia, T., & Azeda, C. (2017). Public policies creating tensions in Montado management models: Insights from farmers' representations. *Land Use Policy*, 64:76–82.

Pinto-Correia, T., N. Guiomar, C. A. Guerra, and S. Carvalho-Ribeiro. 2016. Assessing the ability of rural areas to fulfil multiple societal demands. *Land Use Policy*, 53:86–96.

Pinto-Correia, T., J. Primdahl, and B. Pedroli. 2018. European Landscapes in Transition: *Implications for Policy and Practice*. 1st ed. Cambridge, UK: Cambridge University Press.

Pinto-Correia, T., Muñoz-Rojas, J., Hvarregaard Thorsøe, M., & and E. B. Noe. 2019. Governance Discourses Reflecting Tensions in a Multifunctional Land Use System in Decay; Tradition Versus Modernity in the Portuguese montado. *Sustainability*, 11(12):3363. doi: 10.3390/su11123363

Ploeg J.D. van der (2020) From biomedical to politico-economic crisis: The food system in times of Covid-19. *Journal of Peasant Studies*, 47(3):1–29

Pulido, F., García, E., Obrador, J. J., & Moreno, G. (2010). Multiple pathways for tree regeneration in anthropogenic savannas: Incorporating biotic and abiotic drivers into management schemes. *Journal of Applied Ecology*, 47(6):1272–1281.

Rolo V, Hartel T, Aviron S, Berg, S., Crous-Durán, J., Franca, A., Mirck, J., Palma, J.H.N., Pantera, A., Paulo, J.A., Pulido, F.J., Seddaiu, G., Thenail, C., Varga, A., Viaud, V., Burgess, P.J. & Moreno G. (2020) Challenges and innovations for improving the resilience of European agroforestry systems of high nature and cultural value: A stakeholder perspective. *Sustainability Science*, 15:1301–1315.

Scheidel, A., L. Temper, F. Demaria and J. Martínez-Alier. 2018. Ecological distribution conflicts as forces for sustainability: an overview and conceptual framework. *Sustainability Science*, 13: 585–598

Schröder, C. (2011). Land use dynamics in the dehesas in the Sierra Morena (Spain): the role of diverse management strategies to cope with the drivers of change. *European Countryside*, 3(2):11–28.

Simonson, W. D., Allen, H. D., Parham, E., e Santos, E. D. B., & Hotham, P. (2018). Modelling biodiversity trends in the montado (wood pasture) landscapes of the Alentejo, Portugal. *Landscape Ecology*, 33(5):811–827.

Torralba, M., Fagerholm, N., Hartel, T., Moreno, G., & Plieninger, T. (2018). A social-ecological analysis of ecosystem services supply and trade-offs in European wood-pastures. *Science Advances*, 4(5):eaar2176.

UN. 2015. Sustainable Development Goals. United Nations. Retrieved at http://www.fao.org/state-of-food-security-nutrition/en/.

Varela, E., Pulido, F., Moreno, G., & Zavala, M. A. (2020). Targeted policy proposals for managing spontaneous forest expansion in the Mediterranean. *Journal of Applied Ecology,* 57:2373–2380.

Woestenburg, M. (2018). Heathland farm as new commons? *Landscape Research*, 43(8):1045–1055.

Zavratnik, V., Superina, A., & Stojmenova Duh, E. (2019). Living Labs for Rural Areas: Contextualization of Living Lab Frameworks, *Concepts and Practices. Sustainability*, 11 (14):3797.

Index

Note: *Italicized* folio indicates figures, **bold** indicates tables.

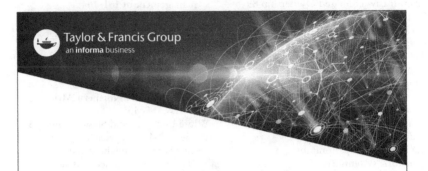

Printed in the United States
by Baker & Taylor Publisher Services